Applied Mathematical Sciences
Volume 123

Springer Science+Business Media, LLC

Applied Mathematical Sciences

(continued following index)

Vy Khoi Le   Klaus Schmitt

Global Bifurcation in Variational Inequalities

Applications to Obstacle and Unilateral Problems

With 22 Illustrations

 Springer

Vy Khoi Le
Department of Mathematics
    and Statistics
University of Missouri-Rolla
Rolla, MO 65409
USA

Klaus Schmitt
Department of Mathematics
University of Utah
Salt Lake City, UT 84112
USA

Editors

J. E. Marsden
Control and
    Dynamical Systems, 104-44
California Institute of Technology
Pasadena, CA 91125
USA

L. Sirovich
Division of
    Applied Mathematics
Brown University
Providence, RI 02912
USA

Mathematics Subject Classification (1991): 73CSO, 73Kxx, 49Rxx, 73Hxx

Library of Congress Cataloging-in-Publication Data
Le, Vy Khoi
    Global bifurcation in variational inequalities : applications to
obstacle and unilateral problems / Vy Khoi Le, Klaus Schmitt.
        p.   cm. -- (Applied Mathematical Sciences ; 123)
    Includes bibliographic references and index.
    ISBN 0-387-94886-4 (alk. paper)
    1. Variational inequalities (Mathematics)   2. Bifurcation theory.
I. Schmitt, Klaus, 1940– . II. Title.   III. Series: Applied
mathematical sciences (Springer-Verlag New York Inc.)   ;   v.   123.
    QA1.A647   vol.   123
    [QA316]
    5110 s—dc20
[515'.36]   96-41133

Printed on acid-free paper.

Production managed by Robert Wexler; manufacturing supervised by Jacqui Ashri.
Photocomposed pages prepared from the authors' TEX file.

9 8 7 6 5 4 3 2 1
ISBN 978-1-4612-7298-4   ISBN 978-1-4612-1820-3 (eBook)
DOI 10.1007/978-1-4612-1820-3

Kính tặng Bố Mẹ   To Claudia, Susan, and Michael

Angesichts von Hindernissen

mag der kürzeste Weg zwischen zwei Punkten

der krumme sein.

– Bertolt Brecht
Leben des Galilei

Contents

List of Figures

List of Symbols

$\mathbb{R}$	The real line.
$\mathbb{R}^N$	N-dimensional Euclidean space.
$\mathbb{R}_*^+$	$= \{x \in \mathbb{R} : x > 0\}$, the set of positive real numbers.
Ω	An open bounded domain in $\mathbb{R}^N$.
X, V	Real reflexive Banach spaces.
V^*	The dual space of V.
$L(X, V)$	The space of bounded linear mappings from X to V.
$B_r(a)$	$= \{x : \|x - a\| < r\}$, the ball of radius r, centered at a.
$\overline{K}$	The closure of a subset K of V.
$\overset{\circ}{K}$	The interior part of K.
∂K	The boundary of K.
∇f	$= (\partial f/\partial x_1, \partial f/\partial x_2, \ldots, \partial f/\partial x_N)$, the gradient of f.
Δf	$= \partial^2 f/\partial x_1^2 + \partial^2 f/\partial x_2^2 + \cdots + \partial^2 f/\partial x_N^2$, the Laplacian of f.
p'	$= p(p-1)^{-1}$, the conjugate exponent of p.
p^*	$= Np(N-p)^{-1}$, the Sobolev conjugate exponent of p.
$\|f\|_{L^p(\Omega)}$	$= \left(\int_\Omega |f|^p dx\right)^{1/p}$, the L^p norm.
$L^p(\Omega)$	The space of p integrable functions (whose L^p norm is bounded).
$\|f\|_{W^{k,p}(\Omega)}$	$= \left(\sum_{|\beta| \leq k} \int_\Omega |D^\beta f|^p dx\right)^{1/p}$, the Sobolev norm.

$W^{k,p}(\Omega)$ The space of functions with bounded $W^{k,p}(\Omega)$ Sobolev norm.

$C_0^\infty(\Omega)$ C^∞ functions with compact support.

$W_0^{k,p}(\Omega)$ The closure in $W^{k,p}(\Omega)$ of $C_0^\infty(\Omega)$.

$H^k(\Omega)$ $= W^{k,2}(\Omega)$.

$H_0^k(\Omega)$ $= W_0^{k,2}(\Omega)$.

$\mathrm{supp} f$ The support of a function f.

1
Introduction

1.1 Variational inequalities and bifurcation

These notes are concerned with global bifurcation problems for variational inequalities, problems which lie at the interface of the theory of variational inequalities and bifurcation theory.

Although the study of variational inequalities dates back to the origins of the calculus of variations, their systematic development began in the sixties with the work of Fichera ([38]) and Stampacchia ([115], [116]), which was motivated by problems in mechanics (obstacle problems in elasticity - the Signorini problem ([114])) and potential theory (the study of the capacity of sets). After the fundamental work of Lions and Stampacchia ([67]), the study of variational inequalities intensified and became an important subject in nonlinear analysis. The rapid growth of the theory, which was made possible by the work of Brézis ([11], [12]), Browder ([15], [16]), Kinderlehrer ([50], [51]), Duvaut and Lions ([36]), Friedman ([39]), Baiocchi and Capelo ([7]), and many others, brought about important contributions to nonlinear analysis, calculus of variations, optimization theory, optimal control, and to many branches of mechanics, mathematical physics, and engineering.

An elementary example of a variational inequality is the following simple deformation problem of a beam constrained by an obstacle. If we consider a homogeneous elastic beam occupying an interval $[a, b]$, subject to a force f, and lying above an obstacle ψ, where ψ is a measurable function, the displacement of the beam is, then, constrained, and the set of admissible displacements is described by the convex set

$$K = \{v : v \geq \psi \text{ on } [a, b]\}.$$

Using the principle of energy minimization, the deflection u of the beam must satisfy the following minimization problem:

$$u \in K : E(u) \leq E(v), \ \forall v \in K, \tag{1.1}$$

where

$$E(v) = \frac{1}{2} \int_a^b (v'')^2 - \int_a^b fv$$

denotes the potential energy. Using the fact that K is a convex set, we must have that

$$(1 - t)u + tv \in K, \ \forall v \in K, \ \forall t \in [0, 1],$$

and, hence, the function

$$i(t) = E((1 - t)u + tv),$$

must have a minimum at $t = 0$, and $i'(0) \geq 0$, i.e.

$$i'(0) = \int_a^b u''(v - u)'' - \int_a^b f(v - u) \geq 0, \ \forall v \in K, \tag{1.2}$$

which can be viewed as the Euler-Lagrange inequality corresponding to (1.1). On the other hand, if u_1 and u_2 both satisfy (1.2), then, after an elementary calculation, we find that

$$- \int_a^b (u_1'' - u_2'')^2 \geq 0$$

or

$$u_1'' = u_2'', \text{ on } [a, b],$$

and, because u_1 and u_2 both satisfy the boundary conditions, we conclude that

$$u_1 = u_2, \text{ on } [a, b].$$

Thus, the minimization problem (1.1) is equivalent to the variational inequality (1.2).

In a similar vein, minimization problems on a more abstract level lead to variational inequalities. For example, if F is a real convex functional of class C^1, defined on a Banach space V, and K is a closed convex subset of V, then, the minimization problem

$$u \in K : F(u) \leq F(v), \ \forall v \in K, \tag{1.3}$$

is equivalent to the variational inequality

$$u \in K : \langle F'(u), v - u \rangle \geq 0, \ \forall v \in K, \tag{1.4}$$

where $\langle \cdot, \cdot \rangle$ stands for the duality pairing between V and its dual space V^*; the argument for the equivalence may be found in ([36]) and is very similar to the one given above.

Detailed presentations and surveys of the theory of variational inequalities and their applications may be found in [51] and [65] (general theory and applications), [7] and [39] (applications to free boundary problems), and [36] and [103] (applications of variational inequalities in physics and mechanics).

Bifurcation theory also has its origin in mechanics, dating to the middle of the eighteenth century with the work of Euler and Bernoulli on equilibria of elastic beams. Bifurcation phenomena, which occur in many parameter-dependent nonlinear problems of mathematical physics, concern the existence and behavior of nontrivial solutions of these problems when parameters are varied. Consider, for example, a functional equation $F(x, \lambda) = 0$ depending on a parameter λ. Suppose that this equation has a known family of (called trivial) solutions $\{x = x(\lambda)\}$. For some particular values λ_0, there may exist, however, for λ arbitrarily near λ_0, solutions different from $x(\lambda)$. It is the purpose of bifurcation theory to locate such bifurcation points for the equation and to describe the nontrivial solution sets bifurcating there. Important contributions to bifurcation theory were made in the latter part of the nineteenth century by Poincaré with his work on celestial mechanics and by Schmidt, Lyapunov, Hammerstein, and Lichtenstein with their work on nonlinear integral equations. Great advances were made in the forties by the Soviet school with Krasnosel'skii, Vainberg, Liusternik, Schnirelman and their colleagues.

Bifurcation in problems in mechanics were investigated intensively since the forties by Antman, Friedrichs, Kolodner, Keller, Kirchgässner, Pimbley, Sather, Stoker, Berger, Fife, and others (see, e.g., a survey in [5] and also [52], [105], and [126]). Motivated by the well-known Krasnosel'skii theorem [53], Crandall and Rabinowitz ([22]) employed Leray–Schauder degree theory to study the global behavior of solutions of nonlinear Sturm-Liouville systems. At about the same time, another important result was proved by Rabinowitz in [98]. Using the topological approach of Krasnosel'skii and exploiting the homotopy invariance property of the Leray–Schauder degree, he showed that the bifurcation occuring in the Krasnosel'skii theorem is actually a global phenomenon (Theorem 1.3, [98] and Theorem 2.4, Chapter 2). There have been several refinements and generalizations of this theorem. Ize ([49]) and Magnus ([68]) considered bifurcation for equations containing Fredholm operators; Dancer ([24] and [23]) treated bifurcation problems for analytic operators and for positive solutions; Stuart, Toland, Alexander and Fitzpatrick ([117], [120], and [2]) studied bifurcation problems for non compact operators; Furi and Vignoli, Nussbaum, McLeod and Turner,

Schmitt and Smith ([40], [85], [69], and [107]) investigated bifurcation for equations containing nonsmooth mappings.

On the other hand, the study of bifurcation from infinity (existence of solutions of arbitrarily large norms) for nonlinear equations seems to begin with the work of Krasnosel'skii (cf. [53] and the references therein), where the author introduced the concept of asymptotically linear operators. In [99], Rabinowitz proved a general global result for bifurcation from infinity by using an inversion technique to convert the problem of bifurcation from infinity to one of bifurcation from trivial solutions.

Another approach to this problem was given by Peitgen and Schmitt ([90], [91], and [106]). Their result relates the existence of global asymptotic bifurcation (in which, one has an alternative similar to the Rabinowitz alternative for bifurcation from trivial solutions) directly to the change of some appropriate degrees defined over large balls, without using intermediate problems of bifurcation from trivial solutions (see also [109]).

A simple example of a bifurcation phenomenon is the buckling problem for an elastic beam with fixed ends points (at a, b), subject to a horizontal force at one end (see Figures 1.1 and 1.2). This problem can be formulated by the following boundary value problem

$$\begin{cases} \int_a^b u''v'' - \lambda \int_a^b \frac{u'}{\sqrt{1+u'^2}} v' = 0, \ \forall v \in V, \\ u \in V, \end{cases} \tag{1.5}$$

where

$$V = \{v : v(a) = v(b) = v'(a) = v'(b) = 0\}$$

is the space of admissible displacements (which satisfy the boundary conditions at a and b) and λ is proportional to the magnitude of the applied force. (1.5) is the Euler-Lagrange equation (weak form) associated with the potential energy of the beam:

$$E(v) = \frac{1}{2} \int_a^b (v'')^2 - \lambda \int_a^b \left[\sqrt{1+u'^2} - 1 \right]. \tag{1.6}$$

1.2 Bifurcation in variational inequalities

It is seen from the example above that the investigation of buckling phenomena for constrained elastic systems leads naturally to a bifurcation problem for variational inequalities.

In fact, if in the example above, the deflection of the beam is restricted by an obstacle ψ (see Figure 1.2), then one obtains a bifurcation problem

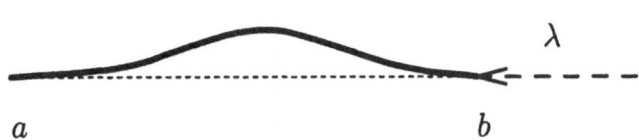

FIGURE 1.1. Buckling of unconstrained beam.

for the following variational inequality:

$$\begin{cases} \int_a^b u''(v-u)'' - \lambda \int_a^b \dfrac{u'}{\sqrt{1+u'^2}}(v-u)' \geq 0, \; \forall v \in K \cap V, \\ u \in K \cap V, \end{cases} \tag{1.7}$$

where K and V are given in (1.1) and (1.5). This inequality is the Euler-Lagrange inequality associated with the minimization problem for the energy E given by (1.6) over the set of admissible displacements K in V.

Bifurcation problems for (1.7) and other variational inequalities will be studied throughout the chapters that follow. In fact, we will consider bifurcation problems for the following general variational inequality:

$$u \in V : \langle A(u) - B(u,\lambda), v - u \rangle + j(v) - j(u) \geq 0, \; \forall v \in V. \tag{1.8}$$

Here V is a reflexive Banach space with dual V^* and dual pairing $\langle \cdot, \cdot \rangle$. $A : V \to V^*$ is a monotone coercive operator (in a sense to be made precise), $B : V \times \mathbb{R} \to V^*$ is a completely continuous mapping, and $j : V \to \mathbb{R} \cup \{\infty\}$ is a convex, lower semicontinuous functional.

Several applications may be modeled as (1.8) in the particular case where V is a Hilbert space, A is the identity mapping (or A is linear), and $j = I_K$ is the indicator function of a closed convex set K:

$$I_K(x) = \begin{cases} 0 & \text{if } x \in K, \\ \infty & \text{if } x \notin K, \end{cases} \tag{1.9}$$

i.e.,

$$\begin{cases} \langle u - B(u,\lambda), v - u \rangle \geq 0, \; \forall v \in K, \\ u \in K. \end{cases} \tag{1.10}$$

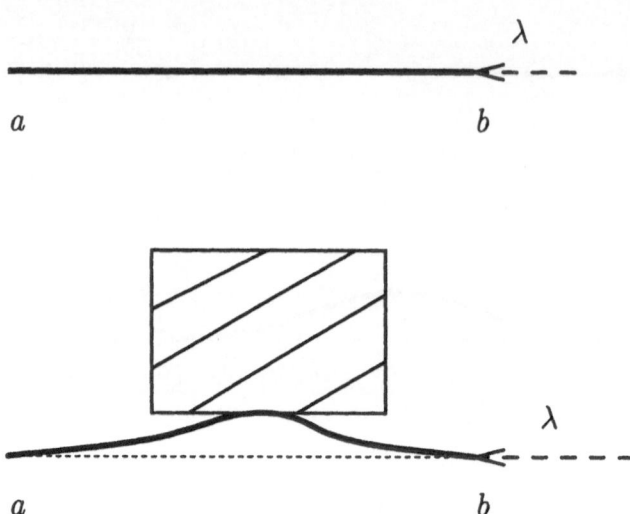

FIGURE 1.2. Buckling of constrained beam.

Because the class of variational inequalities contains nonlinear equations (in their weak formulations), the bifurcation problem for variational inequalities is a natural generalization of the classical problem of bifurcation for nonlinear equations.

Some of the early work on bifurcation for variational inequalities is that of Do ([32] and [33]) and Miersemann ([70], [71], and [72]), who used critical point type arguments to establish the existence of nontrivial solutions and, hence, bifurcation results for such inequalities. This (local) approach to bifurcation is discussed in detail in [126]. Hence, we shall not devote much space to it in this monograph. We simply outline some points of view and topics concerning bifurcation and eigenvalue problems which are different from ours and are not considered in the sequel. We do not attempt to present the most general or recent work, but simply point to some significant results.

As for the bifurcation problem for nonlinear equations, there are two main approaches for the bifurcation problem of variational inequalities, namely, variational and topological approaches. The first approach has been used extensively in recent years. Motivated by the work of Berger and Fife ([9] and [10]), Do ([32] and [33]) applied variational methods to study bifurcation for several unilateral problems for elastic plates under von Kármán's model (see also [20] for the unconstrained case). He proved the existence (and nonexistence) of bifurcation points at the first eigenvalues of some linear operators associated with those problems, formulated as variational inequalities. In a series of papers ([70], [71], [72], [75]), and Miersemann considered the eigenvalue and bifurcation problems for variational inequalities defined on closed convex sets in Hilbert spaces.

For bifurcation of smooth equations, the well-known Krasnosel'skii theorem (cf. [53]) states that bifurcation occurs at each eigenvalue of the associated linear operator. However, for variational inequalities, different scenerios may exist. Simple examples (cf. [74], [73]) show that there are eigenvalues of the corresponding linearized variational inequalities which are not bifurcation values of the original variational inequalities. In [72] (see also [75]), Miersemann showed that, under some general conditions, the first (positive) eigenvalue is also a bifurcation value. The proof is carried out using variational methods based on the Reynolds quotient and Lagrange multiplier approach (see also [55]). Also, a variational method based on an inequality version of the Krasnosel'skii technique for smooth equations ([53]) was used in [74] to establish the existence of higher eigenvalues of variational inequalities. Several applications to beam and plate problems were also considered in [74], [71], [72], and [75]. For further surveys and bibliographical notes concerning this approach, we refer to [75], [126] and the references therein. Parallel to these efforts, in [61], Kučera, Nečas, and Součec employed a version of the Liusternik-Schnirelman theorem and penalty arguments to prove the existence of infinitely many eigenvalues of variational inequalities defined on closed convex cones. In [58], [59], [60], Kučera also used penalty techniques to study eigenvalue and bifurcation problems for variational inequalities defined on convex cones. He was concerned with the existence of boundary eigenvectors, and his method consists of approximating variational inequalities by smooth equations, using some specific penalty functions. Minimax methods based on the Liusternik-Schnirelman theory approach have recently been developed by Schuricht. In [110], [111], [112], and [113], he established some general existence theorems for (higher) eigenvalues and bifurcation points of variational inequalities defined on convex sets (in Hilbert or reflexive Banach spaces). It was proved in [110], [113] that, under some assumptions, the eigenvalues, obtained by minimax procedures on certain appropriate classes of subsets of the space, are also bifurcation values. The results presented in the papers cited are valid for higher eigenvalues and were applied to some buckling problems for beams (see also [55], [56], [83]). Another variational approach was presented in [28], [27], and [88]. The authors used a concept of subdifferentials of nonsmooth functionals and a version of the Liusternik theorem to establish existence results for eigenvalues and bifurcation points of nonsmooth minimization problems related to variational inequalities.

Similar to bifurcation of nonlinear equations, global questions about bifurcation branches of variational inequalities are usually investigated by topological methods. Contrary to local existence theorems, global bifurcation results for variational inequalities seem to be much rarer in the literature and are less systematic. In [118], Szulkin studied the existence, multiplicity, and bifurcation of positive solutions of certain variational inequalities. His tools were index theory and arguments from bifurcation problems for equations containing positive operators. By assuming a change of indices, he established a general result for bifurcation at the least eigenvalue, namely, the existence of an unbounded bifurcation branch of positive solutions. The proof of the change of indices was carried out in a specific obstacle problem of a second-order elliptic operator. Almost at the same time, Quittner ([92] and [95]) employed topological methods, similar to those used by Szulkin in [118], to study bifurcation problems for variational inequalities defined on convex cones in Hilbert spaces. Quittner ([93] and [94]) also used the Leray–Schauder degree to study bifurcation problems for variational inequalities of certain types, including reaction-diffusion systems with unilateral conditions and some variational inequalities of evolution. In [93], by using ideas analogous to those of Szulkin [118], he proved the existence of eigenvalues and bifurcation points of a reaction-diffusion variational inequality by considering situations where there are changes of degrees of some appropriately defined operators. Recently, Saccon ([104]) considered the global bifurcation of an obstacle problem formulated as a second-order elliptic variational inequality on a convex set. The convex set has some specific features due to the relative positions of the obstacles (namely, their zero sets coincide). Employing a topological method, he showed that, at each eigenvalue of odd multiplicity of a linear operator associated with the considered variational inequality, a global bifurcation branch exists that satisfies the alternative in the Krasnosel'skii-Rabinowitz theorem. In [44] and [45], Goeleven, Nguyen, and Théra studied the bifurcation problem for variational inequalities of von Kárman's type, defined on convex cones in Hilbert spaces. Their method is topological and based on index theory. The authors considered conditions where global bifurcation branches exist emanating from the first eigenvalues of those variational inequalities.

Another interesting research area is the numerical analysis and approximation of solutions of bifurcation problems for variational inequalities. Several results were recently obtained by Mittelmann, Conrad et al., Hoppe, and others (see [21], [81], [82], [48], and [76], and the references therein). The main approximation tools used in those papers are continuation procedures, combined with multigrid methods or predictor-corrector schemes. Several obstacle problems for elliptic variational inequalities were studied, and bifurcating solutions, together with bifurcation points and turning points, were numerically analyzed. Further topics in bifurcation of variational inequalities, closely related to numerical approximation, are stability and continuation of solutions of variational inequalities depending on parameters. For results in this direction, we refer to [79], [78], [80], [77], and the references in these papers.

Some of the arguments, theorems, and examples in these notes are related to, improve, or are motivated by the works of Szulkin, Do, Miersemann,

Quittner, and Saccon. Some of our results, obtained independently by a different approach, ameliorate certain results in [44], [45].

Much of the material collected in these notes has its origin in the work of Le [64], [63] which, in turn is based on the thesis [62].

1.3 Discussion of results

In these notes, we shall mainly be concerned with the problem of global bifurcation for variational inequalities, using topological methods. In view of the classical results on bifurcation for nonlinear equations and of the recent works cited on bifurcation for variational inequalities, one notes that there are several essential differences between bifurcation problems for equations containing smooth operators and variational inequalities. These differences are reflected in the following problems.

The first problem that arises concerns linearization of variational inequalities. In bifurcation theory for nonlinear equations (see, e.g., [53] and [100]), the existence of bifurcation points and bifurcating branches is derived from properties of the eigenvalues of the associated linear equations. This follows directly, because, in this case, the solution mapping (i.e., the mapping that associates the right-hand sides with the solutions) is of class C^1, and the solution mapping of the linear equation is the Fréchet derivative of the corresponding solution mapping of the nonlinear equation. Hence one may use properties of the Leray–Schauder degree concerning Fréchet derivatives to study the corresponding bifurcation problems. In the case of variational inequalities, when formulated as fixed-point equations (depending on a parameter), we obtain nonsmooth and nondifferentiable operators because projections on closed convex sets in Hilbert spaces (and, more generally, solution mappings of variational inequalities) are, generally, not differentiable. (For example, (1.10) can be written as $u = P_K[B(u, \lambda)]$, where P_K is the projection onto K. Because P_K is not differentiable, $P_K \circ B$ is also not differentiable, in general.) Hence, it is not, a priori, clear how to relate the original bifurcation problem to a simpler variational inequality, and an important question is how to *linearize* the original bifurcation problem to obtain a simpler problem for which an analysis, similar to the case of equations, may be developed. Moreover, what relationships exist between the topological degrees of the (nondifferentiable) operators in the original variational inequality and those in associated simpler one?

As we shall see, even though the inequalities at hand cannot be linearized, one may associate, with the inequalities, simpler inequalities which have certain homogeneity properties with respect to the dependent variables, i.e. they are homogeneous to some degree. The process of passing from a given inequality to an associated simpler, homogeneous inequality, we call, therefore, *homogenization*. Although homogenization has other uses in the mathematical literature, no confusion should arise by using this term in our context. This process is often also called a *recession* or *asymptotic analysis*.

The second problem is the following. In the case of equations, after linearizing, one obtains a linear equation. The Leray–Schauder result about computation of degrees is usually used to calculate the degree of the linear operators (especially compact perturbations of the identity and similar operators) (cf. [53], [98], and [125]). On the other hand, for variational inequalities, the question must be investigated in which cases we can calculate the degrees of related operators (that may be nonlinear or nondifferentiable) which are analogous to the linearized problem for smooth equations.

The third problem concerns bifurcation from infinity. The usual method used for nonlinear equations is an inversion method ([99] and [125]). By using such an inversion technique, one can transform a problem of bifurcation from infinity to a bifurcation problem from trivial solutions. However, the situation is different for variational inequalities. Simple examples show that, generally, inversion of convex sets no longer yields convex sets. Hence, one cannot indirectly solve the problem of bifurcation from infinity for variational inequalities by using inversion. The question is how to develop a theory for bifurcation from infinity for variational inequalities parallel to that of bifurcation from trivial solutions.

We prove general results (Theorems 3.2 and 6.4) that enable us to reduce the investigation of the bifurcation of a variational inequality of the form (1.8) or (1.10) to that of a simpler, homogeneous, variational inequality. For this homogenized variational inequality, one can generalize the concept of eigenvalue and eigenvector in a natural way. The homogenization process is done by homogenizing the projections on convex sets (or solution mappings). Analogous to bifurcation problems for equations, we prove that bifurcation points of the original variational inequalities correspond to eigenvalues of the homogenized variational inequalities. Moreover, there are relationships between the topological degrees of operators associated with these variational inequalities. These relationships are used in later analysis, permitting us to investigate the global behavior of bifurcation branches of the original variational inequality via the calculation of the degrees of homogeneous operators in the homogenized variational inequality.

In several cases, calculations of the degrees of the homogeneous mappings are used together with the homogenization results above to yield an alternative for global bifurcation branches, similar to the Rabinowitz alternative for nonlinear equations. The degrees can be calculated in the following cases.

(i) The homogenized variational inequality becomes a linear equation of a certain kind. In this case, we use properties of linear equations and calculate the degrees of the corresponding linear mappings by using results about multiplicities of the associated eigenvalues ([53] and [125]). Therefore, we obtain global bifurcation for the original variational inequality at eigenvalues of odd multiplicity of the linearized problem (Corollaries 3.3, 6.5, and 6.6).

(ii) The homogenized variational inequality contains some positive operators (with respect to certain cones, cf. [3] and [85]). Then, by using index theory for positive mappings ([3] and [85]) and a type of Krein-Rutman argument, we can establish a change of index as λ varies over an interval $[0, \lambda_0]$, implying global bifurcation (Theorem 4.3). As an application, we consider a free boundary problem for an integral equation which can be formulated as a complementarity problem. We also prove that, generally, the index and the topological degree of operators at regular points coincide. Hence, the approaches using index and degree are equivalent. Another similar result concerns relationships between bifurcation of variational inequalities of certain kinds and the bifurcation of positive solutions of some corresponding smooth equations (Section 4.2.3).

(iii) The degrees are also computed in the cases where the homogenized variational inequalities and their associated homogenizations are related. Namely, if an eigenvalue of the homogeneous equation is also a simple eigenvalue of the homogenized variational inequality, then, by using arguments based on the Fredholm alternative for linear operators, we show that the degree of the considered mappings is 1 at 0 and is 0 at some point λ near the eigenvalue (Theorems 4.4 and 6.8 (a)). One has a similar result when the simplicity of the eigenvalues is replaced by some asymmetry condition of an appropriate subset of the adjoint eigenspaces at those eigenvalues (Theorem 6.8 (b)).

By using arguments based on some positiveness property of eigenvectors, we also find conditions for global bifurcation at the greatest eigenvalues of the homogenized variational inequality (Theorem 6.9). We prove that, under some conditions, the bifurcation branches are, in fact, unbounded.

Several corollaries are derived for variational inequalities containing symmetric operators (Corollaries 4.6, 6.13, and 6.14) and for cases where one can check the simplicity of the eigenvalues. This may be done by using a result of Miersemann in [72] (Corollary 4.8) or by using the properties of demi-interior points and a lemma of Kučera in [58] (Corollary 4.6).

Based on the work and ideas in [90] and [91], we develop results for bifurcation from infinity of variational inequalities, parallel to those for bifurcation from trivial solutions. We show that a variational inequality of type (1.8), under certain assumptions, can be *linearized* at infinity to a simpler, homogeneous, variational inequality which still has several properties similar to derivatives at infinity for nonlinear operators, as considered in [53]. This is done by using a homogenization process at infinity for the projections onto convex sets and for solution mappings of variational inequalities, in general. Moreover, as in the problem of bifurcation from trivial solutions, we can prove that the asymptotic bifurcation points correspond to eigenvalues of the asymptotically homogenized variational inequality and that there is a relationship between degrees of operators associated with these inequalities (Theorems 5.2 and 7.1). Hence, the investigation of the original bifurcation problem can be reduced to the study of eigenvalues and the corresponding degrees of the simpler, asymptotically homogenized, variational inequality. One significant feature is that the variational inequality, thus obtained, has several properties similar to the one derived in the case of bifurcation from trivial solutions. Hence, as in that case, we prove a number of results for the present problem based on the calculation of the degrees or indices of operators in the homogeneous problem. In the cases where the asymptotically linearized variational inequalities are linear equations, one has global bifurcation from infinity at their eigenvalues of odd multiplicity (Corollaries 5.3, 5.4, and 7.3). If the asymptotic variational inequalities contain positive operators, then, index theory is used to prove global asymptotic bifurcation for the original variational inequalities (Theorem 5.6). We also establish results about global bifurcation from infinity of variational inequalities when the corresponding homogenizations have simple eigenvalues with associated eigenvectors satisfying certain conditions (Theorems 5.7, 7.4, and 7.5, Corollaries 5.9, 5.10, 7.7, and 7.8).

The abstract results considered above are applied to several bifurcation problems with unilateral conditions. In Examples 3.2, 4.2, 4.5, and 6.6, we consider global bifurcation in obstacle problems containing second-order elliptic operators, for instance, the following variational inequality:

$$\begin{cases} \displaystyle\int_\Omega \nabla u \nabla (v - u) dx - \int_\Omega g(x, u, \lambda)(v - u)dx + j(v) - j(u) \geq 0, \ \forall v \in V, \\ u \in V. \end{cases}$$

Here, j represents different convex, lower semicontinuous functionals on $H_0^1(\Omega)$ (Ω is a bounded domain in $\mathbb{R}^N$), j is defined by the obstacles, and g is a certain Carathéodory mapping from $\Omega \times \mathbb{R}^N$ to $\mathbb{R}$. In some examples, $j = I_K$, where K is a closed, convex subset of $H_0^1(\Omega)$. Applications to global bifurcation problems for variational inequalities containing general (nonsymmetric), second-order elliptic operators, with a lower dimensional obstacle on the boundary, or quasilinear elliptic operators are considered in Examples 6.5 and 6.3.

Some of the applications given are for various unilateral buckling problems for plates and beams. An example for the beam problem is the following variational inequality:

$$\begin{cases} \displaystyle\int_0^a u''(v-u)''dx - \lambda \int_0^a \frac{u'}{\sqrt{1+u'^2}}(v-u)'dx + j(v) - j(u) \geq 0, \\ \qquad \forall v \in V, \\ u \in V, \end{cases}$$

and one for the plate problem under von Kárman's model is the following

$$
\begin{cases}
a(u, v - u) + \displaystyle\int_\Omega \sum_{i,j} \sigma_{ij}(u)\, \partial_i u\, \partial_j(v - u) dx \\
\qquad - \lambda \displaystyle\int_\Omega \sum_{i,j} \sigma_{ij}^0\, \partial_i u\, \partial_j(v - u) dx + j(v) - j(u) \geq 0,\ \forall v \in V, \\
u \in V.
\end{cases}
$$

Here $V = H_0^2(0, a)$ $(a > 0)$ in the first example, $V = H_0^2(\Omega)$ or $H^2(\Omega) \cap H_0^1(\Omega)$ in the second example (Ω is a bounded domain in $\mathbb{R}^2$), and j is a convex, lower semicontinuous functional defined on V.

$$
\begin{aligned}
a(u, v) \;=\; & \int_\Omega [(\partial_{11} u\, \partial_{11} v + \partial_{22} u\, \partial_{22} v) + \nu(\partial_{11} u\, \partial_{22} v + \partial_{22} u\, \partial_{11} v) \\
& + 2(1 - \nu)\partial_{12} u\, \partial_{12} v], \quad (0 < \nu < 1/2)
\end{aligned}
$$

is the bilinear form in the theory of plates; and the coefficients $\sigma_{ij}(u), \sigma_{ij}^0$ depend on the data of the plate. The functional j is characterized by unilateral conditions. In the cases of Signorini problems of rigid obstacles (Examples 4.3, 4.4, and 4.6), j is of the form $j = I_K$ where $K = \{u : u \geq \psi\}$ or $K = \{u : \psi_1 \leq u \leq \psi_2\}$ reflects the shapes and positions of the obstacles. Other unilateral conditions, such as contact problems as in the interface model or with elastic obstacles, or problems with different kinds of unilateral conditions on the boundary are represented by various choices of the functional j (section 6.3, Examples 6.7 (a), (b), and (c)). The general theory is also applied to global bifurcation problems for nonlinear variational inequalities containing the p-Laplacian (Example 6.11).

In Chapters 5 and 7, we use the abstract theorems about asymptotic homogenization and calculation of degrees or indices to derive global results for bifurcation from infinity of various concrete variational inequalities. Example 5.1 is devoted to bifurcation from infinity of an integral equation with unilateral conditions. Examples 5.2, 5.3, and 7.2 are concerned with asymptotic bifurcation for variational inequalities containing second- or fourth-order elliptic operators with various forms of the convex functional j. Other examples are for bifurcation from infinity for quasilinear variational inequalities (Example 7.3) and a variational inequality containing the p-Laplacian (Example 7.4).

In addition to new consequences, some results of our analysis are related to, ameliorate, or generalize some theorems in the works of Do, Miersemann, Szulkin, Saccon, McLeod and Turner, Goeleven, Nguyen, Théra, and others. The problems from Section 6.3 are from [32] and [33], where Do proved the existence of bifurcation points at the first eigenvalues of some related linear operators. Corollary 6.7 is concerned with the global behavior of bifurcation branches for those problems and is, therefore, a global result corresponding to the results of Do. Theorem 6.9 is motivated by an index calculation due to Szulkin ([119]).

The result proved by Saccon in [104] is concerned with an obstacle problem for a second-order elliptic variational inequality, where the obstacles are assumed to have the same zero sets. It can be directly verified that, in this particular problem, the corresponding homogenized variational inequality is, in fact, a linear equation. The result in [104] is, therefore, a consequence of the statements in Corollaries 3.3, 6.6, and 6.7. In [45], the authors established global bifurcation results for variational inequalities of von Kármán's type on convex cones, at the first eigenvalues. Parts of their analysis (Theorems 4.2 and Lemma 4.3, [45]) can be proved alternatively using Theorem 6.8. Moreover, the theorems in Chapters 4 and 6 are somewhat more general in nature. Further, Theorem 6.8 and its corollaries are valid for bifurcation from higher eigenvalues and for variational inequalities defined on convex sets (not necessarily cones) or containing convex functionals. In Example 6.6, we prove a global bifurcation result for buckling problems of thin plates resting on foundations, a particular case of which is a global counterpart of the existence results for the same problem considered by McLeod and Turner and Ridell in [69] and [101]. Example 6.3 is for global bifurcation of a variational inequality containing the p-Laplacian. We use the degree calculations due to del Pino, Elgueta, and Manásevich ([30] and [31]), and the result obtained is related to their corresponding results for equations containing the p-Laplacian. Besides giving several new results, our analysis contributes to unifying and putting into a more general setting several results of Do, Saccon, Goeleven et al., and others.

1.4   An outline

The following is a brief outline of the contents of the individual chapters of this monograph. In Chapter 2, we present some basic definitions and tools that will be used later. Chapters 3, 4, and 5 are devoted to bifurcation theory for variational inequalities defined on convex sets in Hilbert spaces, i.e., variational inequalities of the form (1.10). Although a number of results (e.g., Theorems 3.2 and 4.4 in Chapters 3 and 4 and Theorems 5.2 and 5.7 in Chapter 5) in these chapters are considered later in a more general setting, their statements are simpler, and the arguments are more transparent in these particular cases. On the other hand, some results (e.g., Theorems 4.3 and 5.6 and Corollaries 4.8 and 5.10) are considered only on convex sets of Hilbert spaces, because the existence of inner products and the properties of projections on closed convex sets in those spaces are needed.

In Chapter 3, we consider the homogenization process for variational inequalities and the relationships between bifurcation points of the original variational inequalities and the eigenvalues of the associated homogenized variational inequalities (Theorem 3.2 and Corollary 3.3). Chapter 4 is devoted to the calculation of the degrees of the associated mappings, which, together with the results in Chapter 3, yield global bifurcation results for the variational inequality (1.10) (sections 4.1 and 4.2, Theorems 4.3 and 4.4). Applications to concrete problems are considered in Section 4.2.

Bifurcation from infinity for variational inequalities of the form (1.10) is studied in Chapter 5. General results parallel to those in Chapter 3 and section 4.2 are proved in sections 5.1 and 5.2 (Theorems 5.1, 5.6, 5.7, Corollaries 5.9, 5.10), together with many illustrating examples.

Chapters 6 and 7 are devoted to general global bifurcation theories (from trivial solutions or from infinity) for variational inequalities defined on reflexive Banach spaces, containing nonlinear operators and convex functionals which are not necessarily indicator functions on convex sets. Abstract results for bifurcation and asymptotic bifurcation of (1.8), including homogenization and degree calculations, are presented in sections 6.1, 6.2, 6.4, and 7.1 (Theorems 6.4, 6.8, 6.9, 7.2, 7.4, and 7.5). A number of results in previous chapters are generalized in these last two chapters. Several applications to global bifurcation and asymptotic bifurcation of unilateral problems, formulated as various kinds of variational inequalities, are given in sections 6.3, 6.5, and 7.2. These examples include linear, quasilinear, or nonlinear variational inequalities containing different kinds of convex functionals.

1.5   Notation and tools

We shall apply mostly standard notation throughout. Hence, we shall not discuss such here but rather defer it to the point where special notation will be introduced. The results from nonlinear analysis, functional analysis, and the theory of partial differential equations, which will be needed in this monograph, are too numerous to be reproduced here. Hence, we shall provide only those used most frequently (see Chapter 2) and refer to the literature for the others. The texts that we found useful for the work presented here are [1], [14], [57] for properties of Sobolev spaces; [14], [123] for special results from functional analysis; [7], [18], [36], [39], [51], [87], [121], [124] for the theory of variational inequalities; [19], [29], [53], [54], [89], [125] for results from nonlinear analysis; [43], [66] for partial differential equation results; and [6], [18], [20], [36], [47], [66], [103], [126] for applications.

2

Some Auxiliary Results

In this chapter, we present some definitions and theorems that will be used in the sequel. These results concern variational inequalities, degree and index theories, and bifurcation.

2.1 Results on variational inequalities

Before stating some existence and uniqueness results for variational inequalities, one needs some definitions. In what follows, it is usually assumed that V is a real reflexive Banach space with dual V^* and norm $\| \cdot \|$. Let $\langle \cdot, \cdot \rangle$ denote the duality pairing between V and V^*. We always use $\to$ to denote the strong convergence in V or V^* and $\rightharpoonup$ to denote the weak convergence in V or the weak-* convergence in V^*.

Let A be a mapping from V to V^*.

Definition ([13], [51], and [65]) (a) A is called monotone on V if

$$\langle Au - Av, u - v \rangle \geq 0, \ \forall u, v \in V.$$

A monotone mapping A is called strictly monotone if

$$\langle Au - Av, u - v \rangle = 0 \ \text{implies} \ u = v.$$

(b) We say that A is continuous on finite-dimensional subspaces if, for any finite dimensional subspace M of V, the mapping $A|_M$ is weakly continuous, i.e., for all $x \in V$, the mapping

$$u \mapsto \langle Au, x \rangle, \; u \in M,$$

is continuous on M.

(c) A is called bounded if it maps bounded sets of V into bounded sets of V^*.

Let j be a mapping from V to $\mathbb{R} \cup \{\infty\}$.

Definition ([14] and [65]) (a) $D(j) = \{u \in V : j(u) < \infty\}$ is called the effective domain of j. j is called proper if $D(j) \neq \emptyset$ (i.e., $j \not\equiv \infty$).

(b) j is lower semicontinuous (respectively, weakly lower semicontinuous) if $j^{-1}((\lambda, \infty])$ is open (respectively, weakly open) in V for all $\lambda \in \mathbb{R}$.

It is well known ([14]) that j is lower semicontinuous (respectively, weakly lower semicontinuous) if and only if,

$$j(x) \leq \liminf_{y \to x} j(y) \;\; (\text{respectively } j(x) \leq \liminf_{y \to x} j(y)),$$

for all $x \in V$. Further, the weak lower semicontinuity implies the lower semicontinuity of j. The converse is true for convex functionals ([14]):

If j is convex and lower semicontinuous on V, then, it is weakly lower semicontinuous. The latter is a consequence of the fact that convex sets are weakly closed if and only if they are closed, which, in turn, is an easy consequence of the Hahn-Banach theorem.

A particular choice of j is when j is the indicator function of a set $K \subset V$, i.e., j is given by (1.9). Hence, it follows that K is convex and closed if and only if $j = I_K$ is convex and lower semicontinuous.

We have the following existence and uniqueness theorem for variational inequalities:

Theorem 2.1 ([65]) *Let A be a bounded monotone mapping from V to V^* that is continuous on finite-dimensional subspaces. Let j be a convex, lower semicontinuous functional from V to $\mathbb{R} \cup \{\infty\}$, and let A and j satisfy the following coerciveness condition:*

There exists $u_0 \in D(j)$ such that

$$\lim_{\|u\| \to \infty} \frac{\langle Au, u - u_0 \rangle + j(u)}{\|u\|} = \infty. \tag{2.1}$$

Then, for all $f \in V^*$, there exists a solution of the variational inequality

$$\begin{cases} \langle Au - f, v - u \rangle + j(v) - j(u) \geq 0, \; \forall v \in V, \\ u \in V. \end{cases} \tag{2.2}$$

If A is strictly monotone, then, the solution of (2.2) is unique.

If j is given by (1.9), then, (2.1) becomes the following condition:
There exists $u_0 \in K$ such that

$$\lim_{\|u\| \to \infty, u \in K} \frac{\langle Au, u - u_0 \rangle}{\|u\|} = \infty. \tag{2.3}$$

Using Theorem 2.1 in this particular case, one obtains the existence and uniqueness results (Theorem 8.2, [65] and Theorem 1.4, [51]) for the variational inequality:

$$\begin{cases} \langle Au - f, v - u \rangle \geq 0, \ \forall v \in K \\ u \in K. \end{cases} \tag{2.4}$$

An important special case is where V is a Hilbert space and $\langle \cdot, \cdot \rangle$ denotes either the inner product in V or the pairing between V and V^* (it will be clear from the context for which purpose it is being used) and $A = I$ is the identity mapping of V or the isometric isomorphism between V and V^*. From the above results, one sees that, for each $f \in V$, (2.4) has a unique solution $u = P_K f$, which is the orthogonal projection of f on K. $P_K f$ is also characterized as the minimizer of the problem,

$$P_K f \in K : \|f - P_K f\| = \min_{v \in K} \|f - v\|,$$

(Theorem V.2, [14]). Moreover, $P_K : V \to K$ is a nonexpansive mapping, i.e., P_K satisfies a Lipschitz condition with Lipschitz constant 1 (Proposition V.3, [14] and Theorem 2.1, Chapter 4, [103]).

2.2 The Leray–Schauder degree and index

Because many of our bifurcation results depend on arguments using Leray–Schauder degree or index calculations, we provide here the basic properties of these important notions. We first discuss the degree and, then, provide properties of the index.

2.2.1 The Leray–Schauder degree

Some fundamental properties of the Leray–Schauder degree are contained in the following theorem.

Theorem 2.2 ([29] and [125]) *Let $I : V \to V$ be the identity mapping and let*

$$\mathcal{M} = \{(I - F, \Omega, y) : \Omega \subset V \ open, \ bounded, \ F : \overline{\Omega} \to V$$

$$completely \ continuous, \ and \ y \notin (I - F)(\partial\Omega)\}.$$

Then there exists a unique mapping (called the Leray–Schauder degree)

$$d : \mathcal{M} \to \mathbb{Z},$$

$$(I - F, \Omega, y) \mapsto d(I - F, \Omega, y),$$

such that:

(1) (Normalization) $d(I, \Omega, y) = 1$, $\forall y \in \Omega$.

(2) (Additivity) $d(I - F, \Omega, y) = d(I - F, \Omega_1, y) + d(I - F, \Omega_2, y)$, *whenever* Ω_1, $\Omega_2 \subset \Omega$ *are open,* $\Omega_1 \cap \Omega_2 = \emptyset$, *and* $y \notin (I - F)(\overline{\Omega} \setminus (\Omega_1 \cup \Omega_2))$.

(3) (Invariance under homotopy) $d(I - H(t, \cdot), \Omega, y)$ *is independent of* $t \in [0, 1]$, *whenever* $H : [0, 1] \times \overline{\Omega} \to V$ *is completely continuous and* $y \notin (I - H(t, \cdot))(\partial\Omega)$, $\forall t \in [0, 1]$.

(4) (Solution property) $d(I - F, \Omega, y) \neq 0$ *implies* $y \in (I - F)(\Omega)$.

(5) (Excision) $d(I - F, \Omega, y) = d(I - F, \Omega_1, y)$ *if* $\Omega_1 \subset \Omega$ *is open and* $y \notin (I - F)(\overline{\Omega} \setminus \Omega_1)$.

(6) (Continuity) If $(I - F, \Omega, y), (I - G, \Omega, y) \in \mathcal{M}$, *and*

$$\sup\{|F(x) - G(x)| : x \in \partial\Omega\} < \mathrm{dist}\,(y, (I - F)(\partial\Omega)),$$

then, $d(I - F, \Omega, y) = d(I - G, \Omega, y)$.

2.2.2   The fixed-point index

The concept of fixed-point index of mappings defined on retracts is built upon the Leray–Schauder degree (cf. [3], [85], and [126]). Recall that a nonempty subset A of a metric space E is called a retract of E if there exists a continuous mapping $r : A \to E$, a retraction, such that $r(x) = x$, $\forall x \in A$.

It follows from Dugundji's extension theorem ([35]) that every nonempty closed convex subset of a Banach space V is a retract of V. The following result presents some basic properties of the Leray–Schauder index.

Theorem 2.3   ([3], [85], and [126]) *Let*

$$\mathcal{J} = \{(f, U, K) : K \text{ is a retract of } V, \ U \subset K \text{ is open in } K,$$
$$f : \overline{U}^K \to K \text{ is completely continuous, and } x \neq f(x),$$
$$\forall x \in \partial_K U\}$$

($\overline{U}^K$ and $\partial_K U$ denote the closure and boundary of U with respect to the relative topology of K).

There exists a unique mapping (called the Leray–Schauder index or the fixed-point index)

$$\mathrm{ind} : \mathcal{J} \to \mathbb{Z},$$

$$(f, U, K) \mapsto \mathrm{ind}(f, U, K), \ (f, U, K) \in \mathcal{J},$$

such that

(1) (Normalization) If f is a constant mapping from $\overline{U}^K$ to U, then,

$$\text{ind}(f, U, K) = 1.$$

(2) (Additivity) If $(f, U, K) \in \mathcal{J}$, U_1, U_2 are open subsets of U such that $U_1 \cap U_2 = \emptyset$, and f has no fixed points on $\overline{U}^K \setminus (U_1 \cup U_2)$, then, $(f|_{\overline{U_i}^K}, U_i, K) \in \mathcal{J}$, $i = 1, 2$, and

$$\text{ind}(f, U, K) = \text{ind}(f|_{\overline{U_1}^K}, U_1, K) + \text{ind}(f|_{\overline{U_2}^K}, U_2, K).$$

(3) (Invariance under homotopy) If $h : [0, 1] \times \overline{U}^K \to K$ is completely continuous and such that $x \neq h(t, x)$, $\forall (t, x) \in [0, 1] \times \partial_K U$, then, $\text{ind}(h(t, \cdot), U, K)$ is independent of $t \in [0, 1]$.

(4) (Permanence) If $(f, U, K) \in \mathcal{J}$ and $M \subset K$ is a retract of K such that $f(\overline{U}^K) \subset M$, then, $(f|_{\overline{U \cap M}^M}, U \cap M, M) \in \mathcal{J}$ and

$$\text{ind}(f, U, K) = \text{ind}(f|_{\overline{U \cap M}^M}, U \cap M, M).$$

(5) (Solution property) If $(f, U, K) \in \mathcal{J}$, and $\text{ind}(f, U, K) \neq 0$, then, f has at least one fixed point in U.

(6) (Excision) If $V \subset U$ is open such that f has no fixed point in $\overline{U}^K \setminus V$, then $\text{ind}(f, U, K) = \text{ind}(f|_{\overline{V}^K}, V, K)$.

(7) (Relationship with the Leray–Schauder degree) Let $r : V \to K$ be any retraction. Then, $(I - f \circ r, r^{-1}(U), 0) \in \mathcal{M}$, and

$$\text{ind}(f, U, K) = d(I - f \circ r, r^{-1}(U), 0),$$

and $\text{ind}(f, U, K)$ is independent of the particular choice of the retraction r.

2.3   Global bifurcation results

In this section, we shall provide statements of the global bifurcation theorems which will be used in our study.

The classical Krasnosel'skii-Rabinowitz bifurcation theorem is the following:

Theorem 2.4 ([98]) *Let $L : V \to V$ be a compact linear operator, and $F : V \times \mathbb{R} \to V$ be completely continuous, such that $F(u, \lambda) = o(\|u\|)$ as $u \to 0$, uniformly for λ in bounded intervals. Let S be the closure of the nontrivial solution pairs $\{(u, \lambda)\}$, $u \neq 0$ of the equation:*

$$u = \lambda L u + F(u, \lambda) \tag{2.5}$$

in $V \times \mathbb{R}$. If μ is an eigenvalue of odd algebraic multiplicity of L, i.e., the linear equation

$$v = \mu L v, \tag{2.6}$$

has a generalized solution space (eigenspace) of odd dimension, then, S contains a connected component C that contains $(0, \mu)$ and has at least one of the following properties:

(i) C is unbounded, or

(ii) C contains $(0, \nu)$, where ν is another eigenvalue of (2.6).

For mappings, which are not necessarily smooth (as will be the case in our considerations), the following global bifurcation result is valid.

Theorem 2.5   ([100]) *Let $F : V \times \mathbb{R} \to V$ be completely continuous such that $F(0, \lambda) = 0$, $\forall \lambda \in \mathbb{R}$. Let $a, b \in \mathbb{R}$ $(a < b)$ be such that $u = 0$ is an isolated solution of the equation,*

$$u - F(u, \lambda) = 0, \ u \in V, \tag{2.7}$$

for $\lambda = a$ and $\lambda = b$, where $(0, a)$, $(0, b)$ are not bifurcation points of (2.7). Furthermore, assume that

$$\mathrm{d}(I - F(\cdot, a), B_r(0), 0) \neq \mathrm{d}(I - F(\cdot, b), B_r(0), 0),$$

where $B_r(0)$ is an isolating neighborhood of the trivial solution. Let

$$S = \overline{\{(u, \lambda) : (u, \lambda) \ \text{is a solution of (2.7)} \ \text{with} \ u \neq 0\}} \cup (\{0\} \times [a, b]),$$

and let C be the connected component of S containing $\{0\} \times [a, b]$. Then, either

(i) C is unbounded in $V \times \mathbb{R}$, or

(ii) $C \cap [\{0\} \times (\mathbb{R} \setminus [a, b])] \neq \emptyset$.

The following theorem will be used in our analysis of bifurcation from infinity for variational inequalities in Chapters 5 and 7. Here, we employ the following terminology. A set $C = \{(u, \lambda)\} \subset V \times \mathbb{R}$ is said to bifurcate from infinity in the interval $[a, b]$, whenever there exists $\{(u_n, \lambda_n)\} \subset C$ such that $\{\lambda_n\} \subset [a, b]$ and $\|u_n\| \to \infty$, with a similar definition if the set bifurcates from infinity at a point $\bar{\lambda}$.

Theorem 2.6   ([91] and [106]) *Let $F : V \times \mathbb{R} \to V$ be completely continuous, and let $a, b \in \mathbb{R}$ $(a < b)$ be such that the solutions of (2.7) are, a priori, bounded in V for $\lambda = a$ and $\lambda = b$, i.e., there exists an $R > 0$ such that*

$$F(u, a) \neq u \neq F(u, b)$$

for all u with $\|u\| \geq R$. Furthermore, assume that

$$\mathrm{d}(I - F(\cdot, a), B_R(0), 0) \neq \mathrm{d}(I - F(\cdot, b), B_R(0), 0),$$

for $R > 0$ large. Then, there exists a continuum (i.e., a closed connected set) C of solutions of (2.7) that is unbounded in $V \times [a, b]$, and either

(i) C is unbounded in the λ direction, or else

(ii) there exists an interval $[c, d]$ such that $(a, b) \cap (c, d) = \emptyset$ and C bifurcates from infinity in $V \times [c, d]$.

3
Bifurcation in Hilbert Spaces

3.1 Statement of the problem

Let V be a (real) Hilbert space, with inner product $\langle \cdot, \cdot \rangle$ and norm $\| \cdot \|$. Let K be a closed, convex subset of V such that $0 \in K$. Let B be a completely continuous mapping from $V \times \mathbb{R}$ to V.

We consider the following variational inequality

$$\begin{cases} \langle u - B(u, \lambda), v - u \rangle \geq 0, \ \forall v \in K, \\ u \in K, \end{cases} \tag{3.1}$$

under the assumption that $B(0, \lambda) = 0$, $\forall v \in \mathbb{R}$, i.e., (3.1) has the trivial solution $u = 0$ for all $\lambda \in \mathbb{R}$. However, it may be the case that values of λ exist for which (3.1) has solutions other than 0 (we call these nontrivial solutions). The problem to be studied is the following: the existence and local and global behavior of nontrivial solution sets of (3.1), as the parameter λ is varied.

For $\lambda_0 \in \mathbb{R}$, we call $(0, \lambda_0)$ a bifurcation point of (3.1) if there exists a sequence $\{(u_n, \lambda_n)\}$ of solutions of (3.1) such that $\|u_n\| \neq 0$, $\forall n$, and $u_n \to 0$, $\lambda_n \to \lambda_0$ as $n \to \infty$. This chapter is devoted to obtaining and describing necessary conditions for the existence of bifurcation points, and we present, here, a procedure for (3.1) (similar to linearization of smooth mappings) to obtain a simpler, homogeneous variational inequality. Furthermore, we show that the topological degree of the solution mapping of the original variational inequality will be equal to that of the simpler variational inequality. Thus, global bifurcation of the original variational inequality occurs provided there is a change of degree in the homogenized variational inequality, as will follow from Theorem 2.5, Chapter 2.

Let P_K be the projection of V onto K. From properties of projections on closed, convex sets (Chapter 2), we know that $u = P_K f$ if and only if

$$u \in K \text{ and } \| f - u \| = \min\{ \| f - v \| : v \in K \}$$

and that this minimization problem is equivalent to the variational inequality,

$$u \in K \text{ and } \langle u - f, v - u \rangle \geq 0, \ \forall v \in K. \tag{3.2}$$

Hence, it follows from (3.2) that (3.1) is equivalent to the fixed point equation,

$$u = P_K[B(u, \lambda)]. \tag{3.3}$$

Because P_K satisfies a Lipschitz condition (with Lipschitz constant 1) on V (cf. Theorem 2.1, [103] or Chapter 2), it follows that $F(u, \lambda) = P_K[B(u, \lambda)]$ is completely continuous on $V \times \mathbb{R}$. Moreover, $F(0, \lambda) = 0, \ \forall \lambda \in \mathbb{R}$. For $r > 0$,

$$u = P_K[B(u, \lambda)] \text{ and } u \in \overline{B_r(0)}$$

if and only if

$$\langle u - B(u, \lambda), v - u \rangle \geq 0, \ \forall v \in K, \text{ and } u \in K \cap \overline{B_r(0)}.$$

Hence, if 0 is an isolated solution of (3.3) in $\overline{B_r(0)} \cap K$, then, the degree

$$d(I - P_K[B(\cdot, \lambda)], B_r(0), 0)$$

is well defined (I is the identity mapping on V: $I(u) = u, \ \forall u \in V$). Applying Theorem 2.5, Chapter 2 and recalling the equivalence between (3.1) and (3.3), we have the following result:

Theorem 3.1 *Let $a, b \in \mathbb{R}$ $(a < b)$ be such that $u = 0$ is an isolated solution of (3.1) for $\lambda = a$ and $\lambda = b$, where $(0,a)$, $(0,b)$ are not bifurcation points of (3.1). Assume, furthermore, that, for some $r > 0$, small,*

$$d(I - P_K[B(\cdot, a)], B_r(0), 0) \neq d(I - P_K[B(\cdot, b)], B_r(0), 0). \tag{3.4}$$

Let

$$S = \overline{\{(u, \lambda) : (u, \lambda) \text{ is a solution of (3.1) with } u \neq 0\}} \cup (\{0\} \times [a, b]),$$

and let C be the connected component of S containing $\{0\} \times [a, b]$. Then, either

(i) C is unbounded in $V \times \mathbb{R}$, or

(ii) $C \cap (\{0\} \times (\mathbb{R} \setminus [a, b])) \neq \emptyset$.

3.2   Homogenization procedures

In the following theorem, we consider the relationship between the bifurcation points of (3.1) and the eigenvalues of a simpler homogeneous variational inequality. We need the following definitions and assumptions.

We suppose that $B(u, \lambda)$ is differentiable with respect to u at $u = 0$ in the sense that there exists a completely continuous mapping

$$f : V \times \mathbb{R} \to V$$

such that, for all sequences $\{v_n\}, \{\sigma_n\}, \{\lambda_n\}$, satisfying

$$v_n \rightharpoonup v, \ \lambda_n \to \lambda, \ \sigma_n \to 0 \text{ as } n \to \infty, \text{ and } \sigma_n > 0, \ \forall n,$$

$$\frac{1}{\sigma_n} B(\sigma_n v_n, \lambda_n) \to f(v, \lambda) \text{ in } V. \tag{3.5}$$

Then, it follows that $f(\cdot, \lambda)$ is positive homogeneous of degree 1 (i.e., $f(\sigma u, \lambda) = \sigma f(u, \lambda), \ \forall \sigma \geq 0$). Indeed, because $B(0, \lambda) = 0$, we have $f(0, \lambda) = 0, \ \forall \lambda$. For $\sigma > 0$, we have $\sigma / n \to 0$ as $n \to \infty$, and, hence,

$$f(\sigma u, \lambda) = \lim_{n \to \infty} \frac{f(\frac{\sigma}{n} u, u)}{1/n} = \lim_{n \to \infty} \sigma \frac{f(\frac{\sigma}{n} u, u)}{\sigma / n} = \sigma f(u, \lambda),$$

proving the homogeneity of $f(\cdot, \lambda)$.

We denote by K_0 the support cone (tangent cone) of K at 0 (see Figure 3.1):

$$K_0 = \overline{\{tv : t \geq 0, v \in K\}}.$$

It follows from the definition that K_0 is a closed convex cone in V, and moreover, it is the smallest, closed, convex cone in V that contains K.

Next, we consider the following variational inequality, which is a homogenization of (3.1) at 0:

$$\begin{cases} \langle u - f(u, \lambda), v - u \rangle \geq 0, \ \forall v \in K_0, \\ u \in K_0. \end{cases} \tag{3.6}$$

As a consequence of the homogeneity of $f(\cdot, \lambda)$ and the fact that K_0 is a cone, we see that (3.6) is a homogeneous variational inequality, i.e., if u is a solution of (3.6), then, so is tu for all $t \geq 0$.

Similar to the definition of eigenvalues of a linear operator, we call $\lambda \in \mathbb{R}$ an eigenvalue of (3.6) if (3.6) has a solution (u, λ) with $u \in K_0 \setminus \{0\}$, and u is called an eigenvector of (3.6) corresponding to λ. It follows from the above observation that, for all $t > 0$, tu is also an eigenvector corresponding to λ. We have the following relationship between bifurcation points of (3.1) and eigenvalues of (3.6) (the alternatives are illustrated in Figures 3.2 and 3.3):

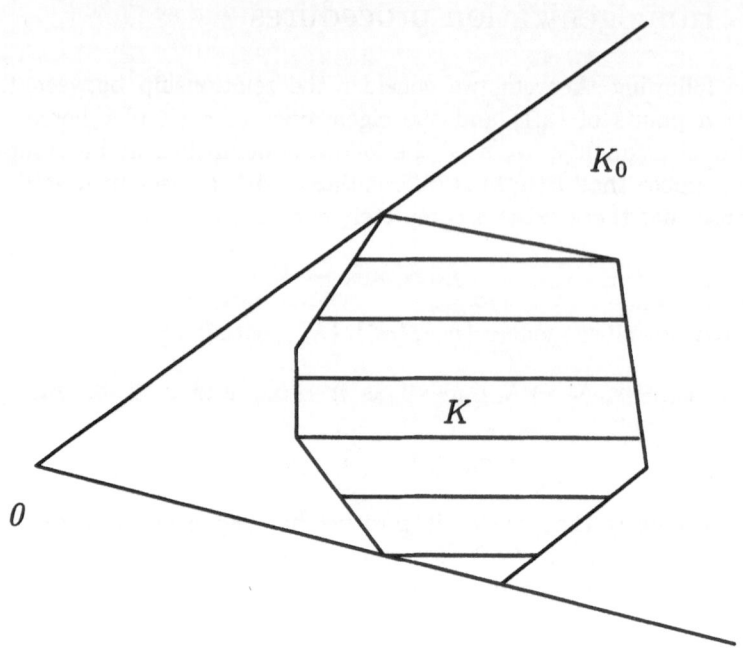

FIGURE 3.1. The support cone K_0 of a convex set K.

Theorem 3.2 *(I) If $(0, \lambda_0)$ is a bifurcation point of (3.1), then, λ_0 is an eigenvalue of (3.6).*
    (II) If a and b $(a < b)$ are not eigenvalues of (3.6) and if

$$d(I - P_{K_0}[f(\cdot, a)], B_r(0), 0) \neq d(I - P_{K_0}[f(\cdot, b)], B_r(0), 0), \qquad (3.7)$$

for some $r > 0$, then, for S, C as in Theorem 3.1, either
    (i) C is unbounded in $V \times \mathbb{R}$, or
    (ii) $(0, \lambda_1) \in C$ for some eigenvalue λ_1 of (3.6), $\lambda_1 \notin [a, b]$.

Proof. We first note that, if a and b are not eigenvalues of (3.6), then, 0 is the unique zero of $I - P_{K_0}[f(\cdot, a)]$ and $I - [P_{K_0}f(\cdot, b)]$ and the degrees in (3.7) are well defined for all $r > 0$ (and do not depend on r). For $\sigma \in [0, 1]$, we define

$$K_\sigma = \begin{cases} \dfrac{1}{\sigma}K = \left\{\dfrac{1}{\sigma}v : v \in K\right\}, & \text{if } \sigma \in (0, 1], \\ \\ K_0, & \text{if } \sigma = 0, \end{cases}$$

and

$$B_\sigma(u, \lambda) = \begin{cases} \dfrac{1}{\sigma}B(\sigma u, \lambda), & \text{if } \sigma \in (0, 1], \\ \\ f(u, \lambda), & \text{if } \sigma = 0. \end{cases}$$

We observe that, for every $\sigma \in [0,1]$, K_σ is a closed, convex subset of V. Next, we show that, if $\{\sigma_n\}$ is a sequence in $[0,1]$ such that $\sigma_n \to \sigma_0$, then,

$$K_{\sigma_n} \to K_{\sigma_0}, \quad \text{as} \quad n \to \infty \tag{3.8}$$

in the Mosco sense (cf. Chapter 3, [103]), i.e.,

- for each $v \in K_{\sigma_0}$, there exists $v_n \in K_{\sigma_n}$, such that $v_n \to v$,

and

- for each subsequence $\{\sigma_{n_k}\} \subset \{\sigma_n\}$, if a sequence $\{v_{n_k}\} \subset V$ satisfies $v_{n_k} \in K_{\sigma_{n_k}}$, $\forall k$, and $v_{n_k} \rightharpoonup v$ in V, then, $v \in K_{\sigma_0}$.

We first prove (3.8) for the case $\sigma_0 > 0$. Let $v \in K_{\sigma_0}$, choosing $v_n = \sigma_0 \sigma_n^{-1} v$, $\forall n \in \mathbb{N}$, then, $v_n \in K_{\sigma_n}$, $\forall n$ and $v_n \to v$ in V as $n \to \infty$. Now, suppose that $v_{n_k} \in K_{\sigma_{n_k}}$, $\forall k$ and $v_{n_k} \rightharpoonup v$ in V. It follows that

$$\sigma_{n_k} v_{n_k} \rightharpoonup \sigma_0 v \quad \text{in} \quad V.$$

We have $\sigma_{n_k} v_{n_k} \in K$, $\forall k$ and, because K is closed and convex, it is weakly closed (Chapter 2). Therefore, $\sigma_0 v \in K$, i.e., $v \in K_{\sigma_0}$, proving (3.8). Now, let $\sigma_0 = 0$, and let $v \in K_0$. By definition, there exist $z_n = t_n w_n$, $t_n > 0$, $w_n \in K$, $n \in \mathbb{N}$ such that $z_n \to v$, as $n \to \infty$. Because $\lim \sigma_n^{-1} = \infty$, we can choose a subsequence $\{\sigma_{n_k}\} \subset \{\sigma_n\}$ with $\sigma_{n_k} \to 0$ such that $\sigma_{n_k}^{-1} \geq t_k$, $\forall k$. For n such that $\sigma_{n_{k+1}} < \sigma_n \leq \sigma_{n_k}$, we choose $v_n = z_{n_k} \in K_{\sigma_{n_k}} \subset K_{\sigma_n}$. Thus, $v_n \in K_{\sigma_n}$, $\forall n \in \mathbb{N}$, and $v_n \to v$ as $n \to \infty$. Therefore, the first condition in (3.8) holds. The second condition is clearly satisfied in this case because $K_\sigma \subset K_0$, $\forall \sigma \in (0,1]$. Therefore, (3.8) holds.

Next, we prove that the mapping

$$(\sigma, v, \lambda) \mapsto P_{K_\sigma}[B_\sigma(v, \lambda)], \quad \sigma \in [0,1], \ v \in V, \ \lambda \in \mathbb{R}, \tag{3.9}$$

is completely continuous in $[0,1] \times V \times \mathbb{R}$.

Indeed, let $\{v_n\} \subset V$, $\{\sigma_n\} \subset [0,1]$, $\{\lambda_n\} \subset \mathbb{R}$ be sequences such that $v_n \rightharpoonup v$ in V and $\sigma_n \to \sigma$, $\lambda_n \to \lambda$ in $\mathbb{R}$. Then, it follows that

$$B_{\sigma_n}(v_n, \lambda_n) \to B_\sigma(v, \lambda) \quad \text{in} \quad V.$$

If $\sigma > 0$, this is a consequence of the complete continuity of B. If $\sigma = 0$, this follows from (3.5). An easy calculation shows that

$$\big\| P_{K_{\sigma_n}}[B_{\sigma_n}(v_n, \lambda_n)] - P_{K_\sigma}[B_\sigma(v, \lambda)] \big\|$$
$$\leq \big\| P_{K_{\sigma_n}}[B_{\sigma_n}(v_n, \lambda_n)] - P_{K_{\sigma_n}}[B_\sigma(v, \lambda)] \big\|$$
$$+ \big\| P_{K_{\sigma_n}}[B_\sigma(v, \lambda)] - P_{K_\sigma}[B_\sigma(v, \lambda)] \big\|$$
$$\leq \big\| B_{\sigma_n}(v_n, \lambda_n) - B_\sigma(v, \lambda) \big\| + \big\| P_{K_{\sigma_n}}[B_\sigma(v, \lambda)] - P_{K_\sigma}[B_\sigma(v, \lambda)] \big\|.$$

$$\tag{3.10}$$

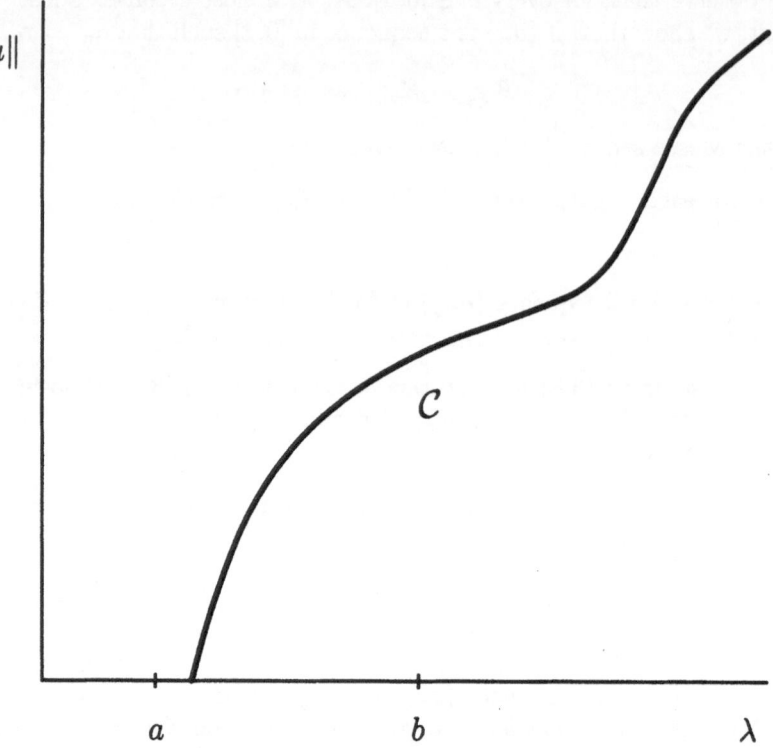

FIGURE 3.2. Unbounded bifurcation branch.

Because K_{σ_n} tends to K_σ in the Mosco sense, by a property of projections in Hilbert spaces (Theorem 4.3, [103]), $P_{K_{\sigma_n}} f \to P_{K_\sigma} f$ in V, as $n \to \infty$, for all $f \in V$. In particular,

$$\lim_{n \to \infty} P_{K_{\sigma_n}} [B_\sigma(v, \lambda)] = P_{K_\sigma} [B_\sigma(v, \lambda)] \text{ in } V.$$

Hence, all terms in the right-hand side of (3.10) tend to 0, as $n \to \infty$. Thus,

$$P_{K_{\sigma_n}} [B_{\sigma_n}(v_n, \lambda_n)] \to P_{K_\sigma} [B_\sigma(v, \lambda)] \text{ in } V,$$

proving the complete continuity of the mapping in (3.9).

Next, we prove (I). Suppose that there exist $u_n \in V$, $\lambda_n \in \mathbb{R}$, $n = 1, 2, \ldots$ such that $u_n \neq 0$, $\forall n$, $\|u_n\| \to 0$, $\lambda_n \to \lambda_0$, as $n \to \infty$ and for all $n \in \mathbb{N}$, (u_n, λ_n) satisfies the variational inequality:

$$\begin{cases} \langle u_n - B(u_n, \lambda_n), v - u_n \rangle \geq 0, \ \forall v \in K, \\ u_n \in K. \end{cases}$$

Dividing both sides of the above inequality by $\|u_n\|^2$,

$$\left\langle \frac{u_n}{\|u_n\|} - \frac{B(u_n, \lambda_n)}{\|u_n\|}, \frac{v}{\|u_n\|} - \frac{u_n}{\|u_n\|} \right\rangle \geq 0, \ \forall v \in K.$$

Setting $v_n = \|u_n\|^{-1}u_n$, $v_n \in K_{\|u_n\|}$, $\forall n$. Because

$$\frac{B(u_n, \lambda_n)}{\|u_n\|} = \frac{1}{\|u_n\|} B(\|u_n\|v_n, \lambda_n),$$

and $v \in K$, if and only if,

$$w = \frac{v}{\|u_n\|} \in \frac{1}{\|u_n\|} K = K_{\|u_n\|},$$

we see that this inequality is equivalent to

$$\begin{cases} \langle v_n - B_{\|u_n\|}(v_n, \lambda_n), w - v_n \rangle \geq 0, \ \forall w \in K_{\|u_n\|}, \\ v_n \in K_{\|u_n\|}. \end{cases}$$

Thus,

$$v_n = P_{K_{\|u_n\|}}\left[B_{\|u_n\|}(v_n, \lambda_n)\right], \ \forall n \in \mathbb{N}. \tag{3.11}$$

Because $\{v_n\}$ is bounded ($\|v_n\| = 1$, $\forall n$), by passing to a subsequence of $\{v_n\}$, if necessary, we can assume, without loss of generality, that $v_n \rightharpoonup v_0$ in V.

Moreover, because $\|u_n\| \to 0$, we have from the complete continuity of the mapping in (3.9),

$$\lim_{n \to \infty} P_{K_{\|u_n\|}}\left[B_{\|u_n\|}(v_n, \lambda_n)\right] = P_{K_0}\left[B_0(v_0, \lambda)\right] = P_{K_0}\left[f(v_0, \lambda)\right] \ \text{in } V. \tag{3.12}$$

Thus, by letting $n \to \infty$ in (3.11), we find from (3.12) that $v_n \to v_0$ in V and $v_0 = P_{K_0}[f(v_0, \lambda_0)]$.

This equation says that (v_0, λ_0) satisfies (3.6). Because $\|v_0\| = \lim \|v_n\| = 1$, v_0 is an eigenvector of (3.6) corresponding to λ_0. This completes the proof of (I).

To prove (II), we first remark that, if a is not an eigenvalue of (3.6), then, 0 is an isolated solution of (3.1), and for $r > 0$, sufficiently small,

$$d(I - P_K[B(\cdot, a)], B_r(0), 0) = d(I - P_{K_0}[f(\cdot, a)], B_r(0), 0). \tag{3.13}$$

Indeed, we claim that there exists $r > 0$, small, such that, for all $\sigma \in [0, 1]$, the equation

$$u - P_{K_\sigma}\left[B_\sigma(u, a)\right] = 0 \tag{3.14}$$

has no nontrivial solution in $\overline{B_r(0)}$.

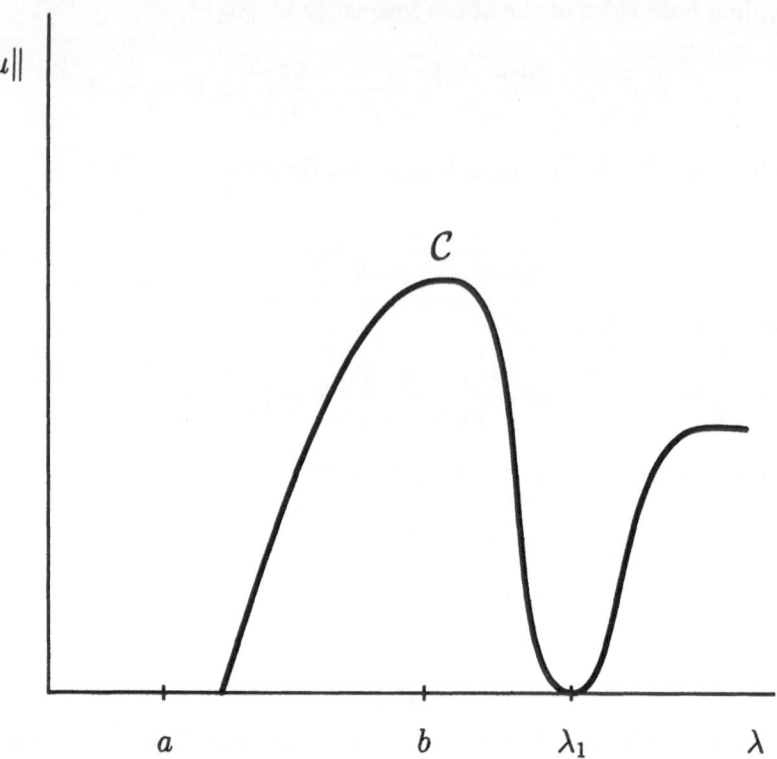

FIGURE 3.3. Bifurcation branch that meets another bifurcation point.

If this is not the case, there exist sequences $\{u_n\} \subset V$, $\{\sigma_n\} \subset [0,1]$ such that $\|u_n\| \neq 0$, $\forall n$, $u_n \to 0$ in V, as $n \to \infty$, and

$$u_n = P_{K_{\sigma_n}}[B_{\sigma_n}(u_n, a)], \ \forall n.$$

This equation can be written in the variational inequality form,

$$\langle u_n - B_{\sigma_n}(u_n, a), v - u_n \rangle \geq 0, \ \forall v \in K_{\sigma_n}.$$

Dividing the inequality by $\|u_n\|^2$ and, again, setting $v_n = \|u_n\|^{-1} u_n$, we get

$$\left\langle v_n - \frac{1}{\|u_n\|} B_{\sigma_n}(\|u_n\| v_n, a), \frac{v}{\|u_n\|} - v_n \right\rangle$$
$$= \left\langle v_n - \frac{1}{\sigma_n \|u_n\|} B(\sigma_n \|u_n\| v_n, a), \frac{v}{\|u_n\|} - v_n \right\rangle$$
$$\geq 0, \ \forall v \in K_{\sigma_n}.$$

Because $u_n \in K_{\sigma_n}$, $v_n \in K_{\sigma_n \|u_n\|}$. Moreover, $v \in K_{\sigma_n}$ if and only if $w = \frac{v}{\|u_n\|} \in K_{\sigma_n \|u_n\|}$. Hence, for all $n \in \mathbb{N}$, v_n satisfies the following variational inequality:

$$\begin{cases} \langle v_n - B_{\|u_n\|\sigma_n}(v_n, a), w - v_n \rangle \geq 0, \ \forall w \in K_{\|u_n\|\sigma_n}, \\ v_n \in K_{\|u_n\|\sigma_n}, \end{cases}$$

whose equivalent operator form is

$$v_n = P_{K_{\|u_n\|\sigma_n}} \left[B_{\|u_n\|\sigma_n}(v_n, a) \right], \ \forall n. \tag{3.15}$$

Again, by passing to a subsequence, if necessary, we may assume that $v_n \rightharpoonup v$. Because $\|u_n\| \to 0$, $\|u_n\|\sigma_n \to 0$, and, by the complete continuity of the mapping in (3.9),

$$P_{K_{\|u_n\|\sigma_n}} \left[B_{\|u_n\|\sigma_n}(v_n, a) \right] \to P_{K_0}[f(v, a)] \text{ in } V.$$

Letting $n \to \infty$ in (3.15), we obtain $v_n \to v$, $\|v\| = 1$ and

$$v = P_{K_0}[f(v, a)].$$

In other words, $v \neq 0$, and v satisfies (3.6) with $\lambda = a$. This contradicts the assumption that a is not an eigenvalue of (3.6).

We have proved that 0 is the unique solution of (3.14) in $\overline{B_r(0)}$ for all r sufficiently small and, in particular, the degree

$$d(I - P_{K_\sigma}[B_\sigma(\cdot, a)], B_r(0), 0)$$

exists for all $\sigma \in [0, 1]$. Consider the following family of mappings from V to V:

$$\{ I - P_{K_\sigma}[B_\sigma(\cdot, a)] : 0 \leq \sigma \leq 1 \}.$$

The complete continuity of the mapping in (3.9) implies that this is a family of completely continuous perturbations of the identity on $\overline{B_r(0)}$. Moreover,

$$u - P_{K_\sigma}[B_\sigma(u, a)] \neq 0, \ \forall u \in \partial B_r(0), \ \forall \sigma \in [0, 1],$$

by the above proof. By the homotopy invariance property of the Leray–Schauder degree (Chapter 2), it follows that

$$\begin{aligned} d(I - P_{K_0}[f(\cdot, a)], B_r(0), 0) &= d(I - P_{K_0}[B_0(\cdot, a)], B_r(0), 0) \\ &= d(I - P_{K_1}[B_1(\cdot, a)], B_r(0), 0) \\ &= d(I - P_K[B(\cdot, a)], B_r(0), 0), \end{aligned}$$

proving (3.13). A similar equality holds for $\lambda = b$. This proves that (3.7) has (3.4) as a consequence. (II) follows from Theorem 1 and (I). ∎

In the next corollary, we consider the case where the support cone K_0 of K is a linear space, i.e.,

$$K_0 = \overline{K_0 - K_0}.$$

By restricting our consideration to $\overline{K_0 - K_0}$ (if necessary), we can assume, further, that $K_0 = V$.

If $B(u, \lambda)$ is a perturbation of linear operators near $u = 0$ (cf. (3.18)), then, $f(u, \lambda)$ is linear with respect to u. In this case, the homogenized variational inequality (3.6) becomes a linear equation containing compact operators.

Thus, we can apply the classical Leray–Schauder calculation of degrees of compact perturbations of the identity, together with Theorem 3.2, to obtain the following result:

Corollary 3.3 *Suppose that $K_0 = V$. Then, the following hold:*
(a) If $(0, \lambda)$ is a bifurcation point of (3.1), then, λ is an eigenvalue of the equation

$$u = f(u, \lambda). \tag{3.16}$$

Conversely, if a and b are not eigenvalues of (3.16) and if

$$d(I - f(\cdot, a), B_r(0), 0) \neq d(I - f(\cdot, b), B_r(0), 0), \tag{3.17}$$

then, we have the conclusion of Theorem 3.2.
(b) Suppose that

$$B(u, \lambda) = \lambda \beta u + N(u, \lambda), \ u \in V, \ \lambda \in \mathbb{R}, \tag{3.18}$$

where $\beta \in L(V, V)$ is a compact linear mapping on V, and N is completely continuous from $V \times \mathbb{R}$ to $\mathbb{R}$ such that

$$\lim_{u \to 0} \frac{1}{\|u\|} N(u, \lambda) = 0 \tag{3.19}$$

uniformly for λ in bounded intervals of $\mathbb{R}$.
   Then, (3.1) has, at most, a countable number of bifurcation points, and, moreover,
   (i) if $(0, \lambda)$ is a bifurcation point of (3.1), then, λ is an eigenvalue of β, and,
   (ii) if λ is an eigenvalue of β of odd (algebraic) multiplicity, then, $(0, \lambda)$ is a bifurcation point of (3.1), and the corresponding global bifurcation branch satisfies the alternative in Theorem 3.2 (II).

Proof. In this case, $P_{K_0} = I$ is the identity mapping in V. The variational inequality (3.6), thus, becomes

$$\begin{cases} \langle u - f(u, \lambda), v - u \rangle \geq 0, \ \forall v \in V, \\ u \in V. \end{cases}$$

For $w \in V$, letting $v = u \pm w$ in this inequality,

$$\langle u - f(u, \lambda), w \rangle = 0, \ \forall w \in V.$$

This is equivalent to (3.16), and (a) is, therefore, a direct consequence of Theorem 3.2.

In case (b), we can show directly that f, given by

$$f(u, \lambda) = \lambda\beta u, \ u \in V, \ \lambda \in \mathbb{R}, \tag{3.20}$$

is the derivative of B with respect to u at 0 in the sense of (3.5). This shows that (3.16), in this case, is the linear equation,

$$(I - \lambda\beta)u = 0, \ u \in V. \tag{3.21}$$

By the classical spectral theory for compact linear operators (Chapter VI, [14]), we know that (3.21) has, at most, a countable number of eigenvalues with the only possible accumulation point at ∞, and, for $\mu \in \mathbb{R}$ not in the spectrum $\sigma(\beta)$ of β,

$$d(I - \mu\beta, B_r(0), 0) = (-1)^{n(\mu)}, \ \forall r > 0, \tag{3.22}$$

where (see [53])

$$n(\mu) = \sum \{m(\lambda) : \lambda \in \sigma(\beta) \cap \mathbb{R}, |\lambda| < |\mu| \text{ and } \lambda \text{ is of the same sign as } \mu\},$$

and $m(\lambda)$ denotes the (algebraic) multiplicity of $\lambda \in \sigma(\beta)$. (i) follows from (a) and the above property.

Now, let $\lambda \in \mathbb{R}$ be an eigenvalue of odd (algebraic) multiplicity of β. For $\epsilon > 0$ sufficiently small, $[\lambda - \epsilon, \lambda + \epsilon] \cap \sigma(\beta) = \{\lambda\}$, and, then, by (3.22),

$$d(I - (\lambda \pm \epsilon)\beta, B_r(0), 0) \in \{-1, 1\},$$

and

$$d(I - (\lambda - \epsilon)\beta, B_r(0), 0) = -d(I - (\lambda + \epsilon)\beta, B_r(0), 0).$$

Applying (3.17) with $a = \lambda - \epsilon < \lambda + \epsilon = b$, we have the result stated in (ii).   ∎

Next, we consider some simple examples to illustrate the results above.

The first example shows, in some sense, that bifurcation for variational inequalities is, in general, different from bifurcation for smooth equations. Even if we place the same obstacle on two equivalent nonlinear equations, the global bifurcation branches of the variational inequalities, thus obtained, may be completely different.

Example 3.1 We consider a one-dimensional variational inequality of the form (3.1) with $V = \mathbb{R}$, $K = (-\infty, 1/2]$, and $B(u, \lambda) = B_1(u, \lambda) = u(u^2 + \lambda^2)$, i.e.

$$\begin{cases} [u - u(u^2 + \lambda^2)] \, (v - u) \geq 0, \ \forall v \leq 1/2, \\ u \leq 1/2. \end{cases} \tag{3.23}$$

From straightforward calculations, it follows that the set of solutions of (3.23) is

$$
\begin{aligned}
\mathcal{S}_1 = \{0\} \times \mathbb{R} \ &\cup \ \{(u,\lambda) : |\lambda| \leq 1 \ \text{and} \ u = -\sqrt{1-\lambda^2}\}, \\
&\cup \ \{(u,\lambda) : \sqrt{3}/2 \leq |\lambda| \leq 1 \ \text{and} \ u = \sqrt{1-\lambda^2}\}, \\
&\cup \ \{(u,\lambda) : |\lambda| \geq \sqrt{3}/2 \ \text{and} \ u = 1/2\}.
\end{aligned}
$$

Now, consider that $B(u,\lambda) = B_2(u,\lambda) = 2(1+u^2+\lambda^2)^{-1}u$. In this case, (3.1) becomes

$$
\begin{cases}
[u(u^2+\lambda^2) - u]\,(v-u) \geq 0, \ \forall v \leq 1/2, \\
u \leq 1/2,
\end{cases}
\tag{3.24}
$$

and the set of solutions of (3.24) is

$$
\begin{aligned}
\mathcal{S}_2 = \{0\} \times \mathbb{R} \ &\cup \ \{(u,\lambda) : |\lambda| \leq 1 \ \text{and} \ u = -\sqrt{1-\lambda^2}\}, \\
&\cup \ \{(u,\lambda) : \sqrt{3}/2 \leq |\lambda| \leq 1 \ \text{and} \ u = \sqrt{1-\lambda^2}\}, \\
&\cup \ \{(u,\lambda) : |\lambda| \leq \sqrt{3}/2 \ \text{and} \ u = 1/2\}.
\end{aligned}
$$

We can check that the conditions in Theorem 3.2 are satisfied at $(0,1)$ and $(0,-1)$ with $f(u,\lambda) = f_1(u,\lambda) = u\lambda^2$ in (3.23) and $f(u,\lambda) = f_2(u,\lambda) = 2u(1+\lambda^2)^{-1}$ in (3.24).

Note that, in the case without constraint, (3.23) and (3.24) yield the equations $u - B_1(u,\lambda) = 0$ and $u - B_2(u,\lambda) = 0$, which are equivalent because the sets of solutions for both equations are the union of the real line $\{u = 0\}$ and the unit circle in $\mathbb{R}^2$. However, in the cases with constraint $u \leq 1/2$, the branches of nontrivial solutions are different.

Remark 3.1 (a) Corollary 3.3 contains the major result in Saccon's paper [104]. There ([104]), the author considers the global bifurcation for the following bi-obstacle variational inequality:

$$
\begin{cases}
a(u, v-u) \geq \displaystyle\int_\Omega [\lambda u + p(x,u,\lambda)](v-u)dx, \ \forall v \in K, \\
u \in K,
\end{cases}
\tag{3.25}
$$

where $a : H_0^1(\Omega) \times H_0^1(\Omega) \to \mathbb{R}$,

$$
a(u,v) = \int_\Omega \sum_{i,j}[(a_{ij}\,\partial_i u + d_j\,u)\,\partial_j v + (b_i\,\partial_i u + cu)v]dx, \ u,v \in H_0^1(\Omega),
$$

is a second-order elliptic bilinear form, assumed to be coercive on the Sobolev space $H_0^1(\Omega)$, and

$$
K = \{u \in H_0^1(\Omega) : \phi_1 \leq u \leq \phi_2 \ \text{on} \ \Omega\}.
$$

Here, ϕ_1 and ϕ_2 are two given measurable functions on Ω such that $\phi_1 \leq \phi_2$ on Ω and

$$\phi_1 = \phi_2 = 0 \text{ on } E, \ \phi_1 < 0 < \phi_2 \text{ on } \Omega \setminus E,$$

for some closed set $E \subset \overline{\Omega}$. Moreover, p is a Carathéodory function satisfying appropriate growth conditions, and $|p(x,s,\lambda)| = o(s)$ as $s \rightarrow 0$ uniformly for $x \in \Omega$ and λ bounded. Saccon ([104]) established the global bifurcation of (3.25) at eigenvalues of odd multiplicity of the following linear equation:

$$\begin{cases} a(u,v) = \int_\Omega \lambda uv dx, \ \forall v \in H_0^1(\Omega \setminus E), \\ u \in H_0^1(\Omega \setminus E). \end{cases} \tag{3.26}$$

We consider on $V = H_0^1(\Omega)$ the norm and inner product generated by a (assumed to be symmetric), and let $\langle B(u,\lambda), v \rangle = \int_\Omega [\lambda u + p(\cdot, u, \lambda)]v$. Then, B satisfies (3.18) with $\langle \beta u, v \rangle = \int_\Omega uv$. Moreover, it can be directly verified that the support cone K_0 of K, in this case, is $K_0 = H_0^1(\Omega \setminus E)$, which is a linear space. Hence, applying the abstract results presented above to this particular problem, we see that (3.26) is the homogenized equation corresponding to (3.25). Corollary 3.3 (b) gives us the result in [104]. The case, where a is not necessarily symmetric, may also be treated by the methods of chapter 6.

(b) We note that, if 0 is an interior point of K, then, $K_0 = V$. In fact, in this case, K_0 is a cone that contains a neighborhood of 0. Thus, $K_0 = V$. However, the condition $0 \in \overset{\circ}{K}$ ($\overset{\circ}{K}$ is the interior part of K) is not necessary for $K_0 = V$, as shown in the next example.

Example 3.2 Suppose $\psi \in H_0^1(0,1)$, $\psi(x) > 0$, $\forall x \in (0,1)$, and that g is a continuous function from $[0,1] \times \mathbb{R}^2$ to $\mathbb{R}$. Moreover, assume that the partial derivative $D_2 g(x, 0, \lambda)$ is continuous with respect to $\lambda \in \mathbb{R}$, $x \in [0,1]$.

We consider the variational inequality

$$\begin{cases} \int_0^1 u'(v-u)' dx \geq \int_0^1 g(x, u(x), \lambda)(v-u)dx, \ \forall v \in K, \\ u \in K, \end{cases} \tag{3.27}$$

where

$$K = \{u \in H_0^1(0,1) : u(x) \leq \psi(x) \text{ on } [0,1]\}.$$

If $g(x, 0, \lambda) \equiv 0$, then, 0 is a trivial solution of (3.27) for all $\lambda \in \mathbb{R}$.

It is clear that K is a closed and convex subset of $V = H_0^1(0,1)$ and $0 \in K$. Now, we check that, in general, $0 \notin \overset{\circ}{K}$. Indeed, let $\psi \in C^1[0,1]$ be such that $\psi(0) = \psi(1) = 0 < \psi(x)$, $\forall x \in (0,1)$. We choose $\alpha > |\psi'(0)|$ and, for $\epsilon > 0$ sufficiently small, set

$$
v_\epsilon(x) = \begin{cases} \alpha x & \text{if } x \in [0, \epsilon], \\ \alpha\epsilon & \text{if } x \in [\epsilon, 1 - \epsilon], \\ \alpha(1 - x) & \text{if } x \in [1 - \epsilon, 1]. \end{cases}
$$

Because v_ϵ is continuous and piecewise smooth on $[0, 1]$, and $v_\epsilon(0) = v_\epsilon(1) = 0$, $v_\epsilon \in H_0^1(0, 1)$, and

$$
\|v_\epsilon\|_{H_0^1(\Omega)} = \left\{ \int_0^\epsilon (\alpha^2 x^2 + \alpha^2) dx + \int_\epsilon^{1-\epsilon} \alpha^2 \epsilon^2 dx \int_{1-\epsilon}^1 [\alpha^2(1-x)^2 + \alpha^2] dx \right\}^{\frac{1}{2}}
$$

$$
= \left[2\alpha^2 \left(\frac{\epsilon^3}{3} + \epsilon \right) + \alpha^2 \epsilon^2 (1 - 2\epsilon) \right]^{\frac{1}{2}}.
$$

Hence, $v_\epsilon \to 0$ in $H_0^1(0, 1)$. On the other hand, because $v_\epsilon(0) = \psi(0) = 0$ and $v_\epsilon'(0) = \alpha > \psi'(0)$, $v_\epsilon(x) > \psi(x)$ for all $x > 0$, sufficiently small.

Thus, $v_\epsilon \notin K$, $\forall \epsilon > 0$, sufficiently small. This shows that $0 \notin \overset{\circ}{K}$.

However, $K_0 = V$. To prove this, we let $v \in C_0^\infty(0, 1)$. Because $\psi \in C[0, 1]$ and $\psi > 0$ on $\text{supp}\, v$ (the support of v),

$$
\psi(x) \geq \inf_{\text{supp}\, v} \psi \equiv m > 0, \ \forall x \in \text{supp}\, v.
$$

We have $tv \leq \psi$ on $[0,1]$ for some $t > 0$, sufficiently small, i.e., $tv \in K$. This means that $C_0^\infty(0, 1) \subset \bigcup_{t>0} tK$. Hence,

$$
H_0^1(0, 1) = \overline{C_0^\infty(0, 1)}^{H_0^1(0,1)} \subset \overline{\bigcup_{t>0} tK}^{H_0^1(0,1)} = K_0.
$$

We have $K_0 = V$. Now, we consider on V the usual H_0^1–inner product induced by Poincaré's inequality: $\langle u, v \rangle = \int_0^1 u'v'$, $u, v \in V$. For $u, v \in V$, $\lambda \in \mathbb{R}$, we set

$$
\langle B(u, \lambda), v \rangle = \int_0^1 g(x, u(x), \lambda) v(x) \, dx.
$$

Then, $B(u, \lambda)$ is defined for all $(u, \lambda) \in V \times \mathbb{R}$, and (3.27) can be written as

$$
\begin{cases} \langle u - B(u, \lambda), v - u \rangle \geq 0, \ \forall v \in K, \\ u \in K. \end{cases}
$$

For $u, \overline{u}, v \in V$, $\lambda, \overline{\lambda} \in \mathbb{R}$,

$$
|\langle B(u, \lambda), v \rangle - \langle B(\overline{u}, \overline{\lambda}), v \rangle| \leq \int_0^1 |g(x, u(x), \lambda) - g(x, \overline{u}(x), \overline{\lambda})| \, |v(x)| \, dx.
$$

Hence,

$$\|B(u, \lambda) - B(\overline{u}, \overline{\lambda})\| \le C \left[\int_0^1 |g(x, u(x), \lambda) - g(x, \overline{u}(x), \overline{\lambda})|^2 \right]^{\frac{1}{2}}. \quad (3.28)$$

Now, let $u_n \rightharpoonup u$ in V, $\lambda_n \to \lambda$ in $\mathbb{R}$. By the compact embedding $H_0^1(0, 1) \hookrightarrow C[0, 1]$,

$$u_n \to u \text{ uniformly on } [0, 1]. \quad (3.29)$$

From (3.28), we see that

$$B(u_n, \lambda_n) \to B(u, \lambda) \text{ in } V,$$

proving the complete continuity of B on $V \times \mathbb{R}$. Now, let

$$\langle f(u, \lambda), v \rangle = \int_0^1 D_2 g(x, 0, \lambda) u(x) v(x) dx, \ u, v \in V, \lambda \in \mathbb{R}.$$

By a similar argument, we can prove that f is completely continuous on $V \times \mathbb{R}$. Moreover,

$$\frac{1}{\sigma_n} B(\sigma_n u_n, \lambda_n) \to f(u, \lambda) \text{ in } V, \text{ as } n \to \infty, \quad (3.30)$$

whenever $\sigma_n \to 0^+$, $\lambda_n \to \lambda$ in $\mathbb{R}$, and $u_n \rightharpoonup u$ in V. In fact, as above,

$$\left\| \frac{1}{\sigma_n} B(\sigma_n u_n, \lambda_n) - f(u, \lambda) \right\|$$

$$\le C \left[\int_0^1 \left| \frac{1}{\sigma_n} g(\cdot, \sigma_n u_n, \lambda_n) - D_2 g(\cdot, 0, \lambda) u \right|^2 \right]^{\frac{1}{2}}.$$

Because

$$g(x, \sigma_n u_n(x), \lambda_n) = \sigma_n D_2 g(x, \theta_n(x) \sigma_n u_n(x), \lambda_n) u_n(x), \ 0 \le \theta_n(x) \le 1,$$

by the mean value theorem, from (3.29),

$$\frac{1}{\sigma_n} g(x, \sigma_n u_n(x), \lambda_n) \to D_2 g(x, 0, \lambda) u(x)$$

uniformly on $[0, 1]$. Therefore, (3.30) results. By Theorem 3.2, to study the bifurcation of (3.27), we can consider the eigenvalue problem for the homogenized variational inequality,

$$\begin{cases} \int_0^1 u'v' = \int_0^1 D_2 g(x, 0, \lambda) u(x) v(x) \, dx, \ \forall v \in H_0^1(0, 1), \\ u \in H_0^1(0, 1), \end{cases}$$

or, equivalently,

$$
\begin{cases}
-u'' = D_2 g(\cdot, 0, \lambda)u \ \ \text{on} \ \ (0,1), \\
u(0) = u(1) = 0.
\end{cases}
$$

This is a boundary value problem for a second-order, linear, ordinary differential equation, which may be studied using classical results.

4

Degree Calculations – The Hilbert Space Case

In this chapter, we consider situations in addition to those in Chapter 3, in which we can calculate the degrees of the operators in the homogenized variational inequalities (3.6). We find conditions that guarantee a change of degree as λ varies. Together with Theorem 3.2, these conditions, then, will imply the existence of global, nontrivial solution continua for (3.1), which bifurcate from the trivial solution.

4.1 Applications of the fixed point index

We shall assume in the sequel that $B(u, 0) = 0$, $\forall u \in V$, from which follows that $f(u, 0) = 0$, $\forall u$. We will need the following lemma.

Lemma 4.1 *Let K_0, f be as in Chapter 3 and assume that a is not an eigenvalue of (3.6). Then, for all $r > 0$, the following equality holds:*

$$d(I - P_{K_0}[f(\cdot, a)], B_r(0), 0) = \text{ind}(P_{K_0}[f(\cdot, a)], B_r(0) \cap K_0, K_0).$$

Proof. Let a not be an eigenvalue of (3.6). We consider the following family of compact perturbations of the identity:

$$\{I - t P_{K_0}[f(\cdot, a) P_{K_0}] - (1 - t) P_{K_0}[f(\cdot, a)] : 0 \leq t \leq 1\}$$

on $\overline{B_r(0)}$. We claim that

$$u - t P_{K_0}[f(P_{K_0} u, a)] - (1 - t) P_{K_0}[f(u, a)] \neq 0, \ \forall u \in \partial B_r(0), \ t \in [0, 1].$$
$$(4.1)$$

In fact, suppose there exist $u \in V \setminus \{0\}$, $t \in [0,1]$ such that

$$u = tP_{K_0}[f(P_{K_0}u, a)] + (1-t)P_{K_0}[f(u, a)].$$

Because K_0 is a convex cone, the right-hand side of this equation is an element of K_0. Hence, $u \in K_0$, and, therefore, $u = P_{K_0}u$. The above equality shows that

$$u = tP_{K_0}[f(u, a)] + (1-t)P_{K_0}[f(u, a)] = P_{K_0}[f(u, a)],$$

and, therefore, (u, a) is a solution of (3.6). Because $u \neq 0$, a is an eigenvalue of (3.6). This contradiction proves (4.1).

Using the homotopy invariance property of the Leray–Schauder degree,

$$d(I - P_{K_0}[f(\cdot, a)], B_r(0), 0) = d(I - P_{K_0}[f(\cdot, a)P_{K_0}], B_r(0), 0), \; \forall r > 0. \tag{4.2}$$

Using the relationship between the Leray–Schauder degree and the fixed-point index (Theorem 2.3, Chapter 2),

$$\begin{aligned} \mathrm{ind}(P_{K_0}[f(\cdot, a)], B_r(0) \cap K_0, K_0) &= d(I - P_{K_0}[f(\cdot, a)P_{K_0}], \\ P_{K_0}^{-1}(B_r(0) \cap K_0), 0). \end{aligned} \tag{4.3}$$

Thus, according to the above (with $t = 0$), 0 is the unique zero of $I - P_{K_0}[f(\cdot, a)P_{K_0}]$ in V. Because P_{K_0} is nonexpansive and $P_{K_0}(0) = 0$,

$$\|P_{K_0}(u)\| \leq \|u\|, \; \forall u \in V.$$

This implies that $P_{K_0}(B_r(0)) \subset B_r(0) \cap K_0$, i.e., $B_r(0) \subset P_{K_0}^{-1}[B_r(0) \cap K_0]$. By the excision property,

$$d(I - P_{K_0}[f(\cdot, a)P_{K_0}], P_{K_0}^{-1}[B_r(0) \cap K_0], 0) = d(I - P_{K_0}[f(\cdot, a)P_{K_0}], B_r(0), 0). \tag{4.4}$$

It follows from (4.2), (4.3), and (4.4) that

$$\mathrm{ind}(P_{K_0}[f(\cdot, a)], B_r(0) \cap K_0, K_0) = d(I - P_{K_0}[f(\cdot, a)], B_r(0), 0)$$

for all $r > 0$, completing the proof. ∎

The following result is a direct consequence of Lemma 4.1 and Theorem 3.2 (II).

Corollary 4.2 *If a, b are not eigenvalues of (3.6) and f is such that $f(K_0 \times \{a, b\}) \subset K_0$, and, further, if*

$$\mathrm{ind}(f(\cdot, a), B_r(0) \cap K_0, K_0) \neq \mathrm{ind}(f(\cdot, b), B_r(0) \cap K_0, K_0),$$

for $r > 0$, then, the conclusion of Theorem 3.2 holds.

To prove this corollary, we simply note that, if $f(K_0 \times \{a\}) \subset K_0$, then, $P_{K_0}[f(u,a)] = f(u,a)$, $\forall u \in K_0$.

To proceed further, we consider the following order relation on V:

$$u \leq v \quad \text{if and only if} \quad v - u \in K_0.$$

We say that $f(u,\lambda)$ is monotone with respect to u (in K_0) and $\lambda \geq 0$ if
(i) $\forall \lambda \geq 0$, $\forall u_1, u_2 \in K_0$, if $u_1 \geq u_2$, then, $f(u_1,\lambda) \geq f(u_2,\lambda)$, and
(ii) $\forall u \in K_0$, $\forall \lambda_1 \geq \lambda_2 \geq 0$, then, $f(u,\lambda_1) \geq f(u,\lambda_2)$.

Let f be monotone with respect to u and λ. Because $f(u,0) = 0$, $\forall \lambda \in \mathbb{R}$, it follows that $f(u,\lambda) \geq 0$, $\forall u \in K_0$, and $\lambda \geq 0$. Hence $f(K_0 \times [0,\infty)) \subset K_0$. We have the following bifurcation result for variational inequalities whose homogenizations contain monotone operators. Its proof is based on Lemma 4.1 and properties of the index and monotone operators on cones.

Theorem 4.3 *Let $f(u,\lambda)$ be monotone with respect to $u \in K_0$ and $\lambda \geq 0$. Suppose that $\lambda_0 > 0$ is an isolated eigenvalue of (3.6) corresponding to an eigenvector $h \in K_0 \setminus (-K_0)$.*

Then, there exists a continuum of nontrivial solutions of (3.1) that emanates from $[0, \lambda_0]$ and satisfies the conclusions of Theorem 2.2.

Proof. First, we note that, for $\lambda \geq 0$, u is an eigenvector of (3.6) corresponding to λ if and only if u is a solution of the equation

$$\begin{cases} u = f(u,\lambda), \\ u \in K_0. \end{cases} \tag{4.5}$$

In fact, for $u \in K_0$, $f(u,\lambda) \in K_0$, and, thus, $f(u,\lambda) = P_{K_0}[f(u,\lambda)]$. Hence, (4.5) is equivalent to

$$u = P_{K_0}[f(u,\lambda)],$$

which is the same as (3.6). Now, because $f(\cdot,0) = 0$ and since the trivial map 0 maps $\overline{B_r(0)} \cap K_0$ into $B_r(0) \cap K_0$, from the normalization property of the fixed-point index (Theorem 2.3),

$$\text{ind}(f(\cdot,0), B_r(0) \cap K_0, K_0) = \text{ind}(0, B_r(0) \cap K_0, K_0) = 1. \tag{4.6}$$

Next, we show that, for $\lambda > \lambda_0$ sufficiently closed to λ_0 and λ not an eigenvalue of f in K_0,

$$x - f(x,\lambda) = \gamma h \tag{4.7}$$

has no solution in K_0 for all $\gamma > 0$. In fact, suppose otherwise that there exists $x \in K_0$ that satisfies (4.7) for some $\gamma > 0$. Letting

$$\tau_0 = \sup\{t \geq 0 : x - th \in K_0\},$$

then, because K_0 is a cone, we easily see that $0 \leq \tau_0 < \infty$. We can choose a sequence $\{t_n\} \subset [0, \tau_0]$ such that $t_n \to \tau_0$ and $x - t_n h \in K_0$, $\forall n$. Because

K_0 is closed, $x - t_n h \to x - \tau_0 h$, and $x + \tau h \notin K_0$, $\forall \tau > \tau_0$, and $x - \tau_0 h \in K_0$, i.e., $x \geq \tau_0 h$. Because $x \geq \tau_0 h$ and $x, \tau_0 h \in K_0$, by (i), $f(x, \lambda) \geq f(\tau_0 h, \lambda)$. Hence,

$$x = f(x, \lambda) + \gamma h \geq f(\tau_0 h, \lambda) + \gamma h. \tag{4.8}$$

On the other hand, by the homogeneity of $f(u, \lambda)$ with respect to u and the monotonicity of f with respect to λ,

$$\tau_0 h = \tau_0 f(h, \lambda_0) = f(\tau_0 h, \lambda_0) \leq f(\tau_0 h, \lambda).$$

Thus,

$$f(\tau_0 h, \lambda) + \gamma h \geq \tau_0 h + \gamma h = (\tau_0 + \gamma)h.$$

Together with (4.8), this implies that

$$x \geq (\tau_0 + \gamma)h.$$

Therefore $x - (\tau_0 + \gamma)h \in K_0$ with $\tau_0 + \gamma > \tau_0$. This contradicts the definition of τ_0. Hence, (4.7) has no solution for all $\gamma > 0$. By the solution property of the Leray–Schauder index,

$$\mathrm{ind}(f(\cdot, \lambda) + \gamma h, B_r(0) \cap K_0, K_0) = 0, \ \forall \gamma > 0. \tag{4.9}$$

Let $r > 0$. Then, because λ is not an eigenvalue of f, it follows that there exists an $\alpha = \alpha(r) > 0$ such that

$$\|x - f(x, \lambda)\| \geq \alpha > 0, \ \forall x \in \partial B_r(0) \cap K_0. \tag{4.10}$$

Now, fix γ in $(0, \alpha \|h\|^{-1})$, and consider the family $\{I - H(\cdot, t) : 0 \leq t \leq 1\}$, with

$$H(x, t) = (1 - t)f(x, \lambda) + t[f(x, \lambda) + \gamma h], \ x \in K_0 \cap \overline{B_r(0)}, \ t \in [0, 1].$$

It is clear that H is completely continuous from $K_0 \times [0, 1]$ to K_0. Moreover, from (4.10),

$$
\begin{aligned}
\|x - H(x, t)\| &= \|x - (1 - t)f(x, \lambda) - t[f(x, \lambda) + \gamma h]\| \\
&= \|x - f(x, \lambda) - t\gamma h\| \\
&\geq \|x - f(x, \lambda)\| - \gamma \|h\| \\
&\geq \alpha - \gamma \|h\| \\
&> 0, \forall x \in K_0, \|x\| = r.
\end{aligned}
$$

Now, (4.9) and the homotopy invariance property of the Leray–Schauder index (Chapter 2) imply that

$$
\begin{aligned}
\mathrm{ind}(f(\cdot, \lambda), B_r(0) \cap K_0, K_0) &= \mathrm{ind}(f(\cdot, \lambda) + \gamma h, B_r(0) \cap K_0, K_0) \\
&= 0.
\end{aligned}
$$

$$\tag{4.11}$$

The conclusion now follows from Corollary 4.2 with $a = 0 < b = \lambda = \lambda_0 + \epsilon$ ($\epsilon > 0$, sufficiently small) and (4.6) and (4.11). ∎

Remark 4.1 (a) If $B(u, \lambda) = \lambda\beta u + N(u, \lambda)$, as in Corollary 2.3 of Theorem 2.2, then, $f(u, \lambda) = \lambda\beta u$ is monotone with respect to u and λ, if β is a positive linear map with respect to K_0, i.e., $\beta(K_0) \subset K_0$. In fact,
   (i) If $\lambda \geq 0$, $u_1, u_2 \in K_0$, $u_1 \geq u_2$, then,

$$f(u_1, \lambda) - f(u_2, \lambda) = \lambda\beta(u_1 - u_2) \in K_0,$$

i.e., $f(u_1, \lambda) \geq f(u_2, \lambda)$.
   (ii) If $u \in K_0$, $\lambda_1 \geq \lambda_2 \geq 0$, then, $\beta u \in K_0$, and, therefore,

$$f(u, \lambda_1) - f(u, \lambda_2) = \lambda_1\beta u - \lambda_2\beta u = (\lambda_1 - \lambda_2)\beta u \in K_0,$$

i.e., $f(u, \lambda_1) \geq f(u, \lambda_2)$.
   (b) Let B be of the form (3.18), and suppose that K_0 is an order cone, i.e., $K_0 \cap (-K_0) = \{0\}$, and β is strongly positive on K_0, i.e., $\beta(K_0\backslash\{0\}) \subset \overset{\circ}{K_0}$. Then, by the strong version of the Krein-Rutman theorem, we see that the bifurcation branch $\mathcal{C}$ in Theorem 4.3 is unbounded in $V \times \mathbb{R}$.

In fact, as noted in the proof of Theorem 4.3, λ_0 is an eigenvalue of (3.6) if and only if λ_0 is an eigenvalue of (4.5). However, the Krein-Rutman theorem (Theorem 85, [54]), implies that (4.5) has a unique eigenvalue $\lambda_0 = \lim \|\beta^n\|^{1/n} > 0$. Hence, by Theorem 3.2 (I) and Theorem 4.3, (3.1) has a unique bifurcation point $(0, \lambda_0)$, and thus, $\mathcal{C}$ must be unbounded in $V \times \mathbb{R}$.

To provide a further example, we apply the above general results to a bifurcation problem for a free boundary value problem of an integral equation.

Example 4.1 We consider the following complementarity problem of a nonlinear integral equation:

$$\begin{cases} u(x) - \lambda \int_\Omega G(x, y, u(y)) dy & \geq \quad 0, \\ u(x) & \geq \quad 0, \\ u(x)\left[u(x) - \lambda \int_\Omega G(x, y, u(y)) dy\right] & = \quad 0, \; \forall x \in \Omega, \end{cases} \qquad (4.12)$$

where Ω is an open, bounded subset of $\mathbb{R}^n$ and

$$G : \overline{\Omega} \times \overline{\Omega} \times \mathbb{R} \to \mathbb{R}$$

is a Carathéodory function such that $G(x, y, 0) = 0$, $\forall x, y \in \Omega$, and G satisfies the following growth condition:

$$|\partial G/\partial u|, |G(x, y, u)| \leq R(x, y)[a + b|u|], \qquad (4.13)$$

for a.e. $x, y \in \Omega$, all $u \in \mathbb{R}$ with $a, b > 0$, and $R \in L^2(\Omega \times \Omega)$.

We show that (4.12) is (formally) equivalent to the following variational inequality:

$$\begin{cases} \left\langle u - \lambda \int_\Omega G(\cdot, y, u(y))dy, v - u \right\rangle \geq 0, \ \forall v \in K, \\ u \in K, \end{cases} \tag{4.14}$$

where $\langle \cdot, \cdot \rangle$ denotes the usual inner product of $L^2(\Omega)$ and

$$K = \{u \in L^2(\Omega) : u \geq 0 \ \text{a.e. on} \ \Omega\}. \tag{4.15}$$

In fact, suppose that u satisfies (4.12). Then, $u \in K$, by the second condition of (4.12). Now, let $v \in K$ and $x \in \Omega$. If $u(x) > 0$, then, by the third condition of (4.12),

$$u(x) - \lambda \int_\Omega G(x, y, u(y))dy = 0.$$

Thus,

$$\left[u(x) - \lambda \int_\Omega G(x, y, u(y))dy \right] [v(x) - u(x)] = 0.$$

If $u(x) = 0$, then, because

$$u(x) - \lambda \int_\Omega G(x, y, u(y))dy \geq 0,$$

$$\left[u(x) - \lambda \int_\Omega G(x, y, u(y))dy \right] [v(x) - u(x)]$$

$$= \left[u(x) - \lambda \int_\Omega G(x, y, u(y))dy \right] v(x) \geq 0.$$

In all cases,

$$\left[u(x) - \lambda \int_\Omega G(x, y, u(y))dy \right] [v(x) - u(x)] \geq 0,$$

for almost all $x \in \Omega$. Integrating this inequality over Ω, one obtains (4.14).

Conversely, if we have (4.14), then $u \geq 0$ on Ω. Suppose that $u(x_0) > 0$ at some $x_0 \in \Omega$. Assuming $u \in C(\Omega)$, it follows that there exists $\rho > 0$ such that $u(x) > 0$ on $B_\rho(x_0)$. Choose $\phi \in C_0^\infty(B_\rho(x_0))$. Because $\text{supp}\,\phi$ is a compact subset of $B_\rho(x_0)$, $u(x) > \inf\limits_{\text{supp}\,\phi} u(x) > 0$, $\forall x \in \text{supp}\,\phi$. Because ϕ is bounded in $B_\rho(x_0)$,

$$u \pm \delta\phi \geq 0 \ \text{on} \ \text{supp}\,\phi,$$

for all $\delta > 0$, sufficiently small. Moreover,

$$u \pm \delta\phi = u \geq 0 \quad \text{on} \ \ \Omega \setminus \operatorname{supp}\phi.$$

Hence, $u \pm \delta\phi \in K$ for all $\delta > 0$, sufficiently small. Inserting $v = u \pm \delta\phi$ in (4.14),

$$\pm \left\langle u - \lambda \int_\Omega G(\cdot, y, u(y))dy, \phi \right\rangle \geq 0,$$

which implies that

$$\left\langle u - \lambda \int_\Omega G(\cdot, y, u(y))dy, \phi \right\rangle = 0.$$

Because this holds for all $\phi \in C_0^\infty(B_\rho(x_0))$,

$$u - \lambda \int_\Omega G(\cdot, y, u(y))dy = 0 \quad \text{on} \ \ B_\rho(x_0).$$

This shows that

$$u - \lambda \int_\Omega G(\cdot, y, u(y))dy = 0$$

on the set

$$\{x \in \Omega : u(x) > 0\}.$$

Thus u satisfies (4.12). This shows that (4.12) is (at least formally) equivalent to (4.14).

Consider now the mappings

$$U : L^2(\Omega) \longrightarrow L^2(\Omega), \ B : L^2(\Omega) \times \mathbb{R} \longrightarrow L^2(\Omega),$$

defined by

$$[U(u)](x) \ = \ \int_\Omega G(x, y, u(y))dy, \ B(u, \lambda) = \lambda U(u),$$

for all $u \in L^2(\Omega)$, $\lambda \in \mathbb{R}$, $x \in \Omega$. Then (4.14) is of the form (3.1). By Urysohn's theorem (Theorem 3.2, Chapter 1, [53]) and (4.13), U is a completely continuous operator from $V = L^2(\Omega)$ into itself. Hence, B is completely continuous from $V \times \mathbb{R}$ to V, and one can apply the results in Chapter 3 and Section 4.1. Suppose that $\dfrac{\partial G}{\partial u}(x, y, 0)$ exists for almost all $x, y \in \Omega$ and that

$$\int_\Omega \int_\Omega \left| \frac{\partial G}{\partial u}(x, y, 0) \right|^2 dx\,dy < \infty.$$

By arguments similar to those in Example 3.2, one can verify that the homogenization f of B is given by $f(u, \lambda) = \lambda\beta u$, $\forall u \in V$, $\lambda \in \mathbb{R}$ with

$$\langle \beta u, v \rangle = \int_\Omega \int_\Omega \frac{\partial G}{\partial u}(x, y, 0)u(y)v(x)\,dy dx, \ u, v \in V,$$

i.e.,

$$[\beta(u)](x) = \int_\Omega \frac{\partial G}{\partial u}(x,y,0)u(y)dy, \ \forall u \in V, \ \text{a.e. } x \in \Omega.$$

It is clear that K, given by (4.15), is a closed convex cone in $L^2(\Omega)$ and $\{u, -u\} \subset K$ if and only if $u = 0$. Hence, $K = K_0$ is an order cone in V.

If $\dfrac{\partial G}{\partial u}(x,y,0) \geq 0$ for a.e. $x, y \in \Omega$, then, β is positive. Indeed, for $u \in K$, $u(y) \geq 0$ for a.e. $y \in \Omega$. Hence,

$$[\beta(u)(x)] = \int_\Omega \frac{\partial G}{\partial u}(x,y,0)u(y)dy$$
$$\geq 0, \ \forall x \in \Omega.$$

Consequently, $\beta u \in K$. Because $B(u,0) = 0$, one can apply Theorem 4.3 to (4.12) and (4.14).

4.2   The case of a simple eigenvalue

This section is concerned with another situation where one can calculate the degree of the operator in (3.6), namely, when the operator in (3.6) has a simple eigenvalue λ_0 and f is symmetric. Using arguments motivated by the Fredholm alternative, one can show that, for $\lambda > \lambda_0$ near λ_0, the corresponding degree is 0. Together with Theorem 3.2, this yields global bifurcation for the original variational inequality (3.1).

Consequences are derived for cases where the simplicity of eigenvalues can be verified. The first case happens at simple eigenvalues of a corresponding linear equation with an eigenvector belonging to the demi-interior part of K_0. Also, in some circumstances, one can prove the simplicity of the first eigenvalue of (3.6) by employing a result of Miersemann.

4.2.1   Some general results

Again, first, some definitions. Consider the homogenized variational inequality (3.6) corresponding to (3.1). As above, we assume that $B(u,0) = 0, \ \forall u \in V$.

An eigenvalue λ of (3.6) is called simple in K_0 if $0 \neq u_1, u_2 \in K_0$ are eigenvectors of (3.6) corresponding to λ, then $u_1 = Cu_2$, for some $C > 0$.

We say that

- $f(u, \lambda)$ is symmetric with respect to u in K_0 if

$$\forall u, v \in K_0, \ \forall \lambda \in \mathbb{R}, \ \langle f(u, \lambda), v \rangle = \langle u, f(v, \lambda) \rangle, \qquad (4.16)$$

and

- f is homogeneous of order γ ($\gamma > 0$), with respect to $\lambda \geq 0$, if

$$f(u, \lambda) = \lambda^\gamma f(u, 1), \ \forall \lambda \geq 0, \ \forall u \in K_0. \tag{4.17}$$

We have the following theorem.

Theorem 4.4 *Suppose that $\lambda_0 > 0$ is a simple eigenvalue of (3.6) and the eigenvectors of (3.6) corresponding to λ_0 are also eigenvectors of the equation*

$$u = f(u, \lambda_0). \tag{4.18}$$

Assume further that f is symmetric with respect to u and homogeneous of order γ with respect to λ. Then, there exists a global bifurcation branch emanating from $\{0\} \times [0, \lambda_0]$ and satisfying the alternative in Theorem 3.2.

Proof. Because $B(u, 0) = 0$, from (3.5), $f(u, 0) = 0$, $\forall u \in V$. Hence,

$$d(I - P_{K_0}[f(\cdot, 0)], B_r(0), 0) = 1. \tag{4.19}$$

We prove that, for $\lambda > \lambda_0$ sufficiently close to λ_0, (3.6) has no nontrivial solution, and

$$d(I - P_{K_0}[f(\cdot, \lambda)], B_r(0), 0) = 0, \tag{4.20}$$

for all $r > 0$.

Let u_0 be the eigenvector of (3.6) corresponding to $\lambda = \lambda_0$ with $\|u_0\| = 1$. u_0 is uniquely determined by the simplicity of λ_0; also, u_0 is a solution of (4.18), i.e.,

$$u_0 - f(u_0, \lambda_0) = 0.$$

Consider the family of compact perturbations of the identity $\{H(t, u, \lambda)\}$ with

$$H(t, u, \lambda) = u - P_{K_0}[(1 - t)f(u, \lambda) + tf(u, \lambda_0) + tu_0],$$

for all $t \in [0, 1]$, $u \in V$, $\lambda \in \mathbb{R}$. We now show that there exist $\lambda_1 > \lambda_0$, and $R_0 > 0$ such that

$$H(t, u, \lambda) \neq 0, \ \forall \lambda \in (\lambda_0, \lambda_1), \ \forall u \in V, \ \|u_0\| \geq R_0, \tag{4.21}$$

$$H(0, u, \lambda) \neq 0, \ \forall u \in V \setminus \{0\}, \ \forall \lambda \in (\lambda_0, \lambda_1), \tag{4.22}$$

and

$$H(1, u, \lambda_0) \neq 0, \ \forall u \in V. \tag{4.23}$$

Suppose that $u_n \in K_0$, $t_n \in [0, 1]$, and $\lambda_n \in \mathbb{R}$ satisfy the following equation:

$$u_n = P_{K_0}[(1 - t_n)f(u_n, \lambda_n) + t_n f(u_n, \lambda_0) + t_n u_0]. \tag{4.24}$$

This equation is equivalent to the variational inequality,

$$\begin{cases} \langle u_n - (1 - t_n)f(u_n, \lambda_n) - t_n f(u_n, \lambda_0) - t_n u_0, v - u_n \rangle \geq 0, \ \forall v \in K_0, \\ u_n \in K_0. \end{cases}$$

Letting $v = u_n + u_0 \in K_0$ in this inequality, one gets the following estimates from (4.16) and (4.17):

$$
\begin{aligned}
0 &\leq \langle u_n - (1 - t_n) f(u_n, \lambda_n) - t_n f(u_n, \lambda_0) - t_n u_0, u_0 \rangle \\
&= \langle u_n, u_0 - (1 - t_n) f(u_0, \lambda_n) - t_n f(u_0, \lambda_0) \rangle - t_n \| u_0 \|^2 \\
&= (1 - t_n) [\langle u_n, f(u_0, \lambda_0) \rangle - \langle u_n, f(u_0, \lambda_n) \rangle] \\
&\quad + \langle u_n, u_0 - f(u_0, \lambda_0) \rangle - t_n \| u_0 \|^2 \\
&= (1 - t_n) \left[\langle u_n, f(u_0, \lambda_0) \rangle - \left(\frac{\lambda_n}{\lambda_0} \right)^\gamma \langle u_n, f(u_0, \lambda_0) \rangle \right] - t_n \| u_0 \|^2 \\
&\quad \left(\text{because } u_0 - f(u_0, \lambda_0) = 0 \text{ and} \right. \\
&\qquad \left. f(u_0, \lambda_n) = f\left(u_0, \left(\frac{\lambda_n}{\lambda_0} \right) \lambda_0 \right) = \left(\frac{\lambda_n}{\lambda_0} \right)^\gamma f(u_0, \lambda_0) \right) \\
&= (1 - t_n) \langle u_n, u_0 \rangle \left[1 - \left(\frac{\lambda_n}{\lambda_0} \right)^\gamma \right] - t_n \| u_0 \|^2. \tag{4.25}
\end{aligned}
$$

We now prove (4.21), (4.22), and (4.23).

If $H(1, u, \lambda_0) = 0$ for some $u \in K_0$, then, (4.24) and (4.25) are satisfied with $u_n = u$, $\lambda_n = \lambda_0$, and $t_n = 1$. From (4.25), one obtains

$$
0 \leq -\| u_0 \|^2 < 0 \quad (\text{because } u_0 \neq 0).
$$

This contradiction proves (4.23).

To prove (4.22), we argue by contradiction. Assume that (4.22) does not hold and there exist sequences $\{u_n\} \subset K_0 \setminus \{0\}$ and $\{\lambda_n\} \subset (\lambda_0, \infty)$ such that $\lambda_n \to \lambda_0^+$, as $n \to \infty$ and $H(0, u_n, \lambda_n) = 0$, $\forall n$. From (4.24) and (4.25) with $t_n = 0$, $\forall n$,

$$
\langle u_n, u_0 \rangle \left[1 - \left(\frac{\lambda_n}{\lambda_0} \right)^\alpha \right] \geq 0.
$$

Because $\dfrac{\lambda_n}{\lambda_0} > 1$, $\langle u_n, u_0 \rangle \leq 0$.

Setting $v_n = u_n / \| u_n \|$,

$$
\langle v_n, u_0 \rangle \leq 0. \tag{4.26}
$$

On the other hand, it follows from (4.24) that

$$
u_n = P_{K_0}[f(u_n, \lambda_n)].
$$

Hence, dividing this equality by $\| u_n \| \neq 0$ and using the homogeneity of $f(\cdot, \lambda_n)$ and of P_{K_0},

$$
v_n = P_{K_0}[f(v_n, \lambda_n)]. \tag{4.27}
$$

Because $\| v_n \| = 1$, $\forall n$, by passing to a subsequence, if necessary, one can suppose that $v_n \rightharpoonup v$ in V. Then, by (4.27) and the complete continuity of $P_{K_0} \circ f$, $v_n \to v$ in V, $v \in K_0$, and

$$
v = P_{K_0}[f(v, \lambda_0)]. \tag{4.28}
$$

Therefore, v is a solution of (3.6) with $\lambda = \lambda_0$. Because λ_0 is a simple eigenvalue, it follows that $v = cu_0$, for some $c > 0$. Because $\|v\| = \|u_0\| = 1$, $c = 1$, and, then, $v = u_0$. On the other hand, by letting $n \to \infty$ in (4.26),

$$\langle v, u_0 \rangle \leq 0, \tag{4.29}$$

which is in contradiction to $u_0 \neq 0$, and, hence, (4.22) holds.

Now, suppose that (4.21) does not hold. One has (4.24) with sequences $\{\lambda_n\} \subset (\lambda_0, \infty)$, $\{u_n\} \subset K_0$, $\{t_n\} \subset [0,1]$ such that $\lambda_n \to \lambda_0^+$, $\|u_n\| \to \infty$, and $t_n \to t_0$.

Let v_n be as above. Dividing (4.24) by $\|u_n\|$,

$$v_n = P_{K_0}\left[(1-t_n)f(v_n, \lambda_n) + t_n f(v_n, \lambda_0) + \frac{t_n u_0}{\|u_n\|}\right], \quad \forall n \in \mathbb{N}. \tag{4.30}$$

Because $\frac{t_n u_0}{\|u_n\|} \to 0$ as $n \to \infty$, by assuming $v_n \to v$ in V, the right-hand side of (4.30) tends to $P_{K_0}[f(v, \lambda_0)]$. Hence, as above, $v_n \to v \in K_0$ and v satisfies (4.28).

Moreover, (4.25) implies (4.26) and, thus, (4.29). Again, one obtains a contradiction. Hence, we must have (4.21).

It follows from (4.23) that the equation $u - P_{K_0}[f(u, \lambda) + u_0] = 0$ has no solution in V. Hence, by the solution property of the Leray–Schauder degree,

$$\mathrm{d}(I - P_{K_0}[f(\cdot, \lambda_0) + u_0], B_R(0), 0) = 0.$$

Using (4.21) and the homotopy invariance property, for $\lambda \in (\lambda_0, \lambda_1)$, $R > R_0$,

$$\begin{aligned}
\mathrm{d}(I - P_{K_0}[f(\cdot, \lambda)], B_R(0), 0) &= \mathrm{d}(H(0, \cdot, \lambda), B_R(0), 0) \\
&= \mathrm{d}(H(1, \cdot, \lambda), B_R(0), 0) \\
&= \mathrm{d}(I - P_{K_0}[f(\cdot, \lambda_0) + u_0], B_R(0), 0) \\
&= 0.
\end{aligned}$$
$$\tag{4.31}$$

Using (4.22), we see that the equation $u - P_{K_0}[f(u, \lambda)] = 0$ has no solution in $B_R(0) \setminus \overline{B_r(0)}$, for $R > r > 0$. Hence, by the excision property,

$$\mathrm{d}(I - P_{K_0}[f(\cdot, \lambda)], B_R(0), 0) = \mathrm{d}(I - P_{K_0}[f(\cdot, \lambda)], B_r(0), 0). \tag{4.32}$$

Therefore, (4.20) is a consequence of (4.31) and (4.32). Our result now follows from Theorem 2.2 with $a = 0, b = \lambda_0 + \epsilon$, and $\epsilon > 0$, sufficiently small. ∎

Some analogs and extensions of Theorem 4.4 are presented in Chapter 6. We consider some consequences. The following definitions and notations are needed.

We call a point $u \in K_0$ a demi-interior point of K_0 (cf. [25]) if the tangent cone of
$$K_0 - u = \{v - u : v \in K_0\}$$
is the whole space V. By K_u, we denote the set
$$K_u = \{v \in V : u + \gamma v \in K_0 \text{ for some } \gamma > 0\}.$$

Let K_0^I be the set of all demi-interior points of K_0. The following property of demi-interior points of K_0 holds:

Lemma 4.5

(a) $u \in K_0^I$ if and only if $\overline{K_u} = V$, if and only if there exists a dense subset D in V such that, for all $w \in D$, there exists an $\epsilon > 0$, such that $u + \epsilon w \in K_0$.

(b) (Similar to Lemma 1.1, [58]) Suppose that the equation (4.18) has a solution $u_0 \in K_0^I$ and that f is symmetric with respect to u. Then, u is a solution of (3.6) if and only if $u \in K_0$ and u is a solution of (4.18).

Proof. (a) First, let $E_u = \bigcup_{t>0} t(K_0 - u)$. Then,

$$
\begin{aligned}
w \in E_u \quad &\Longleftrightarrow \quad w = t(z - u), \text{ for some } z \in K_0, t > 0, \\
&\Longleftrightarrow \quad z = u + \gamma w \in K_0, \text{ for some } \gamma = \frac{1}{t} > 0, \\
&\Longleftrightarrow \quad w \in K_u.
\end{aligned}
$$

Hence, $E_u = K_u$. The support cone of $K_0 - u$ is $\overline{E_u} = \overline{K_u}$. Hence $u \in K_0^I$ if and only if $\overline{K_u} = V$. Now, let $u \in K_0^I$, and choose $D = E_u = K_u$. Then, $\overline{D} = V$, and, for $w \in D$, there exists $\epsilon = \gamma > 0$, such that $u + \epsilon w \in K_0$. Conversely, suppose that $u \in K_0$ is such that there exists a dense set D in V, such that, for all $w \in D$, there is an $\epsilon > 0$ satisfying $u + \epsilon w \in K_0$. One has
$$w \in K_u, \ \forall w \in D,$$
i.e., $D \subset K_u$. Hence, $\overline{K_u} = \overline{D} = V$, and u is a demi-interior point of K_0.

(b) If $u \in K_0$ satisfies (4.18), then,
$$\langle u - f(u, \lambda_0), v - u \rangle = 0, \ \forall v \in K_0,$$
and u is thus a solution of (3.6). Conversely, suppose that u satisfies (3.6) (with $\lambda = \lambda_0$). Let $w \in D$. Because $u_0 \in K_0^I$, we can choose $\epsilon > 0$ such that $u_0 + \epsilon v \in K_0$. By letting $v = u + u_0 + \epsilon w \in K_0$ in (3.6),

$$
\begin{aligned}
0 \ &\leq \ \langle u - f(u, \lambda_0), u_0 + \epsilon w \rangle \\
&= \ \langle u, u_0 - f(u_0, \lambda_0) \rangle + \langle u - f(u, \lambda_0), \epsilon w \rangle \\
&= \ \epsilon \langle u - f(u, \lambda_0), w \rangle.
\end{aligned}
$$

Because this holds for all $w \in D$ and because D is dense in V,

$$0 \leq \langle u - f(u, \lambda_0), w \rangle, \ \forall w \in V.$$

By replacing w by $-w$, $\langle u - f(u, \lambda_0), w \rangle = 0$, $\forall w \in V$, i.e., $u = f(u, \lambda_0)$, or u is a solution of (4.18). ∎

From this lemma, one obtains the following corollary:

Corollary 4.6 *Suppose that f is homogeneous and symmetric with respect to u and homogeneous of order γ with respect to λ. Assume, furthermore, that λ_0 is a simple eigenvalue of (4.18) with an eigenvector in K_0^I. If $K_0 \neq V$, then, we have the conclusion of Theorem 4.4.*

Note that f satisfies the condition in this corollary if $f(u, \lambda) = \lambda \beta u$, with β a symmetric bounded linear mapping on V.

Proof. To apply Theorem 3.2, one needs to prove that, under the assumptions of Corollary 4.6, λ_0 is a simple eigenvalue of (3.6).

Let u_0, $\|u_0\| = 1$, be an eigenvector of (4.18) such that $u_0 \in K_0^I$. According to Lemma 4.5, $u_0 \in K_0$ is also an eigenvector of (3.6), corresponding to $\lambda = \lambda_0$. Let $u \in K_0$ be another eigenvector of (3.6), corresponding to λ_0. By rescaling, we can assume that $\|u\| = 1$. By Lemma 4.5, u is also an eigenvector of (4.18), corresponding to λ_0. By the simplicity of λ_0, as an eigenvalue of (4.18), one has $u = tu_0$ with $t = 1$ or $t = -1$.

Suppose that $t = -1$. Then, $u = -u_0 \in K_0$. Because $u_0 \in K_0^I$, one can find a set D such that $\overline{D} = V$ and, for all $v \in D$, $u_0 + \epsilon v \in K_0$ for some $\epsilon > 0$. Let $v \in D$, and choose ϵ such that this holds. One gets

$$\epsilon v = -u_0 + (u_0 + \epsilon v) \in K_0,$$

i.e., $v \in K_0$. Therefore, $D \subset K_0$. Because K_0 is closed, $V = \overline{D} = K_0$. This contradiction proves the corollary. ∎

Let us consider some examples as applications of the above results.

Example 4.2 This example is concerned with the bifurcation problem with obstacle (3.27) in Chapter 3 with a different closed convex set K:

$$K = \{u \in H_0^1(0, 1) : u \geq 0 \ \text{on} \ [0, a] \ \text{and} \ u \leq l \ \text{on} \ [b, c]\}. \quad (4.33)$$

Here $l > 0$ and $0 < a < b < c < 1$ are given. We claim that the support cone of K, in this case, is given by

$$K_0 = \{u \in H_0^1(0, 1) : u \geq 0 \ \text{on} \ [0, a]\}.$$

In fact, it is clear that K_0 is a closed, convex cone containing K. Therefore, $\overline{\bigcup_{t>0} tK} \subset K_0$. Conversely, let $u \in K_0$, and let $M \equiv \sup_{[0,a]} u < \infty$. It follows that $lM^{-1}u \geq 0$ on $[0, a]$ and $lM^{-1}u \leq l$ on $[b, c]$, i.e., $u \in Ml^{-1}K$. Hence, $u \in \bigcup_{t>0} tK$, $\forall u \in K_0$, showing that

$$K_0 \subset \bigcup_{t>0} tK \subset \overline{\bigcup_{t>0} tK}.$$

Thus, $K_0 = \overline{\bigcup_{t>0} tK}$ is the support cone of K. We now assume that $D_2 g(x, 0, \lambda)$ is homogeneous of order γ with respect to λ, i.e.,

$$D_2 g(x, 0, \lambda) = \lambda^\gamma D_2 g(x, 0, 1), \ \forall \lambda > 0, \ \text{a.e.} \ x \in \Omega.$$

Let $k(x) = D_2 g(x, 0, 1)$, $x \in \Omega$. In this case,

$$\langle f(u, \lambda), v \rangle = \lambda^\gamma \int_0^1 k(x) u(x) v(x) dx, \tag{4.34}$$

for all $u, v \in V$, $\lambda \in \mathbb{R}$. Therefore, the linear equation corresponding to (3.6) $(u = f(u, \lambda))$ is of the form,

$$\begin{cases} \int_0^1 u'v' = \lambda^\gamma \int_0^1 kuv, \ \forall v \in H_0^1(0, 1), \\ u \in H_0^1(0, 1), \end{cases}$$

or, equivalently,

$$\begin{cases} -u'' = \lambda^\gamma k u \ \text{ on } \ (0, 1), \\ u(0) = u(1) = 0. \end{cases} \tag{4.35}$$

From (4.34), we see that $f(u, \lambda)$ is symmetric and linear with respect to u. Suppose that μ^γ is an eigenvalue of (4.35), corresponding to an eigenvalue u_0, such that

$$u_0(x) > 0, \ \forall x \in (0, a]. \tag{4.36}$$

One has $u_0 \in K_0^I$. In fact, consider $D = C_0^\infty(0, 1)$. Then, D is dense in $H_0^1(0, 1)$, and, for $\phi \in D$, $\text{supp} \, \phi \cap [0, a]$ is a compact subset of $(0, a]$. By (4.36),

$$u_0(x) \geq \inf_{\text{supp} \, \phi \cap [0,a]} u_0 \equiv m > 0, \ \forall x \in \text{supp} \, \phi \cap [0, a].$$

For $0 < \epsilon < m \|\phi\|_{L^\infty(0,1)}^{-1}$, for all $x \in \text{supp} \, \phi \cap [0, a]$,

$$\begin{aligned} (u_0 + \epsilon\phi)(x) &\geq u_0(x) - \frac{m}{\|\phi\|_{L^\infty(0,1)}} \|\phi\|_{L^\infty(0,1)} \\ &= u_0(x) - m \\ &\geq 0. \end{aligned}$$

For $x \in [0, a] \setminus \operatorname{supp} \phi$, $(u_0 + \epsilon\phi)(x) = u_0(x) \geq 0$. Hence, $u_0 + \epsilon\phi \geq 0$ on $[0, a]$, i.e., $u_0 + \epsilon\phi \in K_0$, and, thus, $u_0 \in K_0^I$.

Assume that μ^γ is a simple eigenvalue of (4.35), corresponding to an eigenvector u_0 satisfying (4.36). Then, with K given above, all the conditions in Corollary 4.6 are satisfied for (3.27). Applying this corollary, one obtains a global bifurcation result for (3.27) with K given by (4.33). Note that all of the above conditions are satisfied in the particular case where $k = \text{constant}$, and $\mu^\gamma = \lambda^\gamma$ is the first eigenvalue of (4.35).

Example 4.3   In this example, we consider the following fourth-order variational inequality related to the equilibrium problem of beams (cf. [75]):

$$\begin{cases} \displaystyle\int_0^a u''(v - u)'' dx \geq \int_0^a g(x, u', \lambda)(v - u)' dx, \ \forall v \in K, \\ u \in K. \end{cases} \quad (4.37)$$

Here, $a > 0$ and $V = H^2(0, a) \cap H_0^1(0, a)$ with the usual inner product

$$\langle u, v \rangle = \int_0^a u'' v'', \ u, v \in V,$$

and norm $\|u\| = \langle u, u \rangle^{1/2}$ (which is equivalent to the norm induced by that of $H^2(0, a)$);

$$K = \{u \in V : \psi_1(x) \leq u(x) \leq \psi_2(x), \ x \in [0, a]\},$$

where ψ_1, ψ_2 are measurable functions from $[0, a]$ to $[-\infty, +\infty]$, such that $\psi_1 \leq 0 \leq \psi_2$ on $[0, a]$.

For simplicity, we assume that g is a continuous function from $[0, a] \times \mathbb{R}^2$ to $\mathbb{R}$, such that the partial derivative $D_2 g(x, 0, \lambda)$ is also continuous with respect to $\lambda \in \mathbb{R}$ and $x \in [0, a]$.

Let $B : V \times \mathbb{R} \to V$ be given by

$$\langle B(u, \lambda), v \rangle = \int_0^a g(x, u', \lambda) v', \ \forall u, v \in V, \ \lambda \in \mathbb{R}.$$

Then,

$$\|B(u, \lambda) - B(\bar{u}, \bar{\lambda})\| \leq C \left[\int_0^a |g(x, u', \lambda) - g(x, \bar{u}', \bar{\lambda})|^2 \right]^{\frac{1}{2}}, \quad (4.38)$$

$$\forall u, \bar{u} \in V, \ \lambda, \bar{\lambda} \in \mathbb{R}.$$

If $u_n \to u$ in $H^2(0, a) \cap H_0^1(0, a)$, $\lambda_n \to \lambda$, then, $u_n' \to u'$ in $H^1(0, a)$, and, by the compact embedding $H^1(0, a) \hookrightarrow C[0, a]$, $u_n' \to u'$ in $C[0, a]$. Therefore,

$$g(x, u_n'(x), \lambda_n) \to g(x, u'(x), \lambda)$$

uniformly with respect to $x \in [0, a]$. From (4.38), it follows that

$$B(u_n, \lambda_n) \to B(u, \lambda) \text{ in } V.$$

Consider the mapping f given by

$$\langle f(u, \lambda), v \rangle = \int_0^a D_2 g(x, 0, \lambda) u'(x) v'(x) \, dx, \ \forall u, v \in V, \ \lambda \in \mathbb{R}.$$

By arguments similar to those used above, one can verify that f is a completely continuous mapping from $V \times \mathbb{R}$ to V, and, for $u_n \rightharpoonup u$ in V, $\sigma_n \to 0^+$, and $\lambda_n \to \lambda$ in $\mathbb{R}$,

$$\frac{1}{\sigma_n} B(\sigma_n u_n, \lambda_n) \to f(u, \lambda) \text{ in } V.$$

Moreover, f is linear with respect to u. Let K_0 be the support cone of K. With these settings, we see that the homogenized variational inequality (3.6) associated with (4.37) is of the form,

$$\begin{cases} \int_0^a u''(v - u)'' \geq \int_0^a D_2 g(x, 0, \lambda) u'(v - u)', \ \forall v \in K_0, \\ u \in K_0. \end{cases}$$

In what follows, we consider the particular choice, where g is given by

$$g(x, u', \lambda) = \lambda u'(1 + u'^2)^{-1/2}, \ x \in (0, a), \ u', \lambda \in \mathbb{R}, \tag{4.39}$$

i.e., (4.37) is of the form

$$\begin{cases} \int_0^a u''(v - u)'' dx \geq \lambda \int_0^a \frac{u'}{\sqrt{1 + u'^2}} (v - u)' dx, \\ \forall v \in K, \\ u \in K. \end{cases} \tag{4.40}$$

Here, g is independent of x, $\lambda = P/EJ$, P represents the (changing) force, EJ is the stiffness of the beam, and a is the length of the beam. With these assumptions, (4.40) is the variational inequality for the buckling problem of a simply supported beam lying between obstacles (cf. [75], [70]). For g given by (4.39), $D_2 g(x, 0, \lambda) = \lambda$, $\forall x \in (0, a)$, $\lambda \in \mathbb{R}$, and, then,

$$\langle f(u, \lambda), v \rangle = \lambda \int_0^a u'v', \ \forall u, v \in V, \ \lambda \in \mathbb{R}.$$

$f(u, \lambda)$ is linear, symmetric with respect to u and linear with respect to λ. Next, we consider some particular choices of the closed, convex set K,

$K = \{u : \psi_1 \leq u \leq \psi_2\}$, where ψ_1, ψ_2 are functions on $[0, a]$ that represent the obstacles.

(a) Suppose that $\psi_1, \psi_2 \in C[0, a]$, $\psi_1 < 0 < \psi_2$ on $[0, a]$. Then as in Chapter 3, $K_0 = V$. In fact, by the assumptions on ψ_1, ψ_2, (see Figure 4.1)

$$\psi_1(x) \leq \max_{x \in [0,a]} \psi_1(x) \equiv m_1 < 0 < m_2 \equiv \min_{x \in [0,a]} \psi_2(x) \leq \psi_2(x), \; \forall x \in [0, a].$$

ψ_2

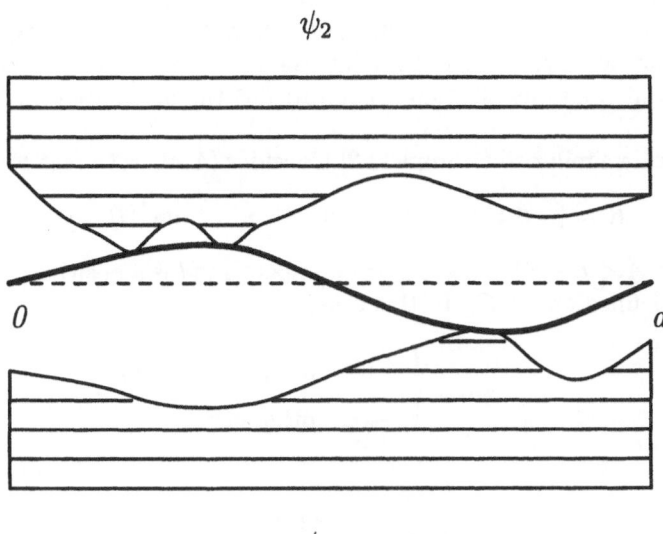

0 a

ψ_1

FIGURE 4.1. Example 4.3 (a).

Let $u \in V$. Because $u \in C[0, a]$, there exists $\epsilon > 0$ sufficiently small, such that

$$m_1 < -\epsilon \|u\|_{L^\infty(0,a)} \leq \epsilon \|u\|_{L^\infty(0,a)} < m_2.$$

Therefore, $\psi_1 \leq m_1 \leq \epsilon u \leq m_2 \leq \psi_2$ on $[0, a]$, i.e., $\epsilon u \in K$. Hence, $V = \bigcup_{t \geq 0} tK = K_0$. The homogenized variational inequality (3.6), in this case, is a linear equation:

$$\begin{cases} \displaystyle\int_0^a u'' v'' = \lambda \int_0^a D_2 g(x, 0, \lambda) \, u' \, v', \; \forall v \in H^2(0, a) \cap H_0^1(0, a), \\ u \in H^2(0, a) \cap H_0^1(0, a), \end{cases} \tag{4.41}$$

or, equivalently,

$$\begin{cases} u^{(4)} + \lambda u'' = 0 \; \text{ on } \; [0, a], \\ u(0) = u(a) = u''(0) = u''(a) = 0. \end{cases} \tag{4.42}$$

The eigenvalues of (4.42) are

$$\lambda = \lambda_k = (k\pi/a)^2, \ k = 1, 2, ...$$

and the eigenfunctions are u_k,

$$u_k(x) = \sin\left(\frac{k\pi x}{a}\right), \ x \in [0, a].$$

In particular, all the eigenvalues are simple. Applying Corollary 3.3, we see that the set of all bifurcation points of (4.40), in this case, is $\{0\} \times \{k^2\pi^2/a^2 : k \in \mathbb{N}\}$, and, at each bifurcation point, there emanates a global continuum of nontrivial solutions of (4.40) that satisfies the alternative in Theorem 3.2.

(b) [A case similar to Example 3.2] Consider (4.40) with (see Figure 4.2)

$$K = \{u \in V : u \geq 0 \text{ on } [0, A], \ u \leq l \text{ on } [B, C]\},$$

where $0 < A < B < C < a$ are given numbers and l is a positive continuous function defined on $[B, C]$. In this case,

$$\psi_1 = \begin{cases} 0 & \text{on } [0, A], \\ -\infty & \text{otherwise,} \end{cases}$$

and

$$\psi_2 = \begin{cases} l & \text{on } [B, C], \\ \infty & \text{otherwise.} \end{cases}$$

As in Example 3.2, the support cone K_0 of K is given by

$$K_0 = \{u \in V : u \geq 0 \text{ on } [0, A]\}.$$

In fact, it is clear that K_0 is a closed, convex cone containing K. Hence, $\bigcup_{t>0} tK \subset K_0$. Conversely, let $u \in K_0$. Because $\inf_{[B,C]} l > 0$, there exists $\epsilon > 0$, such that

$$\epsilon u(x) \leq \epsilon \|u\|_{L^\infty(0,a)} < \inf_{[B,C]} l \leq l(x), \ \forall x \in [B, C].$$

Hence, $\epsilon u \in K$, i.e., $u \in \epsilon^{-1}K$. Therefore, $K_0 \subset \bigcup_{t>0} tK$, and K_0 is the support cone of K, as claimed. Now, the homogenized variational inequality associated with (4.40) is

$$\begin{cases} \int_0^a u''(v - u)'' \geq \lambda \int_0^a u'(v - u)', \ \forall v \in K_0, \\ u \in K_0. \end{cases} \tag{4.43}$$

The equation corresponding to (4.43) is (4.41) or (4.42). The first eigenvalue of (4.42) is $\lambda_1 = \pi^2/a^2$, whose corresponding eigenvector is u_1 with $u_1(x) = \sin\left(\dfrac{\pi x}{a}\right)$, $x \in (0, a)$. Because

$$u_1(x) > 0, \ \forall x \in (0, a),$$

$u_1 \in K_0^I$. This follows from arguments similar to those used in Example 4.2.

Noting that f is symmetric with respect to u and is linear with respect to λ, one can apply Corollary 4.6 of Theorem 4.4 and conclude that there exists a global branch of solutions of (4.40) that bifurcates from $[0, \pi^2/a^2]$ and satisfies the alternative in Theorem 4.4.

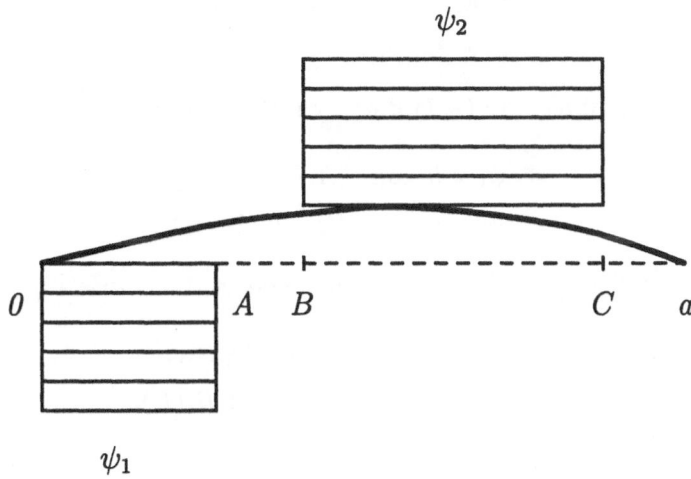

FIGURE 4.2. Example 4.3 (b).

(c) Next, we consider (see also Figure 4.3) the case where both obstacles ψ_1 and ψ_2 vanish on nonempty subsets of $[0, a]$. Let K be of the form,

$$K = \{u \in V : u \geq \psi_1 \ \text{on} \ [A, B], \ u \leq \psi_2 \ \text{on} \ [C, D]\},$$

where $0 \leq A < B < C < D \leq a$, and $\psi_1 : [A, B] \to \mathbb{R}^-, \psi_2 : [C, D] \to \mathbb{R}^+$ are continuous functions such that

$$\begin{cases} I_1 = \{x \in [A, B] : \psi_1(x) = 0\} \neq \emptyset, \\ I_2 = \{x \in [C, D] : \psi_2(x) = 0\} \neq \emptyset. \end{cases}$$

We assume that $0 \notin I_1$ and $a \notin I_2$. The support cone of K is given by

$$K_0 = \{u \in V : u \geq 0 \ \text{on} \ I_1, \ u \leq 0 \ \text{on} \ I_2\}. \tag{4.44}$$

Indeed, K_0 is a closed, convex cone containing K. Then,

$$\bigcup_{t>0} tK \subset K_0.$$

To prove the converse, let $u \in K_0$, and choose $\phi \in C_0^\infty(0,a)$, such that $\phi \geq 1$ on I_1 and $\phi \leq -1$ on I_2. For $\epsilon > 0$ small, we set $u_\epsilon = u + \epsilon\phi \in V$. Then,

$$\begin{cases} u_\epsilon \geq \epsilon\phi \geq \epsilon & \text{on}\quad I_1, \\ u_\epsilon \leq \epsilon\phi \leq -\epsilon & \text{on}\quad I_2. \end{cases}$$

Consequently, there exists $\delta > 0$, sufficiently small, such that

$$\begin{cases} u_\epsilon \geq \epsilon/2 & \text{on}\quad B_\delta(I_1), \\ u_\epsilon \leq -\epsilon/2 & \text{on}\quad B_\delta(I_2), \end{cases}$$

$(B_\delta(I) = \{x \in (0,a) : \text{dist}(x,I) < \delta\})$. On the other hand, because

$$\begin{cases} \psi_1 < 0 & \text{on}\quad [A,B] \setminus B_\delta(I_1), \\ \psi_2 > 0 & \text{on}\quad [C,D] \setminus B_\delta(I_2), \end{cases}$$

and $[A,B] \setminus B_\delta(I_1), [C,D] \setminus B_\delta(I_2)$ are compact, there exists $\eta > 0$, such that

$$\begin{cases} \psi_1 \leq -\eta & \text{on}\quad [A,B] \setminus B_\delta(I_1), \\ \psi_2 \geq \eta & \text{on}\quad [C,D] \setminus B_\delta(I_2). \end{cases}$$

Therefore,

$$\frac{\eta}{\|u_\epsilon\|_{L^\infty(0,a)}} u_\epsilon \geq \frac{\epsilon}{2} \frac{\eta}{\|u_\epsilon\|_{L^\infty(0,a)}} \geq 0 \geq \psi_1 \quad \text{on}\ B_\delta(I_1), \quad \text{and}$$

$$\frac{\eta}{\|u_\epsilon\|_{L^\infty(0,a)}} u_\epsilon \geq -\frac{\eta}{\|u_\epsilon\|_{L^\infty(0,a)}} \|u_\epsilon\|_{L^\infty(0,a)} = -\eta \geq \psi_1 \quad \text{on}\ [A,B] \setminus B_\delta(I_1).$$

These estimates imply that

$$\frac{\eta}{\|u_\epsilon\|_{L^\infty(0,a)}} u_\epsilon \geq \psi_1 \quad \text{on}\ [A,B].$$

Similarly,

$$\frac{\eta}{\|u_\epsilon\|_{L^\infty(0,a)}} u_\epsilon \leq \psi_2 \quad \text{on}\ [C,D].$$

This means that $\frac{\eta}{\|u_\epsilon\|_{L^\infty(0,a)}} u_\epsilon \in K$, and, thus,

$$u_\epsilon \in \frac{\|u_\epsilon\|_{L^\infty(0,a)}}{\eta} K \subset \bigcup_{t>0} tK, \quad \forall \epsilon > 0.$$

On the other hand, because $\|u_\epsilon - u\| = \epsilon\|\phi\| \to 0$, as $\epsilon \to 0$, $u \in \overline{\bigcup_{t>0} tK}$.

Hence, $K_0 = \overline{\bigcup_{t>0} tK}$ is the support cone of K.

The homogenized variational inequality of (4.40), in this case, is (4.43), with the support cone K_0 given by (4.44). We note that $u \in K_0^I$ if $u > 0$ on I_1 and $u < 0$ on I_2. In fact, because I_1, I_2 are compact, there exists $\delta > 0$, such that

$$u > \delta \text{ on } I_1, \ u < -\delta \text{ on } I_2.$$

Let $\phi \in V(\subset C[0,a])$. Then,

$$u + \frac{\delta\phi}{\|\phi\|_{L^\infty(0,a)}} \geq \delta - \frac{\delta|\phi|}{\|\phi\|_{L^\infty(0,a)}} \geq 0 \text{ on } I_1,$$

and similarly,

$$u + \frac{\delta\phi}{\|\phi\|_{L^\infty(0,a)}} \leq -\delta + \frac{\delta|\phi|}{\|\phi\|_{L^\infty(0,a)}} \leq 0 \text{ on } I_2.$$

It follows that $u + \frac{\delta\phi}{\|\phi\|_{L^\infty(0,a)}} \in K$, and, thus, $u \in K_0^I$.

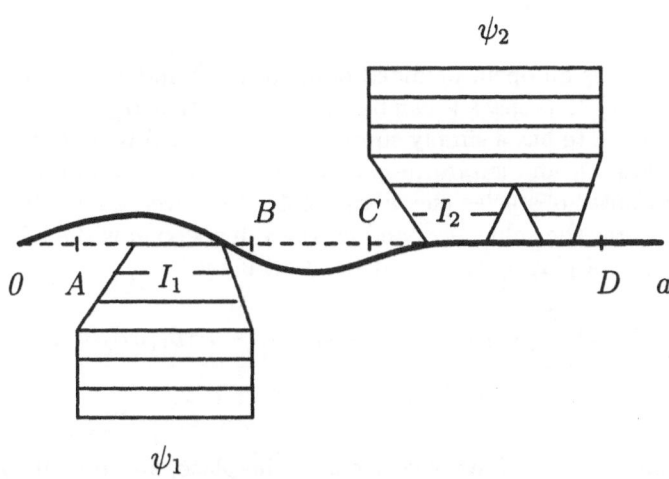

FIGURE 4.3. Example 4.3 (c).

Now, consider the equation (4.42) corresponding to the variational inequality (4.43). In this case, $u_1 \notin K_0$ because $I_2 \neq \emptyset$, and we cannot apply Theorem 3.4 on $[0, \lambda_1]$. However, if $I_1 \subset (0, a/2)$ and $I_2 \subset (a/2, a)$, then,

$$u_2(x) = \sin\left(\frac{2\pi x}{a}\right) = \begin{cases} > 0 & \text{if } x \in (0, a/2), \\ < 0 & \text{if } x \in (a/2, a). \end{cases}$$

Therefore, $u \in K_0^I$ by the above remark. Because λ_2 is simple, we have the following corollary:

Corollary 4.7 *If*

$$\{x : \psi_1(x) = 0\} \subset (0, a/2), \ \{x : \psi_2(x) = 0\} \subset (a/2, a),$$

then, there exists a global bifurcation branch of nontrivial solutions of (4.40) that emanates from $\{0\} \times [0, 4\pi^2/a^2]$ and satisfies the alternative in Theorem 3.2.

Example 4.4 In this example, we consider the global bifurcation problem for variational inequalities in the theory of thin plates. The buckling problem for constrained plates under von Kárman's law is formulated as the following variational inequality (cf. [70], [75], [33]):

$$
\begin{cases}
a(u, v - u) + \displaystyle\int_\Omega \sum_{i,j=1}^2 \sigma_{ij}(u)\, \partial_i u\, \partial_j(v - u) \\
\qquad \geq \lambda \displaystyle\int_\Omega \sum_{i,j=1}^2 \sigma_{ij}^0\, \partial_i u\, \partial_j(v - u), \quad \forall v \in K, \\
u \in K.
\end{cases}
\tag{4.45}
$$

Here, $\Omega \subset \mathbb{R}^2$ is an open, bounded domain in $\mathbb{R}^2$ and $K = \{v \in V : \psi_1 \leq v \leq \psi_2 \text{ on } \Omega\}$. The space V can be $H_0^1(\Omega) \cap H^2(\Omega)$ or $H_0^2(\Omega)$, depending on whether the plate has a simply supported or clamped boundary condition. The functions ψ_1 and ψ_2 are supposed measurable on Ω and represent the upper and lower obstacles (see Figure 4.4). Moreover, $\psi_1 \leq 0 \leq \psi_2$ on Ω, ψ_1 may assume the value $-\infty$, and ψ_2 the value $+\infty$. a is the bilinear form in the theory of plates, which is defined on V by

$$
a(u, v) = D \int_\Omega [(\partial_{11} u\, \partial_{11} v + \partial_{22} u\, \partial_{22} v) + \nu\,(\partial_{11} u\, \partial_{22} v + \partial_{22} u\, \partial_{11} v)
$$
$$
+ (1 - \nu)\, \partial_{12} u\, \partial_{12} v]\, dx, \ \forall u, v \in V.
$$
$$\tag{4.46}$$

$D > 0$ denotes the stiffness coefficient of the plate, and $\nu \in (0, 1/2)$ is the Poisson ratio.

It is known that a is a continuous, symmetric, coercive bilinear form on V (cf. [33], [36]). Consequently, we can consider on V the inner product defined by

$$\langle u, v \rangle = a(u, v), \ \forall u, v \in V,$$

which is equivalent to the usual inner product of $H^2(\Omega)$, restricted to V. The coefficients σ_{ij} and σ_{ij}^0 satisfy the following conditions (cf. [33]):

$$\sigma_{ij} = \sigma_{ji}, \ \sigma_{ij}^0 = \sigma_{ji}^0, \ \forall i, j \in \{1, 2\}.$$

Moreover, for all $i, j \in \{1, 2\}$, $\sigma_{ij}^0 \in L^2(\Omega)$, and σ_{ij} is completely continuous from $H^2(\Omega)$ to $L^2(\Omega)$.

Let $L, C : V \to V$ be defined by

$$\langle Lu, v \rangle = \int_\Omega \sum_{i,j=1}^{2} \sigma_{ij}^0 \, \partial_i u \, \partial_j v \, dx$$

and

$$\langle C(u), v \rangle = \int_\Omega \sum_{i,j=1}^{2} \sigma_{ij}(u) \, \partial_i u \, \partial_j v \, dx,$$

for all $u, v \in V$. From [32] and [33], we know that L is a continuous, self-adjoint, compact linear mapping from V to V, and C is a completely continuous (nonlinear) mapping. Moreover

$$\|C(u)\| \leq C_0 \|u\|^3, \ \forall u \in V. \tag{4.47}$$

Setting $B(u, \lambda) = \lambda Lu - C(u)$, $u \in V, \lambda \in \mathbb{R}$, B is a completely continuous mapping from $V \times \mathbb{R}$ to V.

With these notations, one sees that (4.45) can be written in the operator form as

$$\begin{cases} \langle u, v - u \rangle + \langle C(u), v - u \rangle \geq \lambda \langle Lu, v - u \rangle, \ \forall v \in K, \\ u \in K, \end{cases}$$

or, equivalently,

$$\begin{cases} \langle u - B(u, \lambda), v - u \rangle \geq 0, \ \forall v \in K, \\ u \in K. \end{cases}$$

From (4.47),

$$\lim_{u \to 0} \frac{C(u)}{\|u\|} = 0,$$

which, in turn, implies that the mapping $f : V \times \mathbb{R} \to V$ defined by $f(u, \lambda) = \lambda Lu$, $u \in V$, is the (partial) derivative of $B(u, \lambda)$ at 0 in the sense of (3.5).

Let $Z_i = \{x \in \Omega : \psi_i(x) = 0\}$, $i = 1, 2$. As in the the above examples, one can prove that the support cone K_0 of K is given by

$$K_0 = \{v \in V : u \geq 0 \text{ on } Z_1, \ u \leq 0 \text{ on } Z_2\}.$$

The homogenized variational inequality of (4.45), in this case, is the following:

$$\begin{cases} \langle u, v - u \rangle \geq \lambda \langle Lu, v - u \rangle, \ \forall v \in K_0, \\ u \in K_0. \end{cases}$$

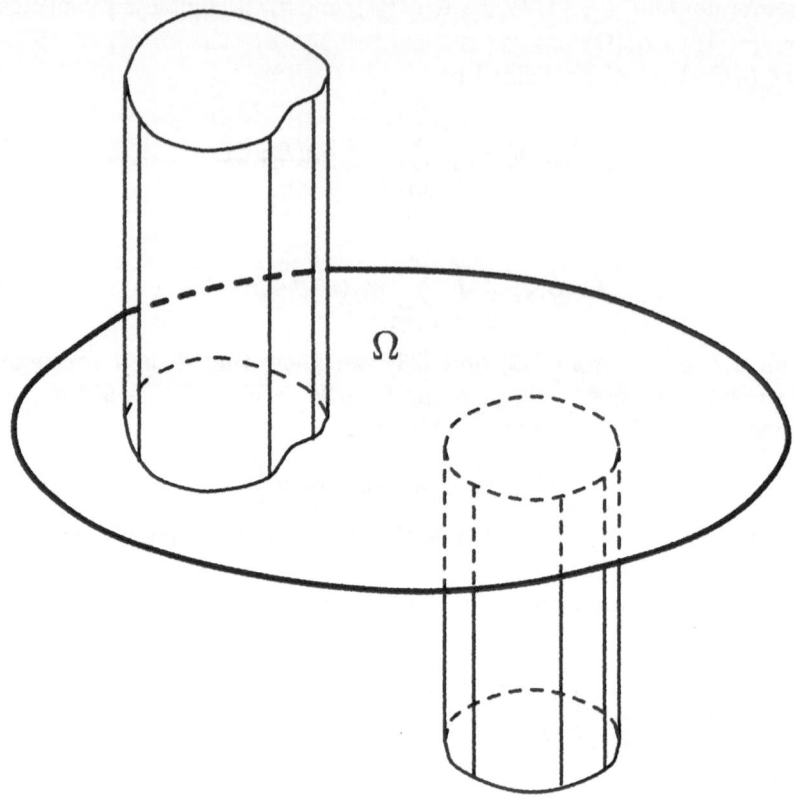

FIGURE 4.4. Example 4.4.

As in Example 4.3, one can apply Theorem 3.2, Corollary 3.3, or Theorem 4.4 to obtain existence results of global bifurcation branches for different obstacles ψ_1 and ψ_2. For example, when $\psi_1 < 0 < \psi_2$ on Ω, $K_0 = V$. Therefore, the following corollary is immediate:

If $(0, \lambda)$ is a bifurcation point of (4.45), then, λ is an eigenvalue of the equation

$$u = \lambda L u, \ u \in V.$$

Conversely, if λ is an eigenvalue of odd (algebraic) multiplicity of the above equation, then, there exists a branch of nontrivial solutions of (4.45) bifurcating from $(0, \lambda)$ and satisfying the alternative in Theorem 3.2.

Similar results are valid for other kinds of obstacles (cf. Example 4.3). More applications for plate and beam problems will be considered in Examples 6.7, 6.9, 6.10, and section 6.3, where the constraints are represented by convex functionals.

4.2.2   A further corollary

In many cases, the smallest (positive) eigenvalue of the homogenized variational inequality (3.6) of (3.1) is also the smallest (positive) eigenvalue of the corresponding equation (4.18) with an eigenvector in K_0. In those cases, we can apply Theorem 4.4 to obtain global properties of the bifurcation branch at this eigenvalue whenever its simplicity can be verified. In this section, we shall employ a finding of Miersemann ([72], [75]) to verify such a result. We assume that B satisfies (3.18) and (3.19). Thus, f is of the form,

$$f(u, \lambda) = \lambda \beta u, \ u \in V, \ \lambda \in \mathbb{R}.$$

We also assume that β is a compact, self-adjoint linear operator on V.

We shall also suppose that there exist two symmetric continuous bilinear forms

$$a_1, a_2 : V \times V \to \mathbb{R},$$

such that $\langle \cdot, \cdot \rangle$ can be decomposed into a sum of a_1 and a_2,

$$\langle u, v \rangle = a_1(u, v) + a_2(u, v),$$

for all $u, v \in V$.

We assume, furthermore, that a_1 is coercive on V and denote by K' the polar cone of K_0 corresponding to the inner product a_1:

$$K' = \{w \in V : a_1(w, v) \leq 0, \ \forall v \in K_0\}.$$

We have the following corollary of Theorem 4.4.

Corollary 4.8 *Assume the following conditions :*
   (a) $-K' \subset K_0$,
   (b) $a_2(v, w) \geq 0$, $\langle \beta v, w \rangle \leq 0$ *for all* $v \in K_0, w \in K'$.
   (c) If u, v *are eigenvectors of (4.18) in* K_0 *corresponding to the smallest eigenvalue*

$$\lambda_V = \left(\max_{v \in V \setminus \{0\}} \frac{\langle \beta v, v \rangle}{\|v\|^2} \right)^{-1},$$

then, $\langle u, v \rangle \neq 0$,
   (d) $K_0 \cap (-K_0) \cap S_{\lambda_V} = \{0\}$, *where* S_{λ_V} *is the set of solutions of (4.18) with* $\lambda_0 = \lambda_V$.
   Then, for the smallest eigenvalue λ_0 *of (3.6),*

$$\lambda_0 = \left(\max_{v \in K_0 \setminus \{0\}} \frac{\langle \beta v, v \rangle}{\|v\|^2} \right)^{-1},$$

$\lambda_V = \lambda_0$, *and, moreover,* $(0, \lambda_0)$ *is a bifurcation point of (3.1) from which a global bifurcation branch emanates which satisfies the alternative in Theorem 3.2.*

Proof. From Theorems 2.1 and 2.2 in [72], we know that the above assumptions (a)-(d) imply that $\lambda_V = \lambda_0$ and that λ_0 is a simple eigenvalue of (4.18) in K_0. We shall prove that λ_0 is a simple eigenvalue of (3.6) in the sense of Theorem 4.4.

Suppose u is an eigenvector of (3.6) corresponding to λ_0. We prove that (u, λ_0) is a solution of (4.18). Setting $v = 0$ and $v = 2u$ in (3.6) with $\lambda = \lambda_0$,

$$\langle u - \lambda_0 \beta u, u \rangle = 0.$$

Hence $\|u\|^2 = \lambda_0 \langle \beta u, u \rangle$, or

$$\lambda_0^{-1} = \frac{\langle \beta u, u \rangle}{\|u\|^2}. \tag{4.48}$$

By the definition of λ_V,

$$\frac{\langle \beta u, u \rangle}{\|u\|^2} \geq \frac{\langle \beta v, v \rangle}{\|v\|^2}, \ \forall v \in V \setminus \{0\}. \tag{4.49}$$

We now employ arguments used in extremum problems for smooth functionals. From (4.49),

$$\langle \beta u, u \rangle \|v\|^2 \geq \langle \beta v, v \rangle \|u\|^2, \ \forall v \in V.$$

Choosing $v = u + tw, w \in V$, for all $t > 0$,

$$(\|u\|^2 + t^2\|w\|^2 + 2t\langle u, w \rangle)\langle \beta u, u \rangle \geq (\langle \beta u, u \rangle + t^2 \langle \beta w, w \rangle + 2t\langle \beta u, w \rangle)\|u\|^2.$$

Thus,

$$t\|w\|^2 \langle \beta u, u \rangle + 2\langle u, w \rangle \langle \beta u, u \rangle \geq t\langle \beta w, w \rangle + 2\langle \beta u, w \rangle \|u\|^2.$$

Letting $t \to 0^+$,

$$\langle u, w \rangle \langle \beta u, u \rangle \geq \langle \beta u, w \rangle \|u\|^2,$$

i.e., $\langle u, w \rangle \langle \beta u, u \rangle / \|u\|^2 \geq \langle \beta u, w \rangle, \ \forall w \in V$.

Therefore, by (4.48), $\langle u, w \rangle \lambda_0^{-1} \geq \langle \beta u, w \rangle, \ \forall w \in V$. Changing w to $-w$, one sees that $\langle u, w \rangle \lambda_0^{-1} = \langle \beta u, w \rangle, \ \forall w \in V$. In other words,

$$\langle u - \lambda_0 \beta u, w \rangle = 0, \ \forall w \in V,$$

i.e., u is an eigenvalue of (4.18) that corresponds to λ_0. We have proved that $u \in S_{\lambda_0}$. Now, let $u_1, u_2 \in K_0 \setminus \{0\}$ be two eigenvectors of (3.6) corresponding to λ_0. By the above proof, one knows that u_1 and u_2 are also eigenvectors of (4.18) in K_0. Because λ_0 is simple as an eigenvalue of (4.18), $u_2 = \mu u_1$ for some $\mu \in \mathbb{R} \setminus \{0\}$, by the quoted results of Miersemann. If $\mu < 0$, then, $u_2 \in K_0$, $-u_2 = (-\mu)u_1 \in K_0$, and $u_2 \in S_{\lambda_0}$. Hence, $u_2 = 0$, by condition (d). This contradiction proves that $\mu > 0$, and that λ_0 is a simple eigenvalue of the variational inequality (3.6). Moreover, as observed before, every eigenvector of (3.6) associated with λ_0 is also an eigenvector of (4.18). Hence, we have verified all conditions of Theorem 4.4, and there exists a global bifurcation branch of (3.1) that bifurcates from $\{0\} \times [0, \lambda_0]$ and satisfies the conclusion of Theorem 3.2. However, because λ_0 is the smallest eigenvalue of (3.6), by Theorem 3.2 (I), we see that the only possible bifurcation point of (3.1) in $\{0\} \times [0, \lambda_0]$ is $(0, \lambda_0)$. ∎

Now, we consider some applications of this corollary.

Example 4.5 [A second-order elliptic variational inequality] Let $\Omega \subset \mathbb{R}^N$ be an open, bounded domain with a smooth boundary. We consider the following variational inequality, which represents an obstacle problem for a membrane stretched over Ω:

$$\begin{cases} \int_\Omega \nabla u \nabla (v - u) \geq \int_\Omega g(x, u, \lambda)(v - u)dx, \ \forall v \in K, \\ u \in K. \end{cases} \tag{4.50}$$

Here, $V = H_0^1(\Omega)$ with the usual inner product $\langle u, v \rangle = \int_\Omega \nabla u \nabla v$, $u, v \in V$, and

$$K = \{v \in H_0^1(\Omega) : v \geq 0 \ \text{on} \ \Omega, \ v \leq \psi \ \text{on} \ \Omega_0\},$$

where Ω_0 is a subdomain of Ω and ψ is an essentially bounded, measurable function on Ω_0 with $\psi(x) \geq \psi_0 > 0$ on Ω_0. We assume, furthermore, that

$$g : \Omega \times \mathbb{R}^2 \to \mathbb{R}$$

is a Carathéodory function that satisfies conditions as in Example 3.2 and that

$$D_2 g(x, 0, \lambda) = \lambda k(x),$$

for $x \in \Omega$, $\lambda > 0$, where k is a bounded, smooth function on Ω, such that $k \not\equiv 0$, $k \geq 0$ on Ω.

As in Examples 3.2, 4.2, by letting

$$\langle B(u, v), v \rangle = \int_\Omega g(x, u(x), \lambda)v(x)dx, \ u, v \in V, \ \lambda \in \mathbb{R},$$

B is completely continuous from $V \times \mathbb{R}$ to V and the derivative f of B at 0 in the sense of (3.5) is given by

$$\langle f(u, \lambda), v \rangle = \lambda \int_\Omega k(x)u(x)v(x)dx, \ u, v \in V, \ \lambda \in \mathbb{R}.$$

On the other hand, as in the above examples, one can check that the support cone of K is given by

$$K_0 = \{v \in H_0^1(\Omega) : v \geq 0 \ \text{on} \ \Omega\}.$$

The homogenized variational inequality of (4.50), therefore, is expressed by

$$\begin{cases} \int_\Omega \nabla u \nabla(v - u) \geq \lambda \int_\Omega k u (v - u) dx, \ \forall v \in K_0, \\ u \in K_0, \end{cases} \tag{4.51}$$

or

$$\begin{cases} \langle u - \lambda \beta u, v - u \rangle \geq 0, \ \forall v \in K_0, \\ u \in K_0, \end{cases}$$

with $\beta : V \to V$, $\langle \beta u, v \rangle = \int_\Omega k u v$. We immediately see that β is compact and symmetric from V to V.

Choosing $a_1(u, v) = \langle u, v \rangle$ and $a_2 = 0$, we check that all conditions in Corollary 4.8 are satisfied.

Let $w \in K' = \{u \in V : \langle u, v \rangle \leq 0, \ \forall v \in K_0\}$. We prove that $w \leq 0$ on Ω. In fact, let $h \in L^2(\Omega)$, $h \geq 0$ on Ω. By the usual existence and regularity results for Poisson's equation ([43]), there exists a unique $v \in H^2(\Omega) \cap H_0^1(\Omega)$, such that

$$-\Delta v = h \ \text{ on } \ \Omega.$$

By the maximum principle for the Laplacian ([43]), $v \geq 0$ on Ω. Hence, $v \in K_0$. Because $w \in K'$,

$$0 \geq a_1(v, w) = \langle v, w \rangle = \int_\Omega \nabla v \nabla w = - \int_\Omega \Delta v . w = \int_\Omega h w.$$

Hence, $\int_\Omega h w \leq 0$, $\forall h \in L^2(\Omega), h \geq 0$ on Ω. It follows that $w \leq 0$ on Ω, i.e., $-w \in K_0$. For $v \in K_0, w \in K'$,

$$\langle \beta v, w \rangle = \int_\Omega k v w \ \leq 0$$

(because $w \leq 0 \leq v$ on Ω). Now assume that $u, v \in K_0 \setminus \{0\}$ satisfy (4.18) (with $\lambda_0 = \lambda_V > 0$), i.e.,

$$u = \lambda_V \beta u, \ v = \lambda_V \beta v. \tag{4.52}$$

Hence $\langle u, v \rangle = \langle \lambda_V \beta u, v \rangle = \lambda_V \langle \beta u, v \rangle = \lambda_V \int_\Omega k u v$. If $\langle u, v \rangle = 0$, then, $\int_\Omega k u v = 0$. Because $u, v \geq 0, k \geq 0$ on Ω, we must have $k u v = 0$ on Ω. On the other hand, from (4.52), u satisfies the boundary value problem:

$$\begin{cases} -\Delta u = \lambda k u \ \text{ in } \ \Omega, \\ u = 0 \ \text{ on } \ \partial \Omega. \end{cases} \tag{4.53}$$

Suppose that $u(x_0) = 0 = \min_\Omega u$ for some $x_0 \in \Omega$. Because $-\Delta u \geq 0$ on Ω, by the strong maximum principle ([43]), $u = $ constant on Ω, i.e.,

$u \equiv 0$ on Ω. This contradiction proves that $u(x) > 0$, $\forall x \in \Omega$. By a similar proof for v, one obtains $u, v > 0$ on Ω. This implies that $k \equiv 0$ on Ω, which contradicts the assumption about k. Hence, assumption (c) in Corollary 4.8 is satisfied. We have checked all conditions of Corollary 4.8 and, therefore, obtain the following result:

Corollary 4.9   *The smallest positive eigenvalue λ_0 of (4.53) is also the smallest eigenvalue of (4.51), and $(0, \lambda_0)$ is a bifurcation point of (4.50). Moreover, at $(0, \lambda_0)$, there bifurcates a global branch of nontrivial solutions of (4.50) that satisfies the alternative in Theorem 3.2.*

Example 4.6   Once more, we consider the von Kármán model for plates, as in Example 4.4. In this example, we assume that Ω is a convex, bounded domain in $\mathbb{R}^2$. Let K be the following convex set (see Figure 4.5):

$$K = \{v \in H_0^1(\Omega) : v \geq 0 \text{ on } \Omega, \ v \leq \psi_2 \text{ on } \Omega_0\},$$

where Ω_0 is a subdomain of $\overline{\Omega}$ (Ω_0 may coincide with Ω) and $\inf_{\Omega_0} \psi_2 > 0$. (This means that, in this case, $\psi_1 \equiv 0$ and $\psi_2 = \infty$ outside of Ω_0, and ψ_2 is strictly positive on Ω_0.) Then, as in the examples in Chapter 3, we can prove that the support cone K_0 is given by

$$K_0 = \{v \in H_0^1(\Omega) \cap H^2(\Omega) : v \geq 0 \text{ on } \Omega\}.$$

The homogenized variational inequality of (4.45), in this case, is expressed by

$$\begin{cases} a(u, v - u) \geq \lambda \int_\Omega \sum_{i,j} \sigma_{ij}^0 \, \partial_i u \, \partial_j (v - u), \ \forall v \in K_0, \\ u \in K_0. \end{cases}$$

We assume that $\sigma_{ij}^0 = \delta_{ij}$, $\forall i, j \in \{1, 2\}$ (δ_{ij} is the Kronecker symbol). The above inequality becomes

$$\begin{cases} a(u, v - u) \geq \lambda \int_\Omega \nabla u \nabla (v - u), \ \forall v \in K_0, \\ u \in K_0. \end{cases}$$

From (4.46) (cf. [36], [72], [75]),

$$a(u, v) = a_1(u, v) + a_2(u, v)$$

with

$$a_1(u, v) = D \int_\Omega \Delta u \Delta v,$$

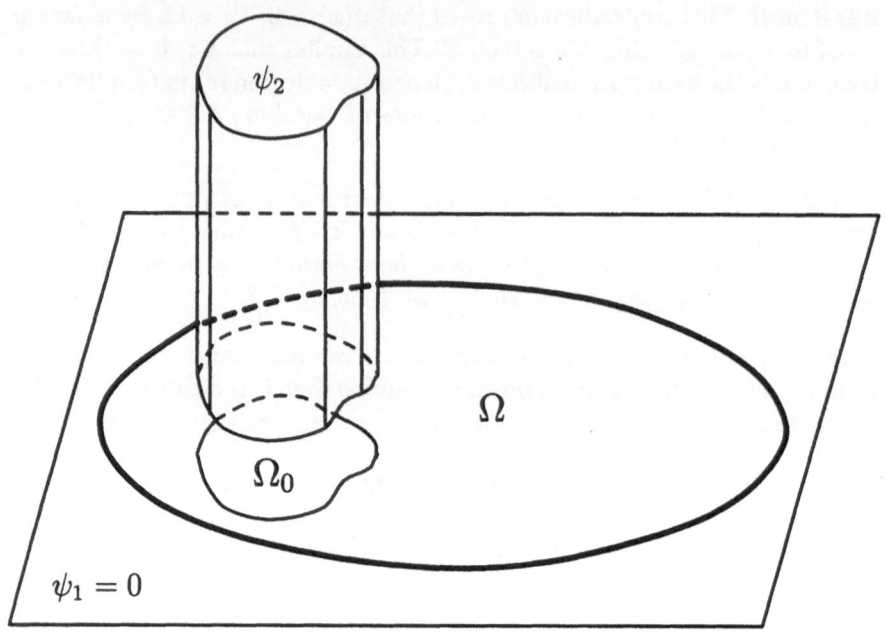

FIGURE 4.5. Example 4.6.

$$a_2(u,v) \;=\; D(1-\nu)\int_\Omega [2\partial_{12}u\,\partial_{12}v - (\partial_{11}u\,\partial_{22}v + \partial_{22}u\,\partial_{11}v)]$$

$$=\; -(1-\nu)\int_\Omega \frac{\partial\theta}{\partial\tau}\frac{\partial u}{\partial n}\frac{\partial v}{\partial n},$$

where, as usual, n and τ denote, respectively, the (unit) outward normal and tangential vectors on $\partial\Omega$ and θ is the angle between the x_1 axis and n. Moreover, in the present case,

$$\langle \beta u, v\rangle = \int_\Omega \nabla u \nabla v, \;\; \forall u,v \in V.$$

Because Ω is convex, it is proved in [72] that the assumptions (a), (b), and (c) in Corollary 4.8 are satisfied by a_1, a_2, and β. Note that, in our case, $K_0 \cap (-K_0) = \{0\}$ and condition (d) in Corollary 4.8 obviously holds. Hence, we can apply that corollary to have a global bifurcation branch for (4.45) emanating from $(0,\lambda_0)$, where λ_0 is the principal eigenvalue of the equation

$$\begin{cases} a(u,v) = \lambda \int_\Omega \nabla u \nabla v, \; \forall v \in H_0^1(\Omega)\cap H^2(\Omega), \\ u \in H_0^1(\Omega) \cap H^2(\Omega). \end{cases}$$

4.2.3   When inequalities are equations

In this short section, we consider some particular cases where bifurcation results from a variational inequality may be deduced from bifurcation of positive solutions of a related smooth equation. For instance, let us consider the variational inequality (3.27) in Example 3.2. We assume that g is a smooth function (with respect to x and u), such that $g(x, 0, \lambda) \equiv 0$, and the closed, convex set K is given by

$$K = \{u \in H_0^1(0,1) : u \leq 0 \text{ on } [A, B], \ u \geq l \text{ on } [C, D]\},$$

with $0 < A < B < C < D < 1$, $l < 0$ on $[C, D]$.

We prove that, if w satisfies the semilinear equation

$$
\begin{cases}
\displaystyle\int_0^1 w'v' dx = \int_0^1 g(x, w(x), \lambda)v dx, \ \forall v \in H_0^1(0, A), \\
w \in H_0^1(0, A),
\end{cases}
\tag{4.54}
$$

and $w \geq 0$ on $[0, A]$, then,

$$
u(x) =
\begin{cases}
w(x) & \text{if } x \in [0, A], \\
0 & \text{if } x \in [A, 1],
\end{cases}
$$

is a solution of (3.27).

In fact, because g is smooth and $w \in H_0^1(0, A) \subset C[0, A]$, by usual regularity properties of second-order elliptic equations and bootstrap arguments ([43]), one can prove that (4.54) is equivalent to

$$
\begin{cases}
w'' = -g(\cdot, w, \lambda) \text{ on } (0, A), \\
w(0) = w(A) = 0.
\end{cases}
\tag{4.55}
$$

Moreover, if w satisfies (4.54), then, $w \in C^2[0, A]$, and w is a classical solution of (4.55). Because $w \geq 0$ on $[0, A]$ and $w(A) = 0$, $w'_-(A) \leq 0$ ($w'_-(A)$ is the left derivative of w at A).

On the other hand, because $w(A) = 0$ and u is continuous on $[0, 1]$, $u \in H_0^1(0, 1)$ (cf. [14]). Let $v \in K$. Because $v(0) = 0$ and $v(A) \leq 0$,

$$
\begin{aligned}
\int_0^1 u'(v - u)' &= \int_0^A u'(v - u)' = \int_0^A w'(v - w)' \\
&= -\int_0^A w''(v - w) + [w'(v - w)]_0^A \\
&= \int_0^A g(\cdot, w, \lambda)(v - w) + w'_-(A)v(A) \\
&\geq \int_0^A g(x, w(x), \lambda)(v - w)(x) dx \\
&= \int_0^1 g(x, u(x), \lambda)(v - u)(x) dx.
\end{aligned}
$$

Hence, u is a solution of (3.27).

We also have a similar proof for the case when w is the negative solution on $[0, A]$ and

$$K = \{u \in H_0^1(0,1) : u \geq 0 \text{ on } [A, B], \; u \leq l \text{ on } [C, D]\}.$$

From this observation, we see that, if $(0, \lambda)$ is a bifurcation point for (4.54) with a bifurcation branch of positive solutions, then, it is also a bifurcation point of (3.27), and the above branch is a bifurcation branch of (3.27). Therefore, we can deduce corresponding properties for bifurcation of variational inequalities from known results for bifurcation of positive solutions of smooth equations.

The following are examples of variational inequalities whose associated equations have known properties of bifurcation branches of positive solutions.

Example 4.7 (a) Let the function g in (3.27) be given by

$$g(x, u, \lambda) = \lambda[a(x)u + F(x, u, \lambda)], \; x \in (0, A), \; u, \lambda \in \mathbb{R},$$

where a and F are in class C^1 and satisfy the following conditions

$$a(x) \geq a_0 > 0 \text{ on } (0, A), \text{ and}$$

$$F(x, u, \lambda) = o(|u|),$$

as $u \to 0$, uniformly for λ in bounded intervals.

The linear equation corresponding to (4.55), in this case, is

$$\begin{cases} w'' = -\lambda a(x)w \text{ on } (0, A), \\ w(0) = w(A) = 0. \end{cases} \tag{4.56}$$

Now, according to Theorem 2.34, [100], there exists a global bifurcation branch of positive solutions of (4.55) emanating from $(0, \lambda_1)$, where λ_1 is the smallest positive eigenvalue of (4.56) given by the Krein-Rutman theorem. Moreover, this bifurcation branch is unbounded.

From the above remark, we see that there exists an unbounded bifurcation branch of solutions of (3.27) emanating from $(0, \lambda_1)$.

(b) Consider the case where g is of the form,

$$g(x, u, \lambda) = \lambda f(x, u), \; x \in (0, A), \; u, \lambda \in \mathbb{R},$$

where $f \in C^1([0, A] \times \mathbb{R})$, $f(x, 0) = 0$ on $[0, A]$, and there exists $x_0 \in [0, A]$, such that $f_u(x_0, 0)(= (\partial f / \partial u)(x_0, 0)) > 0$. In this case, by a result of Hess and Kato (Theorem 1, [46]), we know that the linear equation corresponding to (4.55),

$$\begin{cases} w'' = -\lambda f_u(x,0)w \text{ on } (0,A) \\ w(0) = w(A) = 0, \end{cases} \qquad (4.57)$$

has a unique positive eigenvalue λ_1 corresponding to a positive eigenfunction of (4.57), and, moreover, $(0, \lambda_1)$ is a bifurcation point of (4.54) with a corresponding unbounded branch of positive solutions.

Therefore, by the above remark, $(0, \lambda_1)$ is a bifurcation point of the variational inequality (3.27) where an unbounded branch of positive solutions bifurcates.

We conclude this chapter with two examples of bifurcation problems motivated by some problems in mechanics. The first is a torsion problem for elastic plastic bars, and the second is concerned with variational inequalities associated with the Stokes problem in fluid mechanics.

Example 4.8 In this example we revisit Example 4.5 and change the constraint set K to the following:

$$K = \{v \in H_0^1(\Omega) : |\nabla v| \leq 1 \text{ a.e. on } \Omega\}. \qquad (4.58)$$

The variational inequality, thus, resulting is related to an elastic plastic torsion problem for a cylindrical bar with cross section $\Omega \subset \mathbb{R}^2$ and g measuring the angle of twist of the bar which is not clamped (cf. [18], [39], and [103]). We retain the same hypotheses on g and k as in Example 4.5 and find that K is a closed, convex, and bounded set (as follows from Poincaré's inequality) which contains 0 as a demi-interior point (cf. Lemma 4.5) and the support cone K_0 of K is the whole space $H_0^1(\Omega)$. (In fact, for $w \in C_0^\infty(\Omega)$, because ∇w is bounded, $u \in tK$ for some $t > 0$ large. As $C_0^\infty(\Omega)$ is dense in $H_0^1(\Omega)$, it follows that $K_0 = H_0^1(\Omega)$.) Hence, the homogenized inequality becomes a linear equation containing a spectral parameter λ (an analysis similar to that used in the previous example shows this). Thus, all eigenvalues of odd multiplicity of the linear equation will yield bifurcation points.

Other interesting cases related to this example are when K is chosen in the following way (see [18] again). Assume that there are given open sets (see Figure 4.6),

$$O_1, \cdots, O_n \subset \Omega,$$

whose closures are disjoint. Let

$$K = \{v \in H_0^1(\Omega) : |\nabla v| \leq 1 \text{ a.e. on } \Omega, \ v = c_i \text{ on } O_i\}, \qquad (4.59)$$

where the constants c_i, $i = 1, \cdots, n$ are not given but will depend on the solution. Physically, the sets O_i represent cavities in the bar. Again, it is

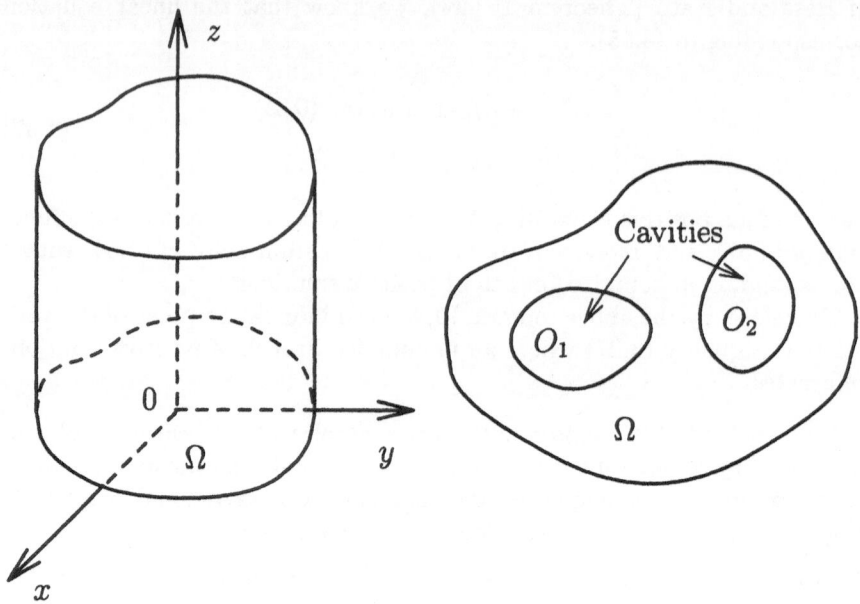

FIGURE 4.6. Example 4.8 - Elastic plastic torsion problem.

easy to see that the support cone K_0 is a linear space,

$$K_0 = \{v \in H_0^1(\Omega) : \ v = c_i \text{ on } O_i\},$$

and, thus, one may determine a linearization at 0.

Other second-order elliptic problems may also be treated by the above types of procedures.

Example 4.9 [Variational inequalities associated with the Stokes problem]

(a) In this example, we consider bifurcation problems related to nonlinear versions of the Stokes equation. We are concerned with the equilibrium state of a steady fluid flow within a reservoir $\Omega \subset \mathbb{R}^3$, under the action of an external force $g = (g_1, g_2, g_3)$. As usual, we assume that Ω is bounded with a smooth boundary.

Let $w = (w_1, w_2, w_3) : \Omega \to \mathbb{R}^3$ be the velocity field of the flow. The functional (to be minimized) associated with the system is given by (see [37])

$$I(w) = \frac{1}{2} \int_\Omega [|Dw|^2 - g \cdot w],$$

where $Dw = (\partial_i w_j)_{1 \le i,j \le 3}$,

$$|Dw| = \left[\sum_1^3 \int_\Omega (\partial_i w_j)^2\right]^{1/2}$$

and $g \cdot w$ is the usual dot product of g and w. We assume that the fluid is incompressible and that the velocity on the boundary is v_0, which, for purposes of illustration, we simply assume equals 0. The set of all admissible velocity fields is a subset K of the space

$$V = \{w \in [H_0^1(\Omega)]^3 : \text{div}\, w = 0 \text{ in } \Omega\}.$$

V is a Hilbert space with the norm and inner product given by those in $[H_0^1(\Omega)]^3$.

In the case without constraints (on the velocity), i.e., $K = V$, the velocity field u of the flow, which is a solution of the minimization problem,

$$u \in V : I(u) = \min_{v \in V} I(v),$$

satisfies the following equation

$$\begin{cases} \int_\Omega Du : Dv - \int_\Omega g \cdot v = 0, \ \forall v \in V \\ u \in V, \end{cases}$$

where $Du : Dv = \sum_{i,j=1}^3 \partial_i u_j \partial_i v_j$. The classical form of this variational equation is the usual Stokes problem:

$$\begin{cases} -\Delta u &=& g - \nabla p, \\ \text{div}\, u &=& 0 \text{ in } \Omega \\ u &=& 0 \text{ on } \partial\Omega. \end{cases}$$

However, if we impose some requirements (constraints) on the velocity, for example, on the speed of the flow:

$$|w(x)| \le c \text{ for a.e } x \in \Omega,$$

(here $c > 0$ is the limit speed), then, the set of admissible velocity fields becomes

$$K = \{w \in [H_0^1(\Omega)]^3 : \text{div}\, w\,(x) = 0, \ |w(x)| \le c \text{ for a.e } x \in \Omega\}.$$

The Euler-Lagrange inequality, corresponding to the minimization problem

$$u \in K : I(u) = \min_{v \in K} I(v),$$

is the following variational inequality:

$$\begin{cases} \int_\Omega Du : D(v-u)dx - \int_\Omega g \cdot (v-u)dx \geq 0, \ \forall v \in K, \\ u \in K. \end{cases} \tag{4.60}$$

Now, we assume that the external force g depends on the velocity (in some nonlinear manner) and also on a real parameter λ, which is usually a measurement of the magnitude of the force. Hence, $g = g(x, u, \lambda)$ is a mapping from $\Omega \times \mathbb{R}^3 \times \mathbb{R}$ to $\mathbb{R}^3$, and (4.60) becomes the following (nonlinear) variational inequality:

$$\begin{cases} \int_\Omega Du : D(v-u)dx - \int_\Omega g(\cdot, u, \lambda) \cdot (v-u)dx \geq 0, \ \forall v \in K \\ u \in K. \end{cases} \tag{4.61}$$

Assuming that $g(x, 0, \lambda) = 0$ (i.e., there is no external force acting on stagnant points of Ω), then, $u = 0$ is obviously a solution of (4.61). This trivial solution corresponds to the state of zero flow. If λ changes, then, we may have nonzero velocity fields, represented by nontrivial solutions of (4.61). This leads to bifurcation problems for (4.61).

To homogenize (4.61) near 0, we assume that g is a Carathéorory function differentiable with respect to u. Moreover, g and $D_u g$ satisfy the usual growth conditions, for example,

$$|g(x, u, \lambda)| \leq A(\lambda) + B(\lambda)|u|^{\frac{s}{s-1}},$$

and

$$|D_u g(x, u, \lambda)| \leq A(\lambda) + B(\lambda)|u|^{\frac{s}{s-2}},$$

for $x \in \Omega, u \in \mathbb{R}^3$, and $\lambda \in \mathbb{R}$. Here $A, B \in L^\infty_{loc}(\mathbb{R})$, and $1 < s < 3 (= 2^*)$.

Under those conditions, we can verify that the mapping $B : V \times \mathbb{R} \to \mathbb{R}$ given by

$$\langle B(u, \lambda), v \rangle = \int_\Omega g(x, u, \lambda) \cdot v \, dx$$

is completely continuous. Moreover, for $\sigma_n \to 0^+$, $u_n \rightharpoonup u$ in V, and $\lambda_n \to \lambda$,

$$\frac{1}{\sigma_n} B(\sigma_n u_n, \lambda_n) \to f(u, \lambda) \quad \text{in} \ \ V.$$

Here, $f(u, \lambda)$ is given by

$$\begin{aligned} \langle f(u, \lambda), v \rangle &= \int_\Omega \sum_{i,j=1}^3 D_{u_i} g_j(x, 0, \lambda) u_i(x) v_j(x) dx \\ &= \int_\Omega [v(x)]^T D_u g(x, 0, \lambda) u(x) dx, \end{aligned}$$

where $D_u g = [D_{u_i} g_j]_{i,j=1,2,3}$. On the other hand, the support cone K_0 of K is the whole space V. In fact, for $w \in [C_0^\infty(\Omega)]^3$ such that $\operatorname{div} w = 0$ in Ω, $tw \in K$ for $t > 0$ sufficiently small. As $[C_0^\infty(\Omega)]^3 \cap V$ is dense in V, $V \subset K_0$, i.e. $V = K_0$.

The above arguments show that the homogenized variational inequality associated with (4.61) is the linear equation

$$\begin{cases} \int_\Omega Du : Dv - \int_\Omega v^T D_u g(\cdot, u, \lambda) u = 0, \ \forall v \in V, \\ u \in V. \end{cases} \tag{4.62}$$

We note, here, that the operator A defined by

$$\langle A(u), v \rangle = \int_\Omega Du : Dv, \ u, v \in V,$$

is linear and coercive on $V(\subset [H_0^1(\Omega)]^3)$. Hence, by using the previous results, one can draw conclusions about relationships between global behavior of bifurcation branches of (4.61) and eigenvalues of (4.62). For example, if we assume that

$$D_u g(x, 0, \lambda) = \lambda k(x), \tag{4.63}$$

where $k = [k_{ij}]_{i,j=1,2,3}$ is a matrix in $[L^\infty(\Omega)]^9$, then, (4.62) is the usual eigenvalue problem for the Stokes equation:

$$\begin{cases} \int_\Omega Du : Dv - \lambda \int_\Omega v^T k u = 0, \ \forall v \in V, \\ u \in V. \end{cases} \tag{4.64}$$

From the general results above, it follows that eigenvalues of odd multiplicity of (4.64) yield global bifurcation branches of (4.60).

(b) In some cases, we need to restrict the flow in one direction for some components of the velocity field. For example, if we assume that the constraints (on the velocity) are on the negative directions of the x and z axes in a subdomain $\Omega_0 \subset \Omega$, i.e., the fluid can flow freely on the y axis and positive directions of the x and z axes at points of Ω_0, then, the set K can be chosen as

$$K = \{w \in V : w_1(x) \geq -c, \ w_3(x) \geq -d \text{ for a.e. } x \in \Omega_0\},$$

where $c, \ d \geq 0$ are the bounds on the x and z components of the the velocity. For example, by choosing $c = d = 0$, we mean that the flow can pass through Ω_0 in only one direction along the x and z axes. Then, the support cone

$$K_0 = \{w \in V : w_1 \geq 0, w_3 \geq 0 \text{ a.e. in } \Omega_0\}$$

is a proper cone in V. The results in Chapters 4 and 6 can be used to obtain conditions for global bifurcation. For instance, by applying Corollary 4.6, one has the following result:

Assume (4.63) holds with a symmetric matrix k. Let $\lambda_0 > 0$ be a simple eigenvalue of (4.64) with an eigenvector u, such that

$$\text{ess inf}_{\Omega_0} u_1 > 0, \ \text{ess inf}_{\Omega_0} u_3 > 0.$$

Then, a global branch of nontrivial solutions bifurcates from the trivial solution in $[0, \lambda_0]$.

Further results (e.g., in the case where the matrix k is not symmetric) can be obtained using the theorems in Chapter 6.

(c) Next, consider the case where some constraints are imposed on the circulation $\text{curl } u$ of the flow. The set K can then be chosen as

$$K = \{u \in V : |(\nabla \times u)(x)| \leq c \text{ for a.e. } x \in \Omega\},$$

if the restriction is placed on the magnitude of the circulation. Note that this is a condition similar to that in the elastic plastic torsion problem (Example 4.8). The support cone coincides with the whole space V, and there is global bifurcation at eigenvalues of odd multiplicity.

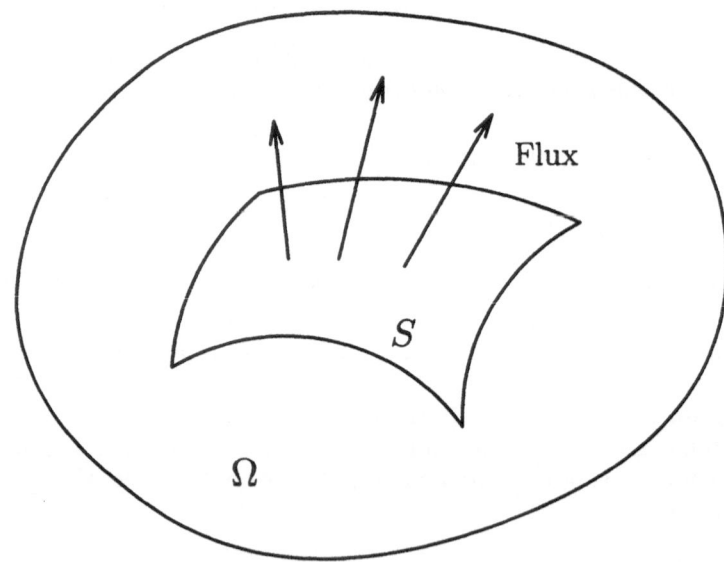

FIGURE 4.7. Example 4.9 (d) - Variational inequality associated with the Stokes problem.

The constraints may also be on the direction of the circulation. One may, for example, require that

$$(\nabla \times u)_1(x) \geq 0, \ (\nabla \times u)_2(x) \geq 0, \ (\nabla \times u)_3(x) \leq 0, \ \text{a.e on } \Omega_0(\subset \Omega).$$

(d) The constraints may also be of a nonlocal nature, e.g. (see Figure 4.7), if we let S be a compact, oriented, smooth surface in Ω. Limiting the flux of the flow across S, then, K is chosen as

$$K = \left\{ u \in V : \left| \int_S u \cdot \nu dS \right| \le c \right\}.$$

Here, ν is the unit normal on S, and $c \ge 0$ is the limit of the flux (if $c = 0$, the fluxes in both normal directions of S balance). In the present case, it is clear that K is a closed, convex subset of V, which has 0 as an interior point, and bifurcation follows from the results of Chapter 3.

If it is required that the average flow passing across S is in the direction of ν, then, the flux is nonnegative. Hence K becomes

$$K = \left\{ u \in V : \int_S u \cdot \nu dS \ge 0 \right\}.$$

Applying the theorems in Chapter 4, we obtain global bifurcation on $[0, \lambda_0]$, if λ_0 is a simple eigenvalue of (4.64) with an eigenvector u, such that

$$\int_S u \cdot \nu dS > 0.$$

5

Bifurcation from Infinity in Hilbert Spaces

In this chapter, we shall consider the problem of bifurcation from infinity of the variational inequality (3.1),

$$
\begin{cases}
\langle u - B(u, \lambda), v - u \rangle \geq 0, \; \forall v \in K, \\
u \in K,
\end{cases}
$$

where K is a closed, convex subset of V (again V is a real Hilbert space with norm $\| \cdot \|$ and inner product $\langle \cdot, \cdot \rangle$, as in Chapter 3), i.e., we consider the problem of the existence of solutions of large norms of (3.1) and, as before, global properties of such solution sets.

Let $\lambda_0 \in \mathbb{R}$. We say that (∞, λ_0) is an asymptotic bifurcation point of (3.1) if there exists a sequence $\{(u_n, \lambda_n)\}$ of solutions of (3.1) such that $\lambda_n \to \lambda_0$ and $\|u_n\| \to \infty$. We say that bifurcation occurs in $\{\infty\} \times [a, b]$ ($a, b \in \mathbb{R}$, $a < b$) if there exists a sequence $\{(u_n, \lambda_n)\}$ of solutions of (3.1) such that $\lambda_n \in [a, b]$, $\forall n$ and $\|u_n\| \to \infty$ as $n \to \infty$.

First, we consider the case where the closed convex set K is a cone. Because K is invariant with respect to inversion with respect to the unit sphere of V, in this case, we can use the usual inversion technique (cf. [99] and [125]) to reduce the problem of bifurcation from infinity to the problem of bifurcation from trivial solutions. Letting $w = u / \|u\|^2$ and

$$
B_1(v, \lambda) = \begin{cases}
\|v\|^2 B(v / \|v\|^2, \lambda) & \text{if } v \neq 0, \\
0 & \text{if } v = 0,
\end{cases}
$$

we see that (3.1) is equivalent to

$$\begin{cases} \langle w - B_1(w, \lambda), v - w \rangle \geq 0, \ \forall v \in K, \\ w \in K. \end{cases} \qquad (5.1)$$

Because $\|w\| \to 0$ if and only if $\|u\| \to \infty$, $(0, \lambda_0)$ is a bifurcation point of (5.1) if and only if (∞, λ_0) is an asymptotic bifurcation point of (3.1). Because $u \mapsto w$ is a homeomorphism of $V \setminus \{0\}$ onto itself, we see that a bifurcation branch of (5.1) emanating from $\{0\} \times [a, b]$ corresponds to a bifurcation branch of (3.1) emanating from $\{\infty\} \times [a, b]$.

We can check that, if B is completely continuous and f satisfies

$$\frac{1}{t} B(tu, \lambda) \to f(u, \lambda), \ \text{as } t \to \infty,$$

then, B_1 is completely continuous and

$$\frac{1}{s} B_1(sv, \lambda) \to f(v, \lambda), \ \text{as } s \to 0^+.$$

Hence, in the case where K is a cone, we can reduce the problem of bifurcation from infinity to that of bifurcation from zero. The result in the previous chapters can, thus, be applied to (5.1) to give corresponding results for bifurcation from infinity for (3.1).

5.1   Asymptotic homogenization

Now, we will consider the more interesting case where K is not necessarily a cone. We remark that the inversion technique used above is not applicable, because, in general, inversion mapping does not transform a convex set into a convex set. This can be seen by the following simple example in $\mathbb{R}^2$. Let K be the strip $[-1, 1] \times \mathbb{R} \subset \mathbb{R}^2$. Through the inversion $x \mapsto x/\|x\|^2$, the straight lines $\{\pm 1\} \times \mathbb{R}$ are transformed into circles centered at $\pm 1/2$ with radii $1/2$:

$$\{(x_1, x_2) : (x_1 \pm 1/2)^2 + x_2^2 = 1/4\}.$$

Hence, K is transformed into the following set

$$\mathbb{R}^2 \setminus [\{(x_1, x_2) \in \mathbb{R}^2 : (x_1 - 1/2)^2 + x_2^2 < 1/4\} \\ \cup \{(x_1, x_2) \in \mathbb{R}^2 : (x_1 + 1/2)^2 + x_2^2 < 1/4\}]$$

which is, clearly, not convex.

However, as shown in the sequel, we can develop results similar to those considered previously for bifurcation from zero. In this chapter, we deal with the Hilbert space case, i.e., the asymptotic bifurcation of (3.1). More general situations will be studied in Chapter 7, where a number of results in this chapter will be revisited in more general settings.

We consider (3.1) with the assumption that $B : V \times \mathbb{R} \to V$ is completely continuous and K is a closed, convex subset of V. By translation, we can reduce the problem to one with $0 \in K$. As before, we see that (3.1) is equivalent to

$$u = P_K[B(u, \lambda)], \qquad (5.2)$$

where P_K is the orthogonal projection of V onto K. Also, $P_K \circ B$ is completely continuous from $V \times \mathbb{R}$ to V.

The general result about global asymptotic bifurcation in [91] (Theorem 2.6) will be crucial for our further analysis. As in the case of bifurcation from trivial solutions, by applying Theorem 2.6 and the equivalence between (3.1) and (5.2), we obtain the following abstract result for bifurcation from infinity of (3.1).

Theorem 5.1 *Suppose that $a < b$ are such that (3.1) with $\lambda = a$ and $\lambda = b$ does not have any solution with large norms (i.e., there exists $R_0 > 0$, such that (3.1) with $\lambda = a, b$ does not have solutions u with $\|u\| \geq R_0$). Furthermore, assume that*

$$\mathrm{d}(I - P_K[B(\cdot, a)], B_R(0), 0) \neq \mathrm{d}(I - P_K[B(\cdot, b)], B_R(0), 0), \qquad (5.3)$$

($R \geq R_0$). Then, there exists an asymptotic bifurcation point (∞, λ) with $a < \lambda < b$, and, moreover, there exists a continuum

$$\mathcal{C} \subset \{(u, \lambda) \in V \times [a, b] : (u, \lambda) \text{ is a solution of (3.1)}\},$$

which is unbounded in $V \times [a, b]$, and either
(i) $\mathcal{C}$ is unbounded in the λ direction, or else
(ii) there exists an interval $[c, d]$, such that $(a, b) \cap (c, d) = \emptyset$ and $\mathcal{C}$ bifurcates from infinity in $V \times [c, d]$.

Proof. Because $B : V \times \mathbb{R} \to V$ is completely continuous and $P_K : V \to V$ ($P_K(V) \subset K$) is continuous, $h(u, \lambda) = P_K[B(u, \lambda)]$ is also completely continuous on $V \times \mathbb{R}$. ∎

Now, we develop a homogenization procedure at infinity for problem (3.1) and obtain a related variational inequality defined on a cone (which is different from the support cone in previous chapters). First, we need some definitions. We denote by rcK the recession cone of K (see Figure 5.1) ([102]):

$$rcK = \bigcap_{t > 0} tK.$$

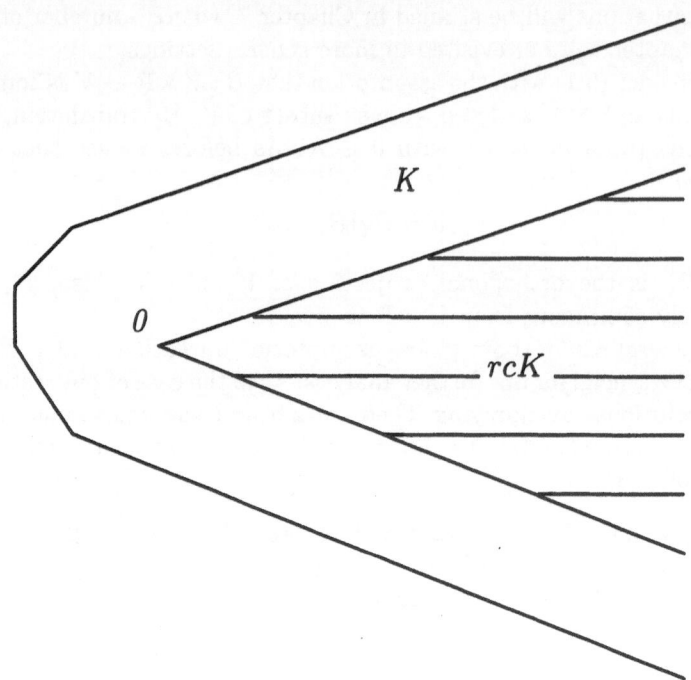

FIGURE 5.1. The recession cone rcK of a convex set K.

It is proved (cf. [102]) that

$$v \in rcK \iff tv \in K, \ \forall t > 0,$$
$$\iff \exists a \in K : a + tv \in K, \ \forall t > 0,$$
$$\iff a + tv \in K, \ \forall a \in K, \ \forall t > 0.$$

We suppose that $B(u, \lambda)$ is differentiable with respect to u at infinity in the sense that there exists a mapping

$$f_\infty : V \times \mathbb{R} \to V,$$

which is completely continuous on $V \times \mathbb{R}$, such that, for all sequences $\{v_n\} \subset V, \{\sigma_n\} \subset [0, \infty), \{\lambda_n\} \subset \mathbb{R}$ satisfying

$$v_n \to v \text{ in } V, \ \lambda_n \to \lambda, \sigma_n \to \infty, \text{ as } n \to \infty,$$

$$\frac{1}{\sigma_n} B(\sigma_n v_n, \lambda_n) \to f_\infty(v, \lambda) \text{ in } V. \tag{5.4}$$

Note that f_∞ is positive homogeneous with respect to $u \in V$, i.e.,

$$f_\infty(\sigma u, \lambda) = \sigma f_\infty(u, \lambda), \ \forall \sigma > 0, \ u \in V, \ \lambda \in \mathbb{R}.$$

Consider the following variational inequality, which is the homogenization of (3.1) at infinity:

$$\begin{cases} \langle u - f_\infty(u, \lambda), v - u \rangle \geq 0, \ \forall v \in rcK, \\ u \in rcK. \end{cases} \tag{5.5}$$

Because rcK is a cone and $f_\infty(\cdot, \lambda)$ is positive homogeneous, we see that, if $u \neq 0$ is a solution of (5.5), then, so is tu, $\forall t > 0$.

The following is an analog of Theorem 3.2 for bifurcation from infinity (see also Figures 5.2 and 5.3).

Theorem 5.2

(I) If (∞, λ) is an asymptotic bifurcation point of (3.1), then, λ is an eigenvalue of (5.5).

(II) Suppose a and b ($a < b$) are not eigenvalues of (5.5) and that

$$d(I - P_{rcK}[f_\infty(\cdot, a)], B_R(0), 0) \neq d(I - P_{rcK}[f_\infty(\cdot, b)], B_R(0), 0) \tag{5.6}$$

for some $R > 0$. Then, there exists a bifurcation point (∞, λ) with $a < \lambda < b$ (λ is an eigenvalue of (5.5)) and a continuum $\mathcal{C}$ of solutions of (3.1), which is unbounded in $V \times [a, b]$. Moreover, either

(i) $\mathcal{C}$ is unbounded in the λ direction, or

(ii) there exists an interval $[c, d]$, such that $(a, b) \cap (c, d) = \emptyset$ and $\mathcal{C}$ bifurcates from infinity in $V \times [c, d]$.

Proof. Because a and b are not eigenvalues of (5.5),

$$u \neq P_{rcK}[f_\infty(u, \lambda)], \ \forall u \neq 0.$$

Hence, the degrees in (5.6) exist and do not depend on $R > 0$.

For $\sigma \in [0, 1]$, define

$$K_\sigma^\infty = \begin{cases} \sigma K = \{\sigma v : v \in K\} & \text{if} \ \sigma \in (0, 1], \\ rcK & \text{if} \ \sigma = 0, \end{cases}$$

and

$$B_\sigma^\infty(u, \lambda) = \begin{cases} \sigma B\left(\frac{u}{\sigma}, \lambda\right) & \text{if} \ \sigma \in (0, 1], \\ f_\infty(u, \lambda) & \text{if} \ \sigma = 0. \end{cases}$$

We observe that, for all $\sigma \in [0, 1]$, K_σ is a closed, convex subset of V. Similar to the case of support cones, we prove that, if $\{\sigma_n\} \subset [0, 1]$ and $\sigma_n \to \sigma_0$, then,

$$K_{\sigma_n} \to K_{\sigma_0} \tag{5.7}$$

in the Mosco sense (i.e., for each $v \in K_{\sigma_0}$, there exists a sequence $\{v_n\}$, such that $v_n \in K_{\sigma_n}$, $\forall n$, and $v_n \to v$ in V, and, for each subsequence

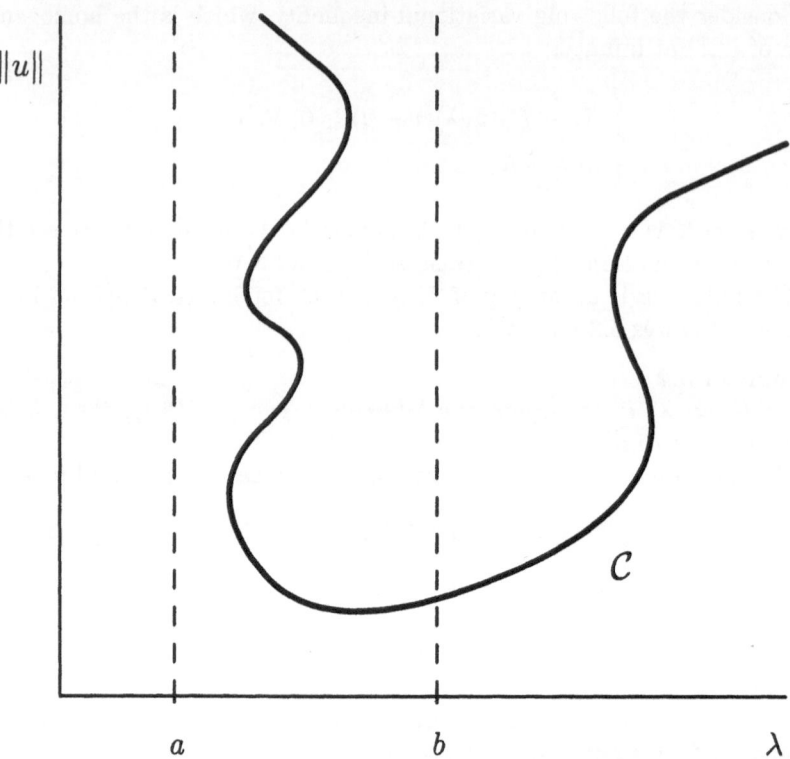

FIGURE 5.2. Asymptotic bifurcation branch unbounded in the λ-direction.

$\{\sigma_{n_k}\} \subset \{\sigma_n\}$, if $\{v_{n_k}\}$ satisfies $v_{n_k} \in K_{\sigma_{n_k}}$, $\forall k$, and if $v_{n_k} \rightharpoonup v$, then, $v \in K_{\sigma_0}$).

In fact, the case where $\sigma_0 > 0$ can be established, as in the proof of Theorem 3.2. Let $v \in K_{\sigma_0}$. Then $\sigma_0^{-1}v \in K$, and, thus, $v_n = \sigma_n \sigma_0^{-1} v \in K_{\sigma_n}$. Moreover, $v_n \to v$ in V. Now, suppose $v_{n_k} \in K_{\sigma_{n_k}}$, $v_{n_k} \rightharpoonup v$ in V. Then,

$$\sigma_0^{-1}v = \lim_k \sigma_{n_k}^{-1}v_{n_k} \in K$$

because $\sigma_{n_k}^{-1}v_{n_k} \in K$, $\forall k$, and K is weakly closed. Hence, $v \in K_{\sigma_0}$.

Now, consider the case $\sigma_0 = 0$. Because $0 \in K$, a direct proof shows that $K_\sigma \subset K_{\sigma'}$ if $\sigma \leq \sigma'$. Moreover, by definition, $rcK = \bigcap_{\sigma>0} K_\sigma = \bigcap_{\sigma \in (0,1]} K_\sigma$.

Hence, for $v \in rcK$, by choosing $v_n = v$, $\forall n$, $v_n \in K_{\sigma_n}$, and $v_n \to v$ in V.

Now suppose $\sigma_{n_k}^{-1}v_{n_k} \in K$ and $v_{n_k} \rightharpoonup v$ with $\sigma_{n_k} \to 0^+$. For $\sigma > 0$, there exists $k_0 \in \mathbb{N}$, such that $\sigma_{n_k} < \sigma$, $\forall k \geq k_0$. Because $0, \sigma n_k^{-1} \in K$ and $0 < \sigma^{-1} < \sigma_{n_k}^{-1}$,

$$\sigma^{-1}v_{n_k} = \frac{\sigma^{-1}}{\sigma_{n_k}^{-1}}(\sigma_{n_k}^{-1}v_{n_k}) + \left(1 - \frac{\sigma^{-1}}{\sigma_{n_k}^{-1}}\right)0 \in K.$$

Because $\sigma^{-1}v_{n_k} \rightharpoonup \sigma^{-1}v$, $\sigma^{-1}v \in K$, i.e., $v \in \sigma K$. Because this holds for every $\sigma > 0$,

$$v \in \bigcap_{\sigma>0} \sigma K = rcK,$$

and (5.7) is verified in all cases.

Now, as in the proof of Theorem 3.2, we prove that the mapping

$$(\sigma, v, \lambda) \mapsto P_{K_\sigma^\infty}[B_\sigma^\infty(v, \lambda)] \qquad (5.8)$$

is completely continuous from $[0,1] \times \mathbb{R} \times V$ to V.

Letting

$$v_n \rightharpoonup v \text{ in } V, \text{ and } \sigma_n \to \sigma, \lambda_n \to \lambda \text{ in } \mathbb{R},$$

we have

$$B_{\sigma_n}^\infty(v_n, \lambda_n) \to B_\sigma^\infty(v, \lambda). \qquad (5.9)$$

In fact, if $\sigma > 0$, (5.9) is a consequence of the complete continuity of B:

$$B_{\sigma_n}^\infty(v_n, \lambda_n) = \sigma_n B\left(\frac{v_n}{\sigma_n}, \lambda_n\right) \to \sigma B\left(\frac{v}{\sigma}, \lambda\right) = B_\sigma^\infty(v, \lambda)$$

(because $v_n/\sigma_n \rightharpoonup v/\sigma$ in V). If $\sigma = 0$, then, $\sigma_n^{-1} \to \infty$, and (5.9) follows from the definition of f_∞.

Now,

$$\left\| P_{K_{\sigma_n}^\infty}\left[B_{\sigma_n}^\infty(v_n, \lambda_n)\right] - P_{K_\sigma^\infty}\left[B_\sigma^\infty(v, \lambda)\right] \right\|$$
$$\leq \left\| B_{\sigma_n}^\infty(v_n, \lambda_n) - B_\sigma^\infty(v, \lambda) \right\| + \left\| P_{K_{\sigma_n}^\infty}\left[B_\sigma^\infty(v, \lambda)\right] - P_{K_\sigma^\infty}\left[B_\sigma^\infty(v, \lambda)\right] \right\|$$

($P_{K_{\sigma_n}^\infty}$ is the orthogonal projection on the convex set $K_{\sigma_n}^\infty$ and is, therefore, nonexpansive).

Moreover, because $K_{\sigma_n}^\infty \to K_\sigma^\infty$ in the Mosco sense,

$$P_{K_{\sigma_n}^\infty}(u) \to P_{K_\sigma^\infty}(u),$$

for all $u \in V$. This and (5.9) give

$$P_{K_{\sigma_n}^\infty}[B_{\sigma_n}^\infty(v_n, \lambda_n)] \to P_{K_\sigma^\infty}[B_\sigma^\infty(v, \lambda)] \text{ in } V,$$

and the complete continuity of the mapping in (5.8) follows.

To prove (I), we suppose that (∞, λ) is an asymptotic bifurcation point of (3.1). This means that there exist u_n, λ_n $(n = 1, 2, ...)$ such that

$$\|u_n\| \to \infty, \lambda_n \to \lambda,$$

and for all n, (u_n, λ_n) satisfies the variational inequality:

$$\begin{cases} \langle u_n - B(u_n, \lambda_n), v - u_n \rangle \geq 0, \ \forall v \in K, \\ u_n \in K. \end{cases}$$

Setting $u_n = \|u_n\|^{-1} u_n$ and dividing both sides of the above inequality by $\|u_n\|^2$,

$$\left\langle v_n - \frac{B(\|u_n\|v_n, \lambda_n)}{\|u_n\|}, \frac{v}{\|u_n\|} - v_n \right\rangle \geq 0, \ \forall v \in K,$$

and v_n is a solution of the variational inequality,

$$\begin{cases} \langle v_n - B^\infty_{1/\|u_n\|}(v_n, \lambda_n), w - v_n \rangle \geq 0, \ \forall w \in K^\infty_{1/\|u_n\|}, \\ v_n \in K^\infty_{1/\|u_n\|}, \end{cases}$$

i.e.,

$$v_n = P_{K^\infty_{1/\|u_n\|}} \left[B^\infty_{1/\|u_n\|}(v_n, \lambda_n) \right], \ \forall n.$$

By passing to a subsequence of $\{v_n\}$, if necessary, we can assume that $v_n \rightharpoonup v_0$ in V.

Because $1/\|u_n\| \to 0$, by the complete continuity of the mapping in (5.8),

$$P_{K^\infty_{1/\|u_n\|}} \left[B^\infty_{1/\|u_n\|}(v_n, \lambda_n) \right] \ \to \ P_{K^\infty_0} [B^\infty_0(v_0, \lambda)]$$
$$= \ P_{rcK} [f_\infty(v_0, \lambda)].$$

Thus, $v_n \to v_0$ in V and $v_0 = P_{rcK} [f(v_0, \lambda)]$. This means that $\|v_0\| = 1$, and v_0 is an eigenvector of (5.5) corresponding to λ.

We now prove (II). Because a is not an eigenvalue of (5.5), we see that (3.1) (respectively, (5.5)) with $\lambda = a$ does not have solutions with large norms (respectively, nontrivial solutions).

Next, we show that

$$d(I - P_K[B(\cdot, a)], B_R(0), 0) = d(I - P_{rcK}[f_\infty(\cdot, a)], B_R(0), 0) \qquad (5.10)$$

for all $R > 0$, sufficiently large. Indeed, there exists $R_0 > 0$, such that, for all $\sigma \in [0, 1]$, the equation

$$u - P_{K^\infty_\sigma} [B^\infty_\sigma(u, a)] = 0 \qquad (5.11)$$

has no solution $u \in V$ with $\|u\| \geq R_0$. This means that (5.11) has no solution on $\partial B_R(0)$ for all $R \geq R_0$. Suppose, otherwise, that there exist sequences $\{u_n\} \subset V$, $\{\sigma_n\} \subset [0, 1]$, such that $\|u_n\| \to \infty$, as $n \to \infty$, and, for all n,

$$u_n = P_{K^\infty_{\sigma_n}} [B^\infty_{\sigma_n}(u_n, a)],$$

i.e.,

$$\begin{cases} \langle u_n - B^\infty_{\sigma_n}(u_n, a), v - u_n \rangle \geq 0, \ \forall v \in K^\infty_{\sigma_n}, \\ u_n \in K^\infty_{\sigma_n}. \end{cases}$$

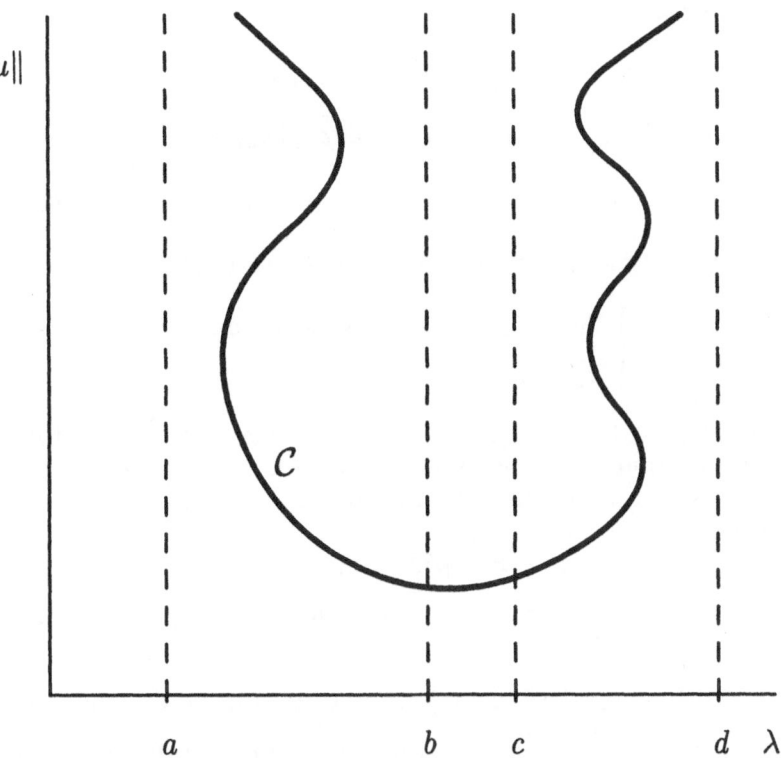

FIGURE 5.3. Asymptotic bifurcation branch bifurcating from infinity in another interval.

As above, dividing both sides of this inequality by $\|u_n\|^2$ and setting $v_n = \|u_n\|^{-1}u_n$,

$$\left\langle \frac{u_n}{\|u_n\|} - \frac{\sigma_n}{\|u_n\|} B\left(\frac{v_n\|u_n\|}{\sigma_n}, a\right), \frac{v}{\|u_n\|} - \frac{u_n}{\|u_n\|}\right\rangle \geq 0, \ \forall v \in K_{\sigma_n}^{\infty},$$

or, by the definitions of B_{σ}^{∞}, K_{σ}^{∞},

$$\begin{cases} \langle v_n - B_{\sigma_n/\|u_n\|}^{\infty}(v_n, a), w - v_n \rangle \geq 0, \ \forall w \in K_{\sigma_n/\|u_n\|}^{\infty}, \\ v_n \in K_{\sigma_n/\|u_n\|}^{\infty}. \end{cases} \tag{5.12}$$

Again, we note that $v \in K_{\sigma_n}^{\infty}$ if and only if

$$w = v/\|u_n\| \in K_{\sigma_n/\|u_n\|}^{\infty}.$$

Because rcK is a cone and $B_0^{\infty}(u, \lambda) = f_{\infty}(u, \lambda)$ is positive homogeneous with respect to u, we see that (5.12) holds for both cases $\sigma_n > 0$ and $\sigma_n = 0$.

(5.12) means that

$$v_n = P_{K^\infty_{\sigma_n/\|u_n\|}}\left[B^\infty_{\sigma_n/\|u_n\|}(v_n, a)\right], \ \forall n \in \mathbf{N}.$$

Assuming $v_n \rightharpoonup v_0$ in V and noting that $\sigma_n/\|u_n\| \to 0$, from (5.8),

$$P_{K^\infty_{\sigma_n/\|u_n\|}}\left[B^\infty_{\sigma_n/\|u_n\|}(v_n, a)\right] \to P_{K^\infty_0}[B^\infty_0(v_0, a)] = P_{rcK}[f_\infty(v_0, a)] \ \text{in} \ V.$$

Hence, $v_n \to v_0$, $\|v_0\| = 1$, and $v_0 = P_{rcK}[f_\infty(v_0, a)]$, i.e.,

$$\begin{cases} \langle v_0 - f_\infty(v_0, a), v - v_0 \rangle \geq 0, \ \forall v \in rcK, \\ v_0 \in rcK. \end{cases}$$

Because $\|v_0\| = 1$, we see that a is an eigenvalue of (5.5) with eigenvector v_0. This contradicts the assumption that a is not an eigenvalue of (5.5) and proves that (5.11) has no solution on $\partial B_R(0)$ for all $R > 0$, sufficiently large. Now,

$$\{I - P_{K^\infty_\sigma}[B^\infty_\sigma(\cdot, a)] : 0 \leq \sigma \leq 1\}$$

is a family of compact perturbations of the identity on $\overline{B_R(0)}$. Moreover, by the above proof,

$$u - P_{K^\infty_\sigma}[B^\infty_\sigma(u, a)] \neq 0, \ \forall u \in \partial B_R(0), \ \forall \sigma \in [0, 1].$$

This implies, via the homotopy invariance property of the Leray–Schauder degree, that

$$\begin{aligned} d(I - P_{rcK}[f_\infty(\cdot, a)], B_R(0), 0) &= d(I - P_{K^\infty_0}[B^\infty_0(\cdot, a)], B_R(0), 0) \\ &= d(I - P_{K^\infty_1}[B^\infty_1(\cdot, a)], B_R(0), 0) \\ &= d(I - P_K[B(\cdot, a)], B_R(0), 0), \end{aligned}$$

proving (5.10). One has a similar equality for $\lambda = b$. These equalities prove that (5.3) and (5.6) are equivalent. Thus, (II) follows from Theorem 5.1 and (I). ∎

From (I) of Theorem 5.2, we immediately have the following nonexistence result:

Corollary 5.3 *If $rcK = \{0\}$, then, (3.1) has no finite asymptotic bifurcation point.*

Now, we consider the case where $rcK = W$ is a vector space. In this case, (5.5) is actually an equation in W:

$$\begin{cases} \langle u - f_\infty(u, \lambda), v \rangle = 0, \ \forall v \in W, \\ u \in W, \end{cases} \tag{5.13}$$

or, equivalently,

$$u = P_W[f_\infty(u, \lambda)],$$

where $P_{rcK} = P_W$ is now the orthogonal projection onto the subspace W, which is a bounded linear mapping from V to W.

We assume, moreover, that B is of the form

$$B(u, \lambda) = \lambda\beta u + G(u, \lambda), \; u \in V, \; \lambda \in \mathbb{R}, \tag{5.14}$$

where $\beta \in L(V, V)$ is linear and compact and

$$G : V \times \mathbb{R} \to \mathbb{R}$$

is completely continuous, such that

$$\frac{\|G(u, \lambda)\|}{\|u\|} \to 0 \; \text{ as } \; \|u\| \to \infty,$$

uniformly for λ in bounded intervals. Note that (5.14) is satisfied in the particular case where G satisfies the following growth condition:

$$\|G(u, \lambda)\| \le C(\lambda)\|u\|^\alpha + D(\lambda), \; u \in V, \; \lambda \in \mathbb{R}, \tag{5.15}$$

with

$$C, D : \mathbb{R} \to \mathbb{R}^+, \; C, D \in L^\infty_{loc}(\mathbb{R}),$$

and $0 \le \alpha < 1$. In fact,

$$\frac{\|G(u, \lambda)\|}{\|u\|} \le C(\lambda)\|u\|^{\alpha-1} + \frac{D(\lambda)}{\|u\|} \to 0$$

as $\|u\| \to \infty$, uniformly for λ belonging to bounded intervals. We see that, if G is bounded, then, G satisfies this condition with $\alpha = 0$ and C and D constants (independent of λ).

Under these conditions, B satisfies (5.4) with f_∞ given by

$$f_\infty(u, \lambda) = \lambda\beta u, \; u \in V, \; \lambda \in \mathbb{R}. \tag{5.16}$$

In fact, suppose $v_n \rightharpoonup v$ in V, $\sigma_n \to \infty$, and $\lambda_n \to \lambda$. Then,

$$\frac{1}{\sigma_n} B(\sigma_n v_n, \lambda_n) = \lambda_n \beta v_n + \frac{G(\sigma_n v_n, \lambda_n)}{\sigma_n}, \; \forall n.$$

We claim that $\frac{1}{\sigma_n}G(\sigma_n v_n, \lambda_n) \to 0$ in V, as $n \to \infty$. If this is not the case, then one can choose a subsequence $\{v_{n_k}\} \subset \{v_n\}$ such that

$$\frac{1}{\sigma_{n_k}}\|G(\sigma_{n_k} v_{n_k}, \lambda_{n_k})\| \ge \epsilon_0 > 0, \; \forall k. \tag{5.17}$$

If the sequence $\{\sigma_{n_k} v_{n_k}\}$ is bounded, then, $\{G(\sigma_{n_k} v_{n_k}, \lambda_{n_k})\}$ is also bounded because G is completely continuous (and, thus, bounded on bounded sets). Therefore,

$$\frac{1}{\sigma_{n_k}} G(\sigma_{n_k} v_{n_k}, \lambda_{n_k}) \to 0 \ \text{ in } \ V,$$

as $k \to \infty$, contradicting (5.17). Hence, $\{\sigma_{n_k} v_{n_k}\}$ is unbounded. Passing to a subsequence, if necessary, we can assume that $\|\sigma_{n_k} v_{n_k}\| \to \infty$ and $v_{n_k} \neq 0$, $\forall k$. Then,

$$\frac{1}{\sigma_{n_k}} \|G(\sigma_{n_k} v_{n_k}, \lambda_{n_k})\| = \frac{\|G(\sigma_{n_k} v_{n_k}, \lambda_{n_k})\|}{\|\sigma_{n_k} v_{n_k}\|} \|v_{n_k}\|.$$

Because $\|\sigma_{n_k} v_{n_k}\| \to \infty$, $\lambda_{n_k} \to \lambda$, by the assumption on G,

$$\lim_{k \to \infty} \frac{\|G(\sigma_{n_k} v_{n_k}, \lambda_{n_k})\|}{\|\sigma_{n_k} v_{n_k}\|} = 0.$$

Moreover, $\{v_{n_k}\}$ is bounded (because $v_{n_k} \rightharpoonup v$ in V). Hence,

$$\frac{1}{\sigma_{n_k}} \|G(\sigma_{n_k} v_{n_k}, \lambda_{n_k})\| \to 0.$$

This contradiction proves that $\dfrac{1}{\sigma_n} G(\sigma_n v_n, \lambda_n) \to 0$ in V, as $n \to \infty$. On the other hand, because β is compact,

$$\lambda_n \beta v_n \to \lambda \beta v \ \text{ in } \ V,$$

and, thus, we have verified (5.4). If f_∞ is given by (5.16), then, (5.13) becomes

$$\begin{cases} \langle u - \lambda \beta u, v \rangle = 0, \ \forall v \in W, \\ u \in W, \end{cases} \tag{5.18}$$

or, in other words,

$$u = \lambda P_W(\beta u).$$

Because $P_W \circ \beta$ is a compact linear operator, (5.18) has, at most, a countable number of isolated eigenvalues λ with ∞ as the only possible accumulation point. Therefore, we have the following result:

Corollary 5.4 *Suppose $rcK = W$ is a subspace of V and that B is given by (5.14). Then, (3.1) has, at most, a countable number of asymptotic bifurcation points. If (∞, λ) is a bifurcation point of (3.1), then, λ is an eigenvalue of (5.18).*

  Conversely, if λ is an eigenvalue of (5.18) of odd (algebraic) multiplicity, then, (∞, λ) is an asymptotic bifurcation point of (3.1) corresponding to an unbounded asymptotic bifurcation branch C that satisfies the alternative in Theorem 5.2.

5.2 Global asymptotic bifurcation

In this section, we consider some analogs of Theorems 4.3 and 4.4 for bifurcation from infinity of the variational inequality (3.1).

5.2.1 Index calculations

We will assume in the sequel that $f_\infty(u, 0) = 0$, $\forall u$ (this holds for instance if $B(u, 0) = 0$, $\forall u$). Assume that B, f_∞, K, and rcK are given as above. By using arguments similar to those in Lemma 4.1, one may prove the following lemma:

Lemma 5.5 *If a is not an eigenvalue of (5.8), then, for $R > 0$, one has*

$$\mathrm{d}(I - P_{rcK}[f_\infty(\cdot, a)], B_R(0), 0) = \mathrm{ind}(P_{rcK}[f_\infty(\cdot, a)], B_R(0) \cap rcK, rcK).$$

Theorem 4.3 has the following counterpart for bifurcation from infinity.

Theorem 5.6 *Suppose that $f_\infty(u, \lambda)$ is monotone with respect to $u \in rcK$ and to $\lambda \geq 0$. If $\lambda_0 > 0$ is an isolated eigenvalue of f_∞ corresponding to an eigenvector $h \in rcK \setminus (-rcK)$, then, there exists an unbounded asymptotic bifurcation branch of solutions of (3.1) that emanates from $\{0\} \times [0, \lambda_0]$ and satisfies the alternative in Theorem 5.2.*

Proof. Using a proof similar to that of Theorem 4.3, for $R > 0$,

$$\begin{cases} \mathrm{ind}(f_\infty(\cdot, 0), B_R(0) \cap rcK, rcK) = 0 \\ \mathrm{ind}(f_\infty(\cdot, \lambda), B_R(0) \cap rcK, rcK) = 1, \end{cases}$$

for all $\lambda > \lambda_0$ near λ_0. Applying the above lemma and Theorem 5.2, we get the result. ■

Let $W = \overline{rcK - rcK}$ be the closed subspace spanned by rcK. Note that, if rcK is an order cone (i.e., $rcK \cap (-rcK) = \{0\}$) and if B is of the form (5.14), where β is a positive operator with respect to the cone rcK, then, by applying the Krein-Rutman theorem, we see that, at the eigenvalues λ_0 of the equation

$$x = \lambda_0 \beta x, \ x \in rcK,$$

(corresponding to (5.5)), all assumptions of Theorem 5.6 are satisfied.

Next, we consider an application of Theorem 5.6.

Example 5.1 Consider the complementarity problem (4.12) in Example 4.1, which can be written in the variational inequality form (4.14). Assume that G satisfies the Urysohn condition in that example:

$$|G(x, y, u)| \leq R(x, y)[A + B|u|]$$

with $A, B \geq 0$, $R \geq 0$, and

$$\int_\Omega \int_\Omega R(x,y)^2 < \infty.$$

Assume, furthermore, that G can be homogenized with respect to u at infinity in the sense that

$$G(x,y,u) = b(x,y)u + h(x,y,u),$$

with

$$|h(x,y,u)| \leq R_0(x,y)[a_0 + b_0|u|^\alpha], \quad x,y \in \Omega, u \in \mathbb{R},$$

where $a_0, b_0 \geq 0$, $0 \leq \alpha < 1$ and b, $R_0 \in L^2(\Omega \times \Omega)$.

Let U and B be as in Example 4.1. We consider the mappings

$$H, \beta : L^2(\Omega) \to L^2(\Omega),$$

given by

$$(\beta u)(x) = \int_\Omega b(x,y)u(y)dy,$$

$$H(u)(x) = \int_\Omega h(x,y,u(y))dy, \quad x \in \Omega.$$

Then H and β are completely continuous on $L^2(\Omega)$, β is linear, and

$$\frac{\|H(u)\|}{\|u\|} \to 0 \quad \text{as} \quad \|u\| \to \infty.$$

In fact, for $x \in \Omega$, by Hölder's inequality,

$$
\begin{aligned}
|H(u)(x)|^2 &\leq \left[\int_\Omega |h(x,y,u(y))|dy\right]^2 \\
&\leq \int_\Omega [R_0(x,y)]^2 \int_\Omega [a_0 + b_0|u(y)|^\alpha]^2 dy.
\end{aligned}
$$

Hence, using Hölder's inequality again, one gets

$$
\begin{aligned}
\|H(u)\|^2 &\leq \int_\Omega \left\{\int_\Omega [R_0(x,y)]^2 \int_\Omega [a_0 + b_0|u(y)|^\alpha]^2 dy\right\} dx \\
&= \left\{\int_\Omega \int_\Omega [R_0(x,y)]^2 dxdy\right\} \left\{\int_\Omega [a_0^2 + b_0^2|u(y)|^{2\alpha} \right. \\
&\qquad \left. + 2a_0 b_0|u(y)|^\alpha] dy\right\} \\
&= C_0 \left(a_0^2|\Omega| + b_0^2 \int_\Omega |u|^{2\alpha} + 2a_0 b_0 \int_\Omega |u|^\alpha\right) \\
&\qquad \left(C_0 = \int_\Omega \int_\Omega R_0(x,y)^2 dxdy \geq 0\right)
\end{aligned}
$$

$$\leq C_0 \left[a_0^2 |\Omega| + b_0^2 \left(\int_\Omega |u|^{2\alpha\left(\frac{1}{\alpha}\right)} \right)^\alpha |\Omega|^{1-\alpha} \right.$$

$$\left. + 2a_0 b_0 \left(\int_\Omega |u|^{\alpha\left(\frac{2}{\alpha}\right)} \right)^{\frac{\alpha}{2}} |\Omega|^{1-\frac{\alpha}{2}} \right]$$

$$= C_0 \left[C_1 + C_2 \|u\|^{2\alpha} + C_3 \|u\|^\alpha \right],$$

where C_0, C_1, C_2, and C_3 are positive constants. From these estimates, it follows that

$$\frac{\|Hu\|^2}{\|u\|^2} \leq C_0 \left[C_1 \|u\|^{-2} + C_2 \|u\|^{2(\alpha-1)} + C_3 \|u\|^{\alpha-2} \right].$$

Because $\alpha < 1$, $\|Hu\|^2/\|u\|^2 \to 0$ as $\|u\| \to \infty$. Therefore,

$$B(u,\lambda) = \lambda\beta u + g(u,\lambda), \ u \in L^2(\Omega), \ \lambda \in \mathbb{R},$$

where $g : (u,\lambda) \mapsto \lambda H(u)$ is completely continuous from $L^2(\Omega) \times \mathbb{R}$ to $L^2(\Omega)$ and satisfies

$$\frac{\|g(u,\lambda)\|}{\|u\|} \to 0 \ \text{ as } \ \|u\| \to \infty,$$

uniformly for λ in bounded intervals.

On the other hand, K is an order cone, and, therefore, $rcK = K$. If $b(x,y) \geq 0$ for a.e. $x, y \in \Omega$, then, $\beta(rcK) \subset rcK$, and, as noted before, we may apply Theorem 5.6.

5.2.2 Some general results

Now we consider an analog of Theorem 4.4 for bifurcation from infinity. As before, we define $W = W_K$ as the closed subspace spanned by rcK:

$$W = \overline{rcK - rcK},$$

which is a Hilbert subspace of V. We have the following theorem:

Theorem 5.7 *Suppose that $\lambda_0 > 0$ is a simple eigenvalue of (5.5) and that the eigenvectors of (5.5) corresponding to λ_0 are also eigenvectors of the equation*

$$\begin{cases} \langle u - f_\infty(u,\lambda_0), v \rangle = 0, \ \forall v \in W, \\ u \in W. \end{cases} \tag{5.19}$$

Assume, furthermore, that f_∞ is symmetric with respect to u and homogeneous of order γ with respect to λ. Then, there exists a global bifurcation branch of solutions of (3.1), which is unbounded in $V \times [0, \lambda_0]$ and satisfies the alternative in Theorem 5.2.

Proof. The proof is rather long, and we present only its main steps here. First, note that because $f_\infty(u, 0) = 0$, $\forall u \in V$,

$$d(I - P_{rcK}[f_\infty(\cdot, 0)], B_R(0), 0) = 1, \ \forall R > 0.$$

Let u_0 ($\|u_0\| = 1$) be the eigenvector of (5.5) corresponding to $\lambda = \lambda_0$. We prove that the family of (compact) perturbations of the identity $\{H(t, u, \lambda): t \in [0, 1]\}$ given by

$$H(t, u, \lambda) = u - P_{rcK}[(1 - t)f_\infty(u, \lambda) + tf_\infty(u, \lambda_0) + tu_0],$$

($t \in [0, 1], u \in V, \lambda \in \mathbb{R}$) does not vanish for $t \in [0, 1], \lambda > \lambda_0$ sufficiently close to λ_0, and $u \in \partial B_R(0)$, $R > 0$, sufficiently large.

Assume, otherwise, that there exist sequences $\{u_n\} \subset rcK, \{\lambda_n\} \subset [0, 1]$, such that $\|u_n\| \to \infty, \lambda_n \to \lambda_0^+$ and

$$H(t_n, u_n, \lambda_n) = 0, \ \forall n \in \mathbb{N},$$

i.e.,

$$\langle u_n - (1 - t_n)f_\infty(u_n, \lambda_n) - t_n f_\infty(u_n, \lambda_0) - t_n u_0, v - u_n \rangle \geq 0, \ \forall v \in rcK, \ \forall n. \tag{5.20}$$

From (5.19) and the symmetry of f_∞,

$$\begin{aligned} 0 &= \langle u_n, u_0 - f_\infty(u_0, \lambda_0) \rangle \\ &= \langle u_0, u_n - f_\infty(u_n, \lambda_0) \rangle. \end{aligned}$$

From this equality, (5.20), and the homogeneity of f_∞,

$$0 \leq (1 - t_n)\left[1 - \left(\frac{\lambda_n}{\lambda_0}\right)^\gamma\right]\langle u_n, u_0 \rangle - t_n \|u_0\|^2. \tag{5.21}$$

On the other hand, letting $v_n = u_n/\|u_n\|$, $n \in \mathbb{N}$, dividing both sides of (5.20) by $\|u_n\|^2$, and converting the variational inequality, thus obtained, into operator form,

$$v_n = P_{rcK}\left[(1 - t_n)f_\infty(v_n, \lambda_n) + t_n f_\infty(v_n, \lambda_0) + \frac{t_n u_0}{\|u_n\|}\right], \ \forall n \in \mathbb{N}.$$

Without loss of generality, we can assume that $v_n \rightharpoonup v \in rcK$ in V. Then, by letting $n \to \infty$ in the above equation, $v_n \to v$ and

$$v = P_{rcK}[f_\infty(v, \lambda_0)].$$

Therefore, by the simplicity of λ_0, $v = \mu u_0$ for some $\mu > 0$. However, from (5.21), $\langle u_n, u_0 \rangle \leq 0$. Hence, $\langle v_n, u_0 \rangle \leq 0$, $\forall n$ and $\mu = \mu\|u_0\|^2 = \langle v, u_0 \rangle \leq 0$. This contradiction shows that there exist $R_0 > 0$ and $\lambda_1 > \lambda_0$ such that

$$H(t, u, \lambda) \neq 0, \ \forall t \in [0, 1], \ u \in V \setminus B_{R_0}(0), \ \text{and} \ \lambda \in (\lambda_0, \lambda_1).$$

Hence, according to the homotopy invariance property of the Leray–Schauder degree, for all $R \geq R_0$,

$$
\begin{aligned}
d(I - P_{rcK}[f_\infty(\cdot, \lambda)], B_R(0), 0) &= d(H(0, \cdot, \lambda), B_R(0), 0) \\
&= d(H(1, \cdot, \lambda), B_R(0), 0) \\
&= d(H(1, \cdot, \lambda_0), B_R(0), 0).
\end{aligned}
$$

By using an argument similar to the previous one,

$$
H(1, u, \lambda_0) \neq 0,
$$

for all $u \in V$. Hence $d(H(1, \cdot, \lambda_0), B_R(0), 0) = 0$.

Consequently,

$$
d(I - P_{rcK}[f_\infty(\cdot, \lambda)], B_R(0), 0) = 0, \ \forall R \geq R_0, \ \forall \lambda \in (\lambda_0, \lambda_1).
$$

Our theorem follows from Theorem 5.2.   ∎

By a proof similar to that of Lemma 4.5, we have the following result:

Lemma 5.8 *(a)* $u \in W$ *is a demi-interior point of rcK with respect to W if and only if there exists a subset $D \subset W$ dense in W, such that, for all $w \in D$, we can choose $\epsilon > 0$, such that $u + \epsilon w \in rcK$.*

(b) Suppose f_∞ is symmetric with respect to u and that (5.19) has a solution u_0, which is a demi-interior point of rcK with respect to W.

Then, u is a solution of (5.5) if and only if $u \in rcK$ and u is a solution of (5.19).

From this lemma, we have the following consequence of Theorem 5.7 (whose proof is similar to that of Corollary 4.6 and is omitted).

Corollary 5.9 *Suppose that f_∞ is homogeneous and symmetric with respect to u and is homogeneous of order γ with respect to $\lambda \geq 0$. Suppose, furthermore, that $\lambda_0 > 0$ is a simple eigenvalue of the equation (5.19) with a corresponding eigenvector u_0 that is a demi-interior point of rcK (with respect to W). Then, the conclusion of Theorem 5.7 holds, provided $rcK \neq W$.*

The remaining part of this section is devoted to some applications of these abstract results. First, we consider an asymptotic bifurcation problem for a second-order elliptic equation with obstacles.

Example 5.2 Consider the variational inequality in Example 3.2:

$$
\begin{cases}
\displaystyle\int_0^1 u'(v - u)' \, dx \geq \int_0^1 g(x, u(x), \lambda)(v - u) \, dx, \ \forall v \in K, \\
u \in K,
\end{cases}
\tag{5.22}
$$

where $K = \{u \in H_0^1(0,1) : \psi_1(x) \le u(x) \le \psi_2(x) \text{ on } [0,1]\}$ and

$$\psi_1, \psi_2 : [0,1] \to [-\infty, \infty]$$

are two, given, measurable barrier functions, such that $\psi_1 \le 0 \le \psi_2$ on $[0,1]$.

Suppose that g has a derivative at infinity with respect to u in the sense that there exists a measurable function

$$F : (0,1) \times \mathbb{R} \times \mathbb{R} \to \mathbb{R}$$

such that g and F satisfy Carathéodory conditions with respect to x and (u, λ), and the following growth condition:

$$|g(x, u, \lambda)|, |F(x, u, \lambda)| \le C(x, \lambda)|u| + D(x, \lambda), \qquad (5.23)$$

with $0 \le C(x, \lambda), D(x, \lambda) \le M(\lambda)P(x)$, where M and P are measurable, $M \in L_{loc}^\infty(\mathbb{R})$, and $P \in L^1(0,1)$. We assume, furthermore, that

$$\frac{g(x, \sigma_n u_n, \lambda_n)}{\sigma_n} \to F(x, u, \lambda), \qquad (5.24)$$

for a.e. $x \in (0,1)$, whenever $\sigma_n \to \infty$, $u_n \to u$, and $\lambda_n \to \lambda$ in $\mathbb{R}$.

Because the embedding $H_0^1(0,1) \hookrightarrow C[0,1]$ is compact, we see that B and f_∞ defined by

$$B, f_\infty : H_0^1(0,1) \times \mathbb{R} \to H_0^1(0,1)(= V),$$

$$\begin{cases} \langle B(u, \lambda), v \rangle = \displaystyle\int_0^1 g(x, u(x), \lambda)v(x)dx, \\ \langle f_\infty(u, \lambda), v \rangle = \displaystyle\int_0^1 F(x, u(x), \lambda)v(x)dx, \ \forall u, v \in V, \ \lambda \in \mathbb{R}, \end{cases}$$

are completely continuous mappings. One verifies that f_∞ is the derivative of B with respect to u at infinity in the sense of (5.4).

Usually, we consider the particular case where

$$g(x, u, \lambda) = \lambda bu + h(x, u, \lambda), \ x \in [0,1], \ u, \lambda \in \mathbb{R}, \qquad (5.25)$$

with $b \in L^1(0,1)$ and h satisfies

$$|h(x, u, \lambda)| \le C(x, \lambda)|u|^\alpha + D(x, \lambda), \qquad (5.26)$$

for a.e. $x \in [0,1]$, $u, \lambda \in \mathbb{R}$, with C and D as in (5.23) and $0 \le \alpha < 1$.

Then g and $F(x, u, \lambda) = \lambda bu$ satisfy (5.23) and (5.24). With these settings, we are now in a position to apply Theorem 5.7 and Corollaries 5.4 and 5.9. Next, we consider some examples with different choices of the convex set K.

(a) Let

$$K = \{u \in H_0^1(0,1) : \psi_1(x) \le u(x) \le \psi_2(x) \text{ on } [A,B]\},$$

where $0 \le A < B \le 1$, and $\psi_1, \psi_2 : [A,B] \to \mathbb{R}$ are two bounded functions such that $\psi_1 \le 0 \le \psi_2$ on $[A,B]$, (i.e., $\psi_1 = -\infty$ and $\psi_2 = \infty$ outside of $[A,B]$).

In this case,

$$rcK = \{u \in H_0^1(0,1) : u(x) = 0 \text{ on } [A,B]\}. \tag{5.27}$$

In fact, if $u(x) = 0$ on $[A,B]$, then, for all $v \in K$, $\psi_1 \le v + tu = v \le \psi_2$ on $[A,B]$. Hence $v + tu \in K$. Conversely, if $u \in rcK$, then, $tu \in K$, $\forall t > 0$. Thus, for $x \in [A,B]$, $tu(x) \le \psi_2(x)$, i.e., $u(x) \le t^{-1}\psi_2(x)$, $\forall t$. Letting $t \to \infty$, $u(x) \le 0$, $x \in [A,B]$.

Similarly, because $u(x) \ge t\psi_1(x)$, $\forall x \in [A,B]$, and $t \ge 0$, it follows that $u \ge 0$ on $[A,B]$ and, therefore, $u = 0$ on $[A,B]$, and (5.27) holds.

Observe that rcK is a linear subspace of V. If g is of the form (5.25), then, the homogenized variational inequality corresponding to (5.22) is the following linear equation:

$$\begin{cases} \int_0^1 u'v' = \lambda \int_0^1 buv, \ \forall v \in rcK, \\ u \in rcK, \end{cases} \tag{5.28}$$

or, equivalently,

$$\begin{cases} -u'' = \lambda bu \ \text{ on } (0,1) \setminus [A,B], \\ u = 0 \ \text{ on } [A,B] \cup \{0,1\}. \end{cases}$$

We can apply Corollary 5.4 to obtain global asymptotic bifurcation branches of (5.22) that bifurcate from the eigenvalues of (5.28).

(b) Next, let K be given by

$$K = \{u \in H_0^1(0,1) : u(x) \ge \psi_1 \text{ on } I\}.$$

Here, I is a closed subset of $[0,1]$, and ψ_1 is a bounded function defined on I. By a proof similar to the above, we see that the recession cone of K, in this case, is given by

$$rcK = \{u \in H_0^1(0,1) : u(x) \ge 0 \text{ on } I\}.$$

Again, if g is given by (5.25), then, the homogenized variational inequality of (5.22) is expressed by

$$\begin{cases} \int_0^1 u'(v-u)' \ge \lambda \int_0^1 bu(v-u), \ \forall v \in rcK, \\ u \in rcK. \end{cases}$$

Note that the closed linear span of rcK, in this case, is the whole space $W = H_0^1(0,1) = V$. The linear equation related to (5.28), therefore, is

$$\begin{cases} \int_0^1 u'v' = \lambda \int_0^1 buv, \ \forall v \in H_0^1(0,1), \\ u \in H_0^1(0,1), \end{cases} \tag{5.29}$$

or, equivalently,

$$\begin{cases} -u'' = \lambda bu \ \text{ on } \ (0,1), \\ u(0) = u(1) = 0. \end{cases}$$

We assume that λ_0 is a simple eigenvalue of (5.29) corresponding to a positive eigenvector u_0, i.e.,

$$u_0(x) > 0, \ \forall x \in (0,1).$$

By a proof similar to that used in Example 4.2, we see that u_0 is a demi-interior point of rcK in V. Moreover, because $\langle f_\infty(u,\lambda), v \rangle = \lambda \int_0^1 buv$, $\forall u$, $v \in V$, all assumptions of Corollary 5.9 are satisfied. We, therefore, have the following corollary:

Let λ_0 be a simple eigenvalue of (5.29) with an associated eigenvector u_0 that is positive on $(0,1)$. Then, there exists a bifurcation branch of (5.22), which is unbounded in $V \times [0, \lambda_0]$ and satisfies the alternative in Theorem 5.2.

Example 5.3 (a) In this example, we consider bifurcation from infinity of the following fourth-order variational inequality modeling deflections of a beam:

$$\begin{cases} \int_0^a u''(v-u)'' \geq \int_0^a g(x,u',\lambda)(v-u)', \ \forall v \in K, \\ u \in K, \end{cases} \tag{5.30}$$

where a and V are as in Example 4.3. We suppose, here, that $g(x,v,\lambda)$ has a derivative at infinity with respect to v in the sense of Example 5.2, i.e., g satisfies the assumption (5.24) or (5.25). By arguments similar to those used in Example 5.2, we see that the mapping $B : V \times \mathbb{R} \to V$ given by

$$\langle B(u,\lambda), v \rangle = \int_0^a g(x,u',\lambda)v', \ u,v \in V, \ \lambda \in \mathbb{R},$$

is completely continuous and satisfies condition (5.4) with

$$\langle f_\infty(u,\lambda), v \rangle = \int_0^a F(x,u',\lambda)v',$$

or, more specifically,

$$\langle f_\infty(u, \lambda), v \rangle = \lambda \int_0^a bu'v', \ \forall u, v \in V, \ \lambda \in \mathbb{R},$$

in the case where g is of the form (5.25). In the particular case where

$$g(x, u', \lambda) = \lambda u'(1 + u'^2)^{-1/2}, \tag{5.31}$$

for $x \in (0, a)$, $u', \lambda \in \mathbb{R}$,

$$\begin{aligned} g(x, u', \lambda) &= \lambda + \lambda \left[\frac{u'}{\sqrt{1 + u'^2}} - 1 \right] \\ &= \lambda + h(u', \lambda), \end{aligned} \tag{5.32}$$

with

$$\begin{aligned} |h(u', \lambda)| &= |\lambda| \left| \frac{u' - \sqrt{1 + u'^2}}{\sqrt{1 + u'^2}} \right| \\ &\leq |\lambda|(1 + |u'|^{1/2}), \ \forall u', \lambda \in \mathbb{R}. \end{aligned}$$

(5.25) and (5.26) hold with $\alpha = 1/2$ and $C = D \equiv 1$.

The asymptotically homogenized variational inequality of (5.30), therefore, is of the following form:

$$\begin{cases} \int_0^a u''(v - u)'' \geq \lambda \int_0^a bu'(v - u)', \ \forall v \in rcK, \\ u \in rcK. \end{cases} \tag{5.33}$$

Using this formulation and Corollary 5.9, we obtain results similar to those in Example 4.3.

Consider, for instance, the case where $b = 1$ and

$$K = \{u \in V : u \geq \psi_1 \text{ on } [A, B], u \leq \psi_2 \text{ on } [C, D]\},$$

where $0 \leq A < B < C < D \leq a$ and $\psi_1 \leq 0, \psi_2 \geq 0$ are two given bounded functions defined, respectively, on $[A, B]$ and $[C, D]$.

Arguing as in the previous example, we obtain

$$rcK = \{u \in V : u \geq 0 \text{ on } [A, B], u \leq 0 \text{ on } [C, D]\}.$$

The linear equation related to (5.33) is:

$$\begin{cases} \int_0^a u''v'' = \lambda \int_0^a u'v', \ \forall v \in H^2(0, a) \cap H_0^1(0, a), \\ u \in H^2(0, a) \cap H_0^1(0, a), \end{cases}$$

which can also be written as

$$\begin{cases} u^{(4)} + \lambda u'' = 0 \text{ on } [0,a], \\ u(0) = u(a) = u''(0) = u''(a). \end{cases} \tag{5.34}$$

Some elementary calculations show that, if $B < a/2 < C$, then, $\lambda_2 = (2\pi/a)^2$ is a simple eigenvalue of (5.34), and the corresponding eigenfunction u_2, given by $u_2(x) = \sin(2\pi x/a)$, $x \in (0,a)$, satisfies

$$u_2(x) > 0 \text{ on } [A, B], \; u_2(x) < 0 \text{ on } [C, D],$$

and is, therefore, an interior point of rcK. We can now apply Corollary 5.9 to obtain the following result:

If $0 \le A < B < a/2 < C < D \le a$, then, (5.30) has an unbounded asymptotic bifurcation branch that bifurcates from $\{\infty\} \times [0, 4\pi^2/a^2]$ and satisfies the alternative in Theorem 5.2.

(b) Similar results about bifurcation from infinity for variational inequalities containing the bilinear form in the plate theory are also valid. Consider the variational inequality (4.45) with V, $\langle \cdot, \cdot \rangle$, L, C, a, σ_{ij}, σ^0_{ij} defined in Example 4.4.

For $u, v \in V$,

$$\begin{aligned} |\langle C(u), v \rangle| &\le \sum_{i,j} \int_\Omega |\sigma_{ij}(u)\, \partial_i u| \, |\partial_j v| \\ &\le \sum_{i,j} \|\sigma_{ij}(u)\, \partial_i u\|_{L^2(\Omega)} \, \|\partial_j v\|_{L^2(\Omega)} \\ &\le \sum_{i,j} \|\sigma_{ij}(u)\, \partial_i u\|_{L^2(\Omega)} \, \|v\|. \end{aligned}$$

Thus,

$$\|C(u)\| \le \sum_{i,j} \|\sigma_{ij}(u)\partial_i u\|_{L^2(\Omega)}, \; \forall u \in V.$$

If $\{\sigma_{ij}\}$ satisfies

$$\lim_{\|u\| \to \infty} \frac{\|\sigma_{ij}(u)\partial_i u\|_{L^2(\Omega)}}{\|u\|} = 0, \tag{5.35}$$

for all $i, j \in \{1, 2\}$, then, $B(u, \lambda) = \lambda Lu - C(u)$, $(u, \lambda) \in V \times \mathbb{R}$, satisfies (5.4) with

$$f_\infty(u, \lambda) = \lambda Lu, \; u \in V, \; \lambda \in \mathbb{R}.$$

The homogenized variational inequality of (43) at infinity is of the form,

$$\begin{cases} \langle u, v - u \rangle \ge \lambda \langle Lu, v - u \rangle, \; \forall v \in rcK \\ u \in rcK. \end{cases} \tag{5.36}$$

Note that (5.35) holds, if there exist $C_{ij} \geq 0$ and $0 \leq \alpha < 1$, such that

$$|\sigma_{ij}(u)\partial_i u| \leq C_{ij}(1 + |\nabla u|^\alpha). \tag{5.37}$$

For example, if $\sigma_{ij}(u) = (1+|\partial_i u|^2+|\partial_j u|^2)^{-1/2}$ or $\sigma_{ij}(u) = (1+|\nabla u|^2)^{-1/2}$, then,

$$|\sigma_{ij}(u)\partial_i u| \leq \frac{|\partial_i u|}{\sqrt{1 + |\nabla u|^2}} \leq 1, \ \forall i,j \in \{1,2\}.$$

Therefore, (5.37) holds with $\alpha = 0$.

5.2.3 A corollary

We conclude this chapter with an analog of Corollary 4.8 of Theorem 4.4 for bifurcation from infinity.

Corollary 5.10 *Suppose that B and F are given by (5.14) and (5.16) and that β is self-adjoint.*

Assume, further, that $\langle \cdot, \cdot \rangle$ can be decomposed into a sum of two symmetric, continuous, bilinear forms a_1 and a_2, where a_1 is coercive on V. Let

$$K' = \{w \in V : a_1(w,v) \leq 0, \ \forall v \in rcK\}$$

be the polar cone of rcK with respect to a_1. In addition, assume the following:

(a) $-K' \subset rcK$,
(b) $a_2(v,w) \geq 0$, $\langle \beta v, w \rangle \leq 0$, $\forall v \in rcK, w \in K'$,
(c) If S_{λ_V} denotes the set of solutions of (4.2) with $\lambda_0 = \lambda_V$,

$$\lambda_V = \left(\max_{V\backslash\{0\}} \frac{\langle \beta v, v \rangle}{\|v\|^2} \right)^{-1},$$

then, $\langle u, v \rangle \neq 0$, $\forall u, v \in S_{\lambda_V}$,
(d) $rcK \cap (-rcK) \cap S_{\lambda_V} = \{0\}$.

Under these assumptions, $\lambda_V = \left(\max_{rcK\backslash\{0\}} \frac{\langle \beta v, v \rangle}{\|v\|^2} \right)^{-1}$ is the smallest eigenvalue of (5.5), and (∞, λ_V) is an asymptotic bifurcation point of (3.1) from which bifurcates an unbounded solution branch that satisfies the alternative in Theorem 5.2.

The proof of this corollary follows along the same lines as those of Corollary 4.8 and is omitted.

Applications of this corollary to the problem of bifurcation from infinity for variational inequalities, similar to those in Example 4.5, are the following. For example, consider the variational inequality (4.51) with the convex set K given by

$$K = \{v \in H_0^1(\Omega) : v \geq \psi_1 \text{ on } \Omega\}, \tag{5.38}$$

where ψ_1 is a bounded function on Ω. Let V and B be as in Example 4.5. We assume that g satisfies (5.25) with $b \geq 0$, $b \not\equiv 0$ on Ω. Hence, one can verify that (5.14) and (5.16) are satisfied, and the asymptotically homogenized mapping f_∞ of B is given by

$$\langle f_\infty(u, \lambda), v \rangle = \lambda \int_\Omega b(x)u(x)v(x)dx, \ u, v \in H_0^1(\Omega).$$

Moreover, in this case, one can show that the recession cone rcK is of the form,

$$rcK = \{v \in H_0^1(\Omega) : v \geq 0 \text{ on } \Omega\}.$$

The homogeneous variational inequality corresponding to (4.50), therefore, is

$$\begin{cases} \int_\Omega \nabla u \nabla (v - u) - \lambda \int_\Omega bu(v - u) \geq 0, \ \forall v \in rcK, \\ u \in rcK. \end{cases} \tag{5.39}$$

Using arguments similar to those in Example 4.5, we see that the conditions (a)-(d) in Corollary 5.10 are all satisfied. Applying that corollary, one obtains the following consequence:

Let λ_0 be the smallest (positive) eigenvalue of (5.39). Then, (∞, λ_0) is an asymptotic bifurcation point of (4.50) (with K given by (5.38)), which corresponds to an unbounded asymptotic bifurcation branch satisfying the alternative in Theorem 5.2.

Similar applications for bifurcation from infinity of variational inequalities containing the plate operator considered in Examples 4.6 and 5.3 may be given.

Additional asymptotic bifurcation problems are considered in Chapter 7 for variational inequalities containing nonlinear operators and general convex functionals.

6

Bifurcation in Banach Spaces

In this chapter, we consider several generalizations and extensions of the results in Chapters 3 and 4 for variational inequalities containing nonlinear operators and convex functionals, defined in reflexive Banach spaces.

6.1 Notation and preparatory results

Let V be a (real) reflexive Banach space with norm $\| \cdot \|$ and dual V^*. The dual norm on V^* is also denoted by $\| \cdot \|$, and $\langle \cdot, \cdot \rangle$ denotes the pairing between V^* and V.

Let $A : V \to V^*$ be a (nonlinear) operator satisfying the following properties:

(A1) A is continuous and bounded (in the sense that A maps bounded sets in V into bounded subsets of V^*), and $A(0) = 0$,

(A2) A is strictly monotone on V and coercive, i.e., there exist constants $p > 1$ and $C > 0$, such that

$$\langle A(u), u \rangle \geq C\|u\|^p, \ \forall u \in V, \tag{6.1}$$

$$\langle A(u) - A(v), u - v \rangle > 0, \ \forall u, v \in V, \ u \neq v, \tag{6.2}$$

and A is of class (S) in V ([17], [89]), i.e., for all sequences $\{u_n\} \subset V$, such that

$$u_n \rightharpoonup u \ \text{in} \ V,$$

and

$$\lim_{n\to\infty} \langle A(u_n), u_n - u \rangle = 0, \qquad (6.3)$$

$u_n \to u$ in V.

Because V is reflexive, we know, by the Lindenstrauss-Asplund-Trojanski theorem ([65], [122]), that there exists a norm on V equivalent to $\|\cdot\|$, such that V is locally uniformly convex with this new norm. Hence, in such cases, we can assume, without loss of generality, that V has this property. A useful property of locally uniformly convex spaces is the following ([17] and [89]):

$$\text{If } u_n \rightharpoonup u \text{ in } V, \text{ and } \|u_n\| \to \|u\|, \text{ then } u_n \to u. \qquad (6.4)$$

We next provide a particular criterion for an operator to be in class (S):

Assume that V is locally uniformly convex and that A satisfies the following monotonicity condition:

$$\langle A(u) - A(v), u - v \rangle \ge g(\|u\|, \|v\|), \ \forall u, \quad v \in V, \qquad (6.5)$$

where

$$g : \mathbb{R}^+ \times \mathbb{R}^+ \to \mathbb{R}^+$$

is such that, for any sequence $\{(x_n, y_n)\} \subset \mathbb{R}^+ \times \mathbb{R}^+$ satisfying

$$g(x_n, y_n) \to 0 \quad \text{and} \quad x_n \to a \in \mathbb{R}^+, \qquad (6.6)$$

$y_n \to a$. Then, A belongs to class (S).

To prove this, we suppose that $u_n \rightharpoonup u$ in V and that (6.3) is satisfied. We have $\lim \langle A(u), u_n - u \rangle = 0$, and, hence,

$$\lim \langle A(u_n) - A(u), u_n - u \rangle = 0.$$

By (6.5), this implies that

$$g(\|u_n\|, \|u\|) \to 0, \text{ as } n \to \infty. \qquad (6.7)$$

Letting $x_n = \|u\|$ and $y_n = \|u_n\|$, $\lim_{n\to\infty} x_n = \|u\|$. Hence, by (6.7),

$$y_n = \|u_n\| \to \|u\|.$$

Using (6.7), we obtain $u_n \to u$, proving that A is in class (S).

Next, let $B : V \times \mathbb{R} \to V^*$ be a completely continuous mapping, such that

$$B(0, \lambda) = 0, \ \forall \lambda \in \mathbb{R},$$

and

$$j : V \to \mathbb{R} \cup \{\infty\}$$

is a proper, convex, lower semicontinuous functional, such that

$$j(v) \ge 0, \ \forall v \in V, \text{ and } j(0) = 0.$$

With these settings, we consider the following variational inequality: Find $u \in V$, such that

$$\langle A(u) - B(u, \lambda), v - u \rangle + j(v) - j(u) \geq 0, \ \forall v \in V. \tag{6.8}$$

From the above assumptions, we see that, for all $\lambda \in \mathbb{R}$, 0 is a (trivial) solution of (6.8). We are concerned, here, with the bifurcation problem for (6.8), i.e., as before, with the existence and properties of nontrivial solution sets of (6.8).

As in the case of bifurcation for variational inequalities defined on convex sets in Hilbert spaces (variational inequalities of the form (3.1)), we call $(0, \lambda)$ $(\lambda \in \mathbb{R})$ a bifurcation point of (6.8) if there exists a sequence $\{(u_n, \lambda_n)\}$ of solutions of (6.8), such that

$$\|u_n\| \neq 0, \ \forall n, \quad \text{and} \quad u_n \to 0, \lambda_n \to \lambda \ (n \to \infty).$$

Remark 6.1 (a) We see that (6.1)-(6.2) and, hence, (A2) are satisfied if A satisfies the following coerciveness condition:

$$\langle A(u) - A(v), u - v \rangle \geq C\|u - v\|^p, \ \forall u, v \in V. \tag{6.9}$$

(b) Some other examples of operators satisfying (6.1), (6.5), and (6.6) are the following:

1. A is linear and coercive in the sense that

$$\langle Au, u \rangle \geq \|u\|^2, \ u \in V.$$

In this case, (6.9) follows from the linearity of A.

2. Let A be the following second-order, quasilinear, elliptic operator on $H_0^1(\Omega)$:

$$\langle Au, v \rangle = \int_\Omega \left[\sum_{i=1}^N a_i(x, u(x), \nabla u(x)) \partial_i v(x) + a_0(x, u(x), \nabla u(x)) v(x) \right] dx,$$

where Ω is a bounded domain in $\mathbb{R}^N$ $(N \geq 1)$ with smooth boundary and, for $0 \leq i \leq N$,

$$a_i : \Omega \times \mathbb{R}^{N+1} \to \mathbb{R}$$

are Carathéodory functions satisfying

$$a_i(x, 0, 0) = 0, \ x \in \Omega,$$

and the following growth condition:

$$|a_i(x, u, \xi)| \leq M(|u| + |\xi|)$$

for a.e. $x \in \Omega$, all $u \in \mathbb{R}$, and $\xi \in \mathbb{R}^N$, where M is a fixed positive constant. We also assume that a_i satisfies the usual uniform monotonicity condition:

$$\sum_{i=1}^{N} [a_i(x, u, \xi) - a_i(x, v, \eta)](\xi_i - \eta_i) + [a_0(x, u, \xi) - a_0(x, v, \eta)](u - v)$$

$$\geq C|\xi - \eta|^2$$

for a.e. $x \in \Omega$, all $u, v \in \mathbb{R}$, $\xi, \eta \in \mathbb{R}^N$, where $C > 0$ is a fixed constant. From this condition,

$$
\begin{aligned}
\langle Au - Av, u - v \rangle \; &= \; \int_{\Omega} \{ \sum_{i=1}^{N} [a_i(x, u, \nabla u) - a_i(x, v, \nabla v)] \partial_i (u - v) \\
&\quad + [a_0(x, u, \nabla u) - a_0(x, v, \nabla v)](u - v) \} \, dx \\
&\geq \; C \|u - v\|^2,
\end{aligned}
$$

for all $u, v \in H_0^1(\Omega)$. Hence, (6.9) holds.

3. Let $V = W_0^{1,p}(\Omega)$ (Ω is as above) with the usual equivalent norm derived from Poincaré's inequality:

$$\|u\| = \|u\|_{W_0^{1,p}(\Omega)} = \left(\int_{\Omega} |\nabla u|^p \right)^{1/p}.$$

We consider the p-Laplacian

$$A : W_0^{1,p}(\Omega) \to W^{-1,p'}(\Omega) (\equiv [W_0^{1,p}(\Omega)]^*),$$

given by

$$\langle Au, v \rangle = \int_{\Omega} |\nabla u|^{p-2} \nabla u \nabla v, \; u, v \in W_0^{1,p}(\Omega).$$

First, consider the case $N = 1$, $p \geq 2$, and, for example, $\Omega = (0, 1)$. Because of the inequality

$$(|x|^{p-2}x - |y|^{p-2}y)(x - y) \geq C(p)|x - y|^p, \; \forall x, y \in \mathbb{R},$$

($C(p)$ is a positive constant depending only on p), for $u, v \in W_0^{1,p}(0, 1)$,

$$
\begin{aligned}
\langle Au - Av, u - v \rangle &= \int_0^1 (|u'|^{p-2}u' - |v'|^{p-2}v')(u' - v') \\
&\geq c(p) \int_0^1 |u' - v'|^p \\
&= c(p) \|u - v\|_{W_0^{1,p}(0,1)}^p .
\end{aligned}
$$

Hence, (6.9) holds.

For the p-Laplacian in the general case, we no longer have (6.9). However, we still have (6.1), (6.5), and (6.6).

In fact, for $u, v \in W_0^{1,p}(\Omega)$,

$$\langle Au - Av, u - v \rangle = \int_\Omega \left[|\nabla u|^p + |\nabla v|^p - \left(|\nabla u|^{p-2} + |\nabla v|^{p-2} \right) \nabla u \nabla v \right]$$

$$\geq \|u\|^p + \|v\|^p - \int_\Omega |\nabla u|^{p-1} |\nabla v| - \int_\Omega |\nabla v|^{p-1} |\nabla u| \quad (6.10)$$

However,

$$\int_\Omega |\nabla u|^{p-1} |\nabla v| \leq \left(\int_\Omega |\nabla u|^p \right)^{\frac{p-1}{p}} \left(\int_\Omega |\nabla v|^p \right)^{\frac{1}{p}}$$

$$= \|u\|^{p-1} \|v\|, \quad (6.11)$$

by Hölder's inequality. Using similar estimates in the second integral in the right-hand side of (6.10), one obtains

$$\langle Au - Av, u - v \rangle \geq (\|u\|^{p-1} - \|v\|^{p-1})(\|u\| - \|v\|).$$

Hence, (6.5) follows with

$$g(x, y) = (x - y)(x^{p-1} - y^{p-1}), \quad x, y \geq 0.$$

Because $p > 1$, $g(x,y) \geq 0$, $\forall x, y \geq 0$. Now, suppose that, for sequences $\{x_n\}$, $\{y_n\}$, $x_n \geq 0$, $y_n \geq 0$, $\forall n$,

$$\begin{cases} g(x_n, y_n) = (x_n - y_n)(x_n^{p-1} - y_n^{p-1}) \to 0, \ n \to \infty \\ x_n \to a. \end{cases}$$

If $y_n \not\to a$, then, there exists a subsequence $\{y_{n_k}\} \subset \{y_n\}$, such that $y_{n_k} \to \infty$ or $y_{n_k} \to b \neq a$. Hence,

$$g(x_{n_k}, y_{n_k}) \to \infty \quad \text{or} \quad g(x_{n_k}, y_{n_k}) \to (a - b)(a^{p-1} - b^{p-1}) \neq 0.$$

This contradiction shows that g satisfies (6.6). Because $W_0^{1,p}(\Omega)$ is uniformly convex ([1]), A belongs to class (S) by the above remarks.

Now, with $v = 0$,

$$\langle Au, u \rangle = \|u\|^p, \ \forall u \in W_0^{1,p}(\Omega).$$

Hence, (6.1) is satisfied. Moreover, A is strictly monotone on $W_0^{1,p}(\Omega)$. In fact, if $\langle Au - Av, u - v \rangle = 0$, then, we have equality signs in (6.10) and (6.11). Hence, $\|u\| = \|v\|$, and

$$\nabla u \nabla v = \|\nabla u\| \, \|\nabla v\| \quad \text{a.e. in } \Omega,$$

i.e.,

$$\nabla u(x) = a(x)\nabla v(x), \ a(x) \geq 0, \ x \in \Omega,$$

and from (6.11), $\|\nabla u\|^p = c\|\nabla v\|^p$ a.e. in Ω with $c = $ const. Hence $a^p = c$, i.e., $a = $ const. Because

$$\int_\Omega \|\nabla u\|^p = \int_\Omega \|\nabla v\|^p,$$

$a \equiv 1$. Hence, $\nabla u = \nabla v$ a.e. in Ω. Because $u, v \in W_0^{1,p}(\Omega)$, $u = v$, proving the strict monotonicity of A.

4. Next, we consider the following anisotropic operator corresponding to the p-Laplacian, $p > 1$:

$$A : W_0^{1,p}(\Omega) \to W^{-1,p'}(\Omega),$$

given by

$$\langle Au, v \rangle = \sum_{i=1}^N \int_\Omega a_i(x)|\partial_i u(x)|^{p-2}\partial_i u(x)\partial_i v(x)dx, \qquad (6.12)$$

where $a_i \in L^\infty(\Omega)$, $a_i \geq a > 0$. For this example, we will consider on $W_0^{1,p}(\Omega)$ the norm

$$\|u\|_0 = \left(\sum_{i=1}^N \int_\Omega |\partial_i u|^p dx\right)^{1/p} = \left(\sum_{i=1}^N \|\partial_i u\|_{L^p(\Omega)}^p\right)^{1/p}.$$

It is known (cf. [1] and [89]) that $\|\cdot\|_0$ is equivalent to the usual norm of $\|\cdot\|_{W_0^{1,p}(\Omega)}$ and $W_0^{1,p}(\Omega)$ is also uniformly convex with respect to $\|\cdot\|_0$.

If $p \geq 2$, then, (cf. [18])

$$\langle Au - Av, u - v \rangle = \sum_{i=1}^N \int_\Omega a_i(x)(|\partial_i u|^{p-2}\partial_i u - |\partial_i v|^{p-2}\partial_i v)(\partial_i u - \partial_i v)\, dx$$

$$\geq ac(p) \sum_{i=1}^N \int_\Omega |\partial_i u - \partial_i v|^p dx$$

$$= \|u - v\|_0^p.$$

Hence A satisfies (6.9) and, therefore, (A2). On the other hand, if $1 < p \leq 2$, then, because

$$[|\partial_i u(x)|^{p-2}\partial_i u(x) - |\partial_i v(x)|^{p-2}\partial_i v(x)][\partial_i u(x) - \partial_i v(x)] \geq 0$$

in Ω,

$$\langle Au - Av, u - v \rangle = \sum_{i=1}^N a\int_\Omega (|\partial_i u|^{p-2}\partial_i u - |\partial_i v|^{p-2}\partial_i v)(\partial_i u - \partial_i v)dx$$

$$\geq a \sum_{i=1}^N \int_\Omega (|\partial_i u|^p + |\partial_i v|^p - |\partial_i u|^{p-1}|\partial_i v| - |\partial_i u||\partial_i v|^{p-1}).$$

As above, $\int_\Omega |\partial_i u|^{p-1}|\partial_i v| \leq \|\partial_i u\|_{L^p(\Omega)}^{p-1}\|\partial_i v\|_{L^p(\Omega)}$. Hence,

$$\langle Au - Av, u - v\rangle$$

$$\geq a\sum_{i=1}^{N}(\|\partial_i u\|_{L^p(\Omega)}^{p} + \|\partial_i v\|_{L^p(\Omega)}^{p} - \|\partial_i u\|_{L^p(\Omega)}^{p-1}\|\partial_i v\|_{L^p(\Omega)}$$

$$- \|\partial_i v\|_{L^p(\Omega)}^{p-1}\|\partial_i u\|_{L^p(\Omega)})$$

$$\geq a\left[\sum_{i=1}^{N}\|\partial_i u\|_{L^p(\Omega)}^{p} + \sum_{i=1}^{N}\|\partial_i v\|_{L^p(\Omega)}^{p}\right.$$

$$- \left(\sum_{i=1}^{N}\|\partial_i u\|_{L^p(\Omega)}^{p}\right)^{\frac{p-1}{p}}\left(\sum_{i=1}^{N}\|\partial_i v\|_{L^p(\Omega)}^{p}\right)^{\frac{1}{p}}$$

$$- \left(\sum_{i=1}^{N}\|\partial_i v\|_{L^p(\Omega)}^{p}\right)^{\frac{p-1}{p}}\left(\sum_{i=1}^{N}\|\partial_i u\|_{L^p(\Omega)}^{p}\right)^{\frac{1}{p}}\right]$$

$$= a(\|u\|_0^{p-1} - \|v\|_0^{p-1})(\|u\|_0 - \|v\|_0).$$

Hence, A satisfies (6.5) with $g(x,y) = a(x^{p-1} - y^{p-1})(x - y)$, which also satisfies (6.6). It is clear that (6.1) holds. Also, by arguments similar to those used in 3, one may prove that A is strictly monotone, and, therefore, A satisfies (A2).

For $f \in V^*$, consider the variational inequality:

$$\begin{cases} \langle A(u) - f, v - u\rangle + j(v) - j(u) \geq 0, \ \forall v \in V, \\ u \in V. \end{cases} \tag{6.13}$$

From (A2)-(A4), we see that A is strictly monotone, bounded, continuous, and coercive on V in the sense that

$$\lim_{\|u\|\to\infty} \frac{\langle A(u), u\rangle}{\|u\|} = \infty.$$

Moreover, j is a proper, lower semicontinuous, convex functional on V. Thus, by classical results about existence and uniqueness of solutions of variational inequalities (Chapter 2, [65] and Theorem 2.1, Chapter 2), we know that, for each $f \in V^*$, (6.13) has a unique solution $u = u_f \in V$.
Let

$$P = P_{A,j} : V^* \to V$$

be the mapping that associates with each $f \in V^*$ the (unique) solution u_f of (6.13):

$$u_f = P_{A,j}(f) = Pf, \ f \in V^*. \tag{6.14}$$

With this definition, we see that (6.8) is equivalent to

$$u = P[B(u, \lambda)].$$ (6.15)

Before proceeding further, we need a stability result for $P_{A,j}(f)$.

Let A, A_n, $n = 1, 2, \ldots$ be continuous, bounded, and strictly monotone operators from V to V^* of class (S), such that $A_n(0) = A(0) = 0$, $\forall n$. Also, let j, j_n, $n = 1, 2, \ldots$ be proper, convex, nonnegative functionals from V to $\mathbb{R} \cup \{\infty\}$, such that $j_n(0) = j(0) = 0$, $\forall n$, and let $f, f_n \in V^*$, $n = 1, 2, \ldots$.

We assume that

$$A_n \to A, \; j_n \to j, \; f_n \to f, \;\; \text{as} \;\; n \to \infty$$

in the following sense:

(A3) $f_n \to f$ in V^* (in the norm topology).

(A4) (a) For all $v \in V$, all subsequences $\{n_k\} \subset \mathbb{N}$, there exists a sequence $\{v_{n_k}\} \subset V$, such that

$$\begin{cases} v_{n_k} \to v \text{ in } V, \text{ and} \\ j_{n_k}(v_{n_k}) \to j(v) \text{ in } \mathbb{R} \cup \{\infty\}, \; k \to \infty. \end{cases}$$ (6.16)

(b) If $\{n_k\}$ is a subsequence of $\mathbb{N}$ and if

$$v_{n_k} \rightharpoonup v \text{ in } V,$$

then,

$$j(v) \leq \liminf_{k \to \infty} j_{n_k}(v_{n_k}).$$ (6.17)

(A5) (a) If $\{n_k\} \subset \mathbb{N}$ and $v_{n_k} \to v$ in V, then,

$$A_{n_k}(v_{n_k}) \to A(v) \text{ in } V^*.$$ (6.18)

(b) A_n satisfies (6.1) with the same constant C, and A_n belongs to class (S) uniformly, i.e., if $\{n_k\} \subset \mathbb{N}$ and $v_{n_k} \rightharpoonup v, w_{n_k} \to v$ in V, such that

$$\lim \langle A_{n_k}(v_{n_k}), v_{n_k} - w_{n_k} \rangle = 0,$$ (6.19)

then,

$$v_{n_k} \to v \text{ in } V.$$ (6.20)

Before proving the main result of this section (Lemma 6.1) about the stability of solutions of the variational inequality (6.8) with respect to A, f, and j, some remarks on assumption (A5) are in order.

- If $A_n = A$ for all n, then, (A5) is immediately satisfied because A is bounded, continuous on V, and belongs to class (S). In fact, let

$\{n_k\}, \{v_{n_k}\}$, and $\{w_{n_k}\}$ satisfy (6.19). Because $\{v_{n_k}\}$ is bounded in V, $\{A(v_{n_k})\}$ is also bounded in V^*. Hence,

$$\lim \langle A(v_{n_k}), w_{n_k} - v \rangle = 0.$$

This and (6.19) (with $A_{n_k} = A$) imply that

$$\lim \langle A(v_{n_k}), v_{n_k} - v \rangle = 0.$$

Because A is in class (S), we have (6.20).

- Assume that V is locally uniformly convex. Then, condition (b) in (A5) is satisfied if A_n satisfies condition (6.5) uniformly, i.e., we have (6.5) for all A_n with the same function g.

  In fact, assume that $v_{n_k} \rightharpoonup v$ in V, such that (6.19) holds. From (6.18), we see that

  $$A_{n_k}(w_{n_k}) \to A(v) \quad \text{in } V^*. \tag{6.21}$$

  Hence, because $v_{n_k} - w_{n_k} \rightharpoonup 0$,

  $$\langle A_{n_k}(w_{n_k}), v_{n_k} - w_{n_k} \rangle \to \langle A(v), 0 \rangle = 0.$$

  Together with (6.19), this gives

  $$\lim_{n \to \infty} \langle A_{n_k}(v_{n_k}) - A_{n_k}(w_{n_k}), v_{n_k} - w_{n_k} \rangle = 0. \tag{6.22}$$

  Hence, by (6.5),
  $$g(\|v_{n_k}\|, \|w_{n_k}\|) \to 0$$
  as $k \to \infty$. Letting $x_k = \|w_k\| \to a = \|v\|$, $y_k = \|v_{n_k}\|$, from (6.6), $\|v_{n_k}\| \to \|v\|$. This and (6.4) imply that $v_{n_k} \to v$.

  Hence, A_n is in class (S) uniformly.

- If A_n satisfies (6.9) with the same constant C, then, A_n belongs to class (S) uniformly.

  In fact, if $v_{n_k} \rightharpoonup v$ and if (6.19) is satisfied, then, arguing as above, we have (6.21) and (6.22). However,

  $$C\|v_{n_k} - w_{n_k}\|^p \leq \langle A_{n_k}(v_{n_k}) - A_{n_k}(w_{n_k}), v_{n_k} - w_{n_k} \rangle, \quad \forall k \in \mathbb{N}.$$

  Hence, $\|v_{n_k} - w_{n_k}\| \to 0$, and, therefore, (6.20) holds because $w_{n_k} \to v$.

Next, we prove the following lemma about the stability of solutions of (6.8) with respect to the given data.

Lemma 6.1 *If (A3), (A4), and (A5) are satisfied, then,*

$$P_{A_n,j_n}(f_n) \to P_{A,j}(f) \quad \text{in } V.$$

Proof. Let $u = P_{A,j}(f)$ and $u_n = P_{A_n,j_n}(f_n)$ (u_n, u exist and are unique by the above remarks). We have to prove that $u_n \to u$ in V. For $n \in \mathbb{N}$,

$$\langle A_n(u_n) - f_n, v - u_n \rangle + j_n(v) - j_n(u_n) \geq 0, \ \forall v \in V, \tag{6.23}$$

and

$$\langle A(u) - f, v - u \rangle + j(v) - j(u) \geq 0, \ \forall v \in V. \tag{6.24}$$

We first prove that $\{u_n\}$ is bounded in V. Letting $v = 0$ in (6.23),

$$-\langle A_n(u_n) - f_n, u_n \rangle - j_n(u_n) \geq 0,$$

and, thus,

$$-\langle A_n(u_n), u_n \rangle + \langle f_n, u_n \rangle \geq j_n(u_n) \geq 0.$$

It follows from (6.1) that

$$\begin{aligned} C\|u_n\|^p &\leq \langle A_n(u_n), u_n \rangle \\ &\leq \langle f_n, u_n \rangle \\ &\leq \|f_n\| \, \|u_n\|. \end{aligned}$$

Hence,

$$\|u_n\|^{p-1} \leq C^{-1}\|f_n\|, \ \forall n.$$

Because $\{\|f_n\|\}$ is bounded, so is $\{\|u_n\|\}$. Now, because V is reflexive, we can choose a subsequence $\{u_\eta\} \subset \{u_n\}$, such that

$$u_\eta \rightharpoonup w \quad \text{in } V. \tag{6.25}$$

Then, w is a solution of (6.24). In fact, let $v \in V$. By (6.16), we can choose a sequence $\{v_\eta\}$ in V, such that

$$\begin{cases} v_\eta \to v & \text{in} \quad V \\ j_\eta(v_\eta) \to j(v) & \text{in} \quad \mathbb{R} \cup \{\infty\}. \end{cases} \tag{6.26}$$

Because A_n and A are monotone in V, we find from Minty's lemma ([51]) that (6.23) and (6.24) are equivalent to

$$\langle A_n(v) - f_n, v - u_n \rangle + j_n(v) - j_n(u_n) \geq 0, \ \forall v \in V \tag{6.27}$$

and

$$\langle A(v) - f, v - u \rangle + j(v) - j(u) \geq 0, \ \forall v \in V. \tag{6.28}$$

Now, letting $n = \eta$ and $v = v_\eta$ in (6.27),

$$\langle A_\eta(v_\eta) - f_\eta, v_\eta - u_\eta \rangle + j_\eta(v_\eta) - j_\eta(u_\eta) \geq 0, \ \forall \eta. \qquad (6.29)$$

From (6.25) and (6.17),

$$j(w) \leq \liminf j_\eta(u_\eta). \qquad (6.30)$$

Hence, passing to the limit in (6.29), and recalling (6.18),

$$\begin{cases} A_\eta(v_\eta) - f_\eta \to A(v) - f & \text{in} \quad V^* \\ v_\eta - u_\eta \rightharpoonup v - w & \text{in} \quad V. \end{cases}$$

Thus,

$$\langle A_\eta(v_\eta) - f_\eta, v_\eta - u_\eta \rangle \to \langle A(v) - f, v - u \rangle.$$

Letting $\eta \to \infty$ in (6.29), it follows from (6.26) and (6.30) that

$$\langle A(v) - f, v - w \rangle + j(v) - j(w)$$
$$\geq \quad \lim \langle A(v_\eta) - f_\eta, v_\eta - u_\eta \rangle + \lim j_\eta(v_\eta) - \liminf j_\eta(u_\eta)$$
$$\geq \quad \liminf \left[\langle A(v_\eta) - f_\eta, v_\eta - u_\eta \rangle + j_\eta(v_\eta) - j_\eta(u_\eta) \right]$$
$$\geq \quad 0.$$

Hence, w is a solution of (6.28) and, therefore, of (6.24). By the uniqueness of the solution of (6.24), we must have $w = u$, i.e.,

$$u_\eta \rightharpoonup u \ \text{in} \ V. \qquad (6.31)$$

Next, we prove that this convergence is, in fact, a strong convergence in V.
Again, applying (6.16), we can choose a sequence $\{w_\eta\}$ in V, such that

$$\begin{cases} w_\eta \to u & \text{in} \quad V, \\ j_\eta(w_\eta) \to j(u) & \text{in} \quad \mathbb{R} \cup \{\infty\}. \end{cases} \qquad (6.32)$$

Letting $n = \eta$ and $v = w_\eta$ in (6.23),

$$\langle A_\eta(u_\eta) - f_\eta, w_\eta - u_\eta \rangle + j_\eta(w_\eta) - j_\eta(u_\eta) \geq 0. \qquad (6.33)$$

From (6.31) and (6.32),

$$w_\eta - u_\eta \rightharpoonup 0 \ \text{in} \ V. \qquad (6.34)$$

Now, we have

$$\langle A_\eta(w_\eta) - A_\eta(u_\eta), w_\eta - u_\eta \rangle$$
$$= \quad \langle A_\eta(w_\eta) - f_\eta, w_\eta - u_\eta \rangle - \langle A_\eta(u_\eta) - f_\eta, w_\eta - u_\eta \rangle$$
$$\leq \quad \langle A_\eta(w_\eta) - f_\eta, w_\eta - u_\eta \rangle + j_\eta(w_\eta) - j_\eta(u_\eta) \quad \text{(by (6.33))}$$
$$= \quad \langle A_\eta(w_\eta) - f_\eta, w_\eta - u_\eta \rangle + [j_\eta(w_\eta) - j(u)] + [j(u) - j_\eta(u_\eta)].$$

By (6.32), (6.34), and (6.18),

$$A_\eta(w_\eta) - f_\eta \to A(u) - f,$$

and, then,

$$\langle A_\eta(w_\eta) - f_\eta, w_\eta - u_\eta \rangle \to 0.$$

Therefore, by (6.32) and (6.30),

$$
\begin{aligned}
\limsup &\langle A_\eta(u_\eta) - A_\eta(w_\eta), u_\eta - w_\eta \rangle \\
&\leq \limsup\{[\langle A_\eta(w_\eta) - f_\eta, w_\eta - u_\eta \rangle] + [j_\eta(w_\eta) - j(u)] + [j(u) - j_\eta(u_\eta)]\} \\
&= \lim\langle A_\eta(w_\eta) - f_\eta, w_\eta - u_\eta \rangle + \lim[j_\eta(w_\eta) - j(u)] + \limsup[j(u) \\
&\quad - j_\eta(u_\eta)] \\
&= \limsup[j(u) - j_\eta(u_\eta)] \\
&= j(u) - \liminf j_\eta(u_\eta) \\
&\leq 0.
\end{aligned}
$$

Because A_η is monotone,

$$\liminf\langle A_\eta(u_\eta) - A_\eta(w_\eta), u_\eta - w_\eta \rangle \geq 0.$$

It follows that

$$\lim\langle A_\eta(u_\eta) - A_\eta(w_\eta), u_\eta - w_\eta \rangle = 0.$$

On the other hand, because $A_\eta(w_\eta) \to A(u)$ in V^*, also from (6.34),

$$\langle A_\eta(w_\eta), u_\eta - w_\eta \rangle \to \langle A(u), 0 \rangle = 0.$$

Therefore,

$$\lim\langle A_\eta(u_\eta), u_\eta - w_\eta \rangle = 0. \tag{6.35}$$

From (6.31) and (6.32), we have (6.19) with $\{A_{n_k}\} = \{A_\eta\}$, $\{w_{n_k}\} = \{w_\eta\}$, $\{v_{n_k}\} = \{u_\eta\}$, and $u = v$. Because A_η belong to class (S) uniformly, from (6.31) and (6.20),

$$u_\eta \to u \text{ in } V.$$

Now, suppose that $u_n \not\to u$, as $n \to \infty$. We can, therefore, choose a subsequence $\{u_{n_k}\} \subset \{u_n\}$, such that

$$\|u_{n_k} - u\| \geq \epsilon_0 > 0, \ \forall k.$$

By the above proof, we can choose a subsequence $\{u_\eta\} \subset \{u_{n_k}\}$, such that

$$u_\eta \to u \text{ in } V.$$

This contradicts the choice of $\{u_{n_k}\}$ and completes the proof of Lemma 6.1.   ∎

As an immediate consequence of this lemma, we have the following corollary:

Corollary 6.2 *The mapping P defined by (6.14) is continuous from V^* to V.*

Because B is completely continuous from $V \times \mathbb{R}$ to V^*, it follows from this corollary that $P \circ B$ is completely continuous from $V \times \mathbb{R}$ to V. Hence, the topological degree $\mathrm{d}(I - P[B(\cdot, \lambda)], U, 0)$ is defined for all open, bounded subsets $U \ni 0$ of V provided

$$u \neq P[B(u, \lambda)], \ \forall u \in \partial U.$$

Applying Theorem 2.5, we have, as in Chapter 3, the following result:

Theorem 6.3 *Let $a, b \in \mathbb{R}, (a < b)$ be such that $u = 0$ is an isolated solution of (6.8) for $\lambda = a$ and $\lambda = b$, where $(0, a), (0, b)$ are not bifurcation points of (6.8).*
  Assume, furthermore, that, for some $r > 0$, small,

$$\mathrm{d}(I - P[B(\cdot, a)], B_r(0), 0) \neq \mathrm{d}(I - P[B(\cdot, b)], B_r(0), 0). \tag{6.36}$$

Let

$$S = \overline{\{(u, \lambda) : (u, \lambda) \ is \ a \ solution \ of \ (6.8) \ with \ u \neq 0\}} \cup (\{0\} \times [a, b]),$$

and let C be the connected component of S containing $\{0\} \times [a, b]$. Then, either
  (i) C is unbounded in $V \times \mathbb{R}$, or
  (ii) $C \cap (\{0\} \times (\mathbb{R} \setminus [a, b])) \neq \emptyset$.

6.2 Homogenization procedures

In this section, we shall establish homogenization procedures for the variational inequality (6.8) to obtain a homogeneous variational inequality (in the sense that if u is one of its solutions, then, so is tu for all $t \geq 0$). We also consider relationships between bifurcation points of (6.8) and eigenvalues of the homogenized variational inequality.
  We assume that A and B are differentiable of order $p - 1$ with respect to u at $u = 0$ in the following sense:
  There exist

$$\alpha : V \to V^*, \ f : V \times \mathbb{R} \to V^*$$

such that
  (A6) (a) α satisfies (A1) and (A2), and, for all sequences $\{v_n\} \subset V$, $\{\sigma_n\} \subset \mathbb{R}_*^+$ satisfying

$$v_n \to v \ \text{in} \ V, \ \sigma_n \to 0^+,$$

$$\frac{1}{\sigma_n^{p-1}} A(\sigma_n v_n) \to \alpha(v) \quad \text{in} \quad V^*. \tag{6.37}$$

(b) For $\sigma > 0$, let

$$A_\sigma(v) = \frac{1}{\sigma^{p-1}} A(\sigma v), \quad v \in V. \tag{6.38}$$

Then, A_{σ_n} satisfies (6.1) uniformly, and A_{σ_n} belongs to class (S) uniformly in the sense of (A5) (b).

(A7) f is completely continuous, such that, for all sequences $\{v_n\} \subset V$, $\{\sigma_n\} \subset \mathbb{R}_*^+$, $\{\lambda_n\} \subset \mathbb{R}$ satisfying

$$v_n \to v \quad \text{in} \quad V, \quad \lambda_n \to \lambda, \quad \sigma_n \to 0^+,$$

$$\frac{1}{\sigma_n^{p-1}} B(\sigma_n v_n, \lambda_n) \to f(v, \lambda) \quad \text{in} \quad V^*. \tag{6.39}$$

We see that α and f are uniquely determined by (6.37) and (6.39). In fact,

$$\alpha(v) = \lim_{t \to 0^+} \frac{A(tv)}{t^{p-1}}, \tag{6.40}$$

and

$$f(v, \lambda) = \lim_{t \to 0^+} \frac{B(tv, \lambda)}{t^{p-1}}, \tag{6.41}$$

for all $v \in V, \lambda \in \mathbb{R}$.

Moreover, α and $f(\cdot, \lambda)$ are positive homogeneous of degree $p - 1$, i.e.,

$$\alpha(\sigma v) = \sigma^{p-1}\alpha(v), \quad f(\sigma v, \lambda) = \sigma^{p-1}f(v, \lambda) \tag{6.42}$$

for all $\sigma \geq 0, v \in V, \lambda \in \mathbb{R}$. The proof of these properties is straightforward.

We note that (A6) (b) is satisfied automatically in either of the following cases:

• A satisfies (6.9), or

• V is locally uniformly convex, and A satisfies (6.5) with g homogeneous of degree p with respect to x and y in the sense that

$$g(tx, ty) = t^p g(x, y), \quad \forall x, y, t \geq 0. \tag{6.43}$$

In fact, in the first case, for all $n \in \mathbb{N}$, $u, v \in V$,

$$
\begin{aligned}
\langle A_{\sigma_n}(u) - A_{\sigma_n}(v), u - v \rangle &= \frac{1}{\sigma_n^p} \langle A(\sigma_n u) - A(\sigma_n v), \sigma_n u - \sigma_n v \rangle \\
&\geq \frac{1}{\sigma_n^p} C \|\sigma_n u - \sigma_n v\|^p \\
&= C\|u - v\|^p.
\end{aligned}
$$

Hence, $A_{\sigma_n}(n = 1, 2, \dots)$ satisfies (6.9) with the same constant C. Thus, A_{σ_n} belongs to class (S) uniformly, as follows from the remark after (6.18), (6.19), and (6.20).

The proof for the second case is similar. As above,

$$
\begin{aligned}
\langle A_{\sigma_n}(u), u \rangle &= \frac{1}{\sigma_n^p} \langle A(\sigma_n u), \sigma_n u \rangle \\
&\geq \frac{1}{\sigma_n^p} C \| \sigma_n u \|^p \\
&= C \| u \|^p,
\end{aligned}
$$

and, moreover,

$$
\begin{aligned}
\langle A_{\sigma_n}(u) - A_{\sigma_n}(v), u - v \rangle &= \frac{1}{\sigma_n^p} \langle A(\sigma_n u) - A(\sigma_n v), \sigma_n u - \sigma_n v \rangle \\
&\geq \frac{1}{\sigma_n^p} g(\sigma_n \| u \|, \sigma_n \| v \|) \\
&= g(\| u \|, \| v \|),
\end{aligned}
$$

for all $n \in \mathbb{N}$, all $u, v \in V$. Hence, A_{σ_n} satisfies (6.5) uniformly, and our conclusion follows, as in the first case.

Note that, in the case A is the p-Laplacian, then, g given by

$$
g(x, y) = (x - y)(x^{p-1} - y^{p-1}), \quad x, y \geq 0,
$$

satisfies the homogeneity condition (6.43).

Now, we consider the homogenization of the convex functional j at 0. For $\sigma > 0$, we denote by j_σ the functional from V to $[0, \infty]$ defined by

$$
j_\sigma(v) = \frac{1}{\sigma^p} j(\sigma v), \quad v \in V. \tag{6.44}
$$

We assume that there exists a proper, convex, lower semicontinuous functional $J : V \to [0, \infty]$, such that j_σ tends to J (as $\sigma \to 0^+$) in the following sense:

(A8) (a) If

$$
v_n \rightharpoonup v \text{ in } V \text{ and } \sigma_n \to 0^+ \ (\sigma_n > 0, \ \forall n),
$$

then,

$$
J(v) \leq \liminf j_{\sigma_n}(v_n). \tag{6.45}
$$

(b) For each $v \in V$, each sequence $\{\sigma_n\} \subset \mathbb{R}_*^+$, such that

$$
\sigma_n \to 0^+, \text{ as } n \to \infty,
$$

we can choose a sequence $\{v_n\} \subset V$, such that

$$
\begin{cases}
v_n \to v \text{ in } V, \text{ and} \\
j_{\sigma_n}(v_n) \to J(v).
\end{cases} \tag{6.46}
$$

From (6.45), we see that $J(0) = 0$. We also note that J, if it exists, is uniquely determined by (6.45) and (6.46). In fact, let J_1 be another functional that satisfies (6.45) and (6.46). Let $v \in V$, and choose a sequence $\{v_n\}$ satisfying (6.46) (with respect to J). Because J_1 also satisfies (6.45) (and because $v_n \rightarrow v$),

$$
\begin{aligned}
J_1(v) &\leq \liminf j_{\sigma_n}(v_n) \\
&= \lim j_{\sigma_n}(v_n) \\
&= J(v).
\end{aligned}
$$

Because this holds for all $v \in V$, we must have $J_1 \leq J$. Similarly, $J \leq J_1$, and therefore, $J = J_1$.

A useful property of J is that it is positive homogeneous of degree p on V, i.e.,

$$J(\sigma u) = \sigma^p J(u), \quad \forall u \in V, \sigma \geq 0. \tag{6.47}$$

This is obviously true for $\sigma = 0$. For $\sigma > 0$ fixed, we define

$$J_1 : V \rightarrow [0, \infty], \quad J_1(v) = \sigma^{-p} J(\sigma v), \quad v \in V.$$

Then, J_1 is a proper, convex, lower semicontinuous functional on V. We check that (6.45), (6.46) are satisfied for J_1. In fact, let $v_n \rightarrow v$, $\sigma_n \rightarrow 0^+$. Then, $\sigma_n/\sigma \rightarrow 0^+$, and $\sigma v_n \rightarrow \sigma v$ in V. From (6.45),

$$
\begin{aligned}
J(\sigma v) &\leq \liminf \frac{j\left(\frac{\sigma_n}{\sigma}(\sigma v_n)\right)}{(\sigma_n/\sigma)^p} \\
&= \sigma^p \liminf \frac{j(\sigma_n v_n)}{\sigma_n^p},
\end{aligned}
$$

i.e.,

$$J_1(v) = \frac{J(\sigma v)}{\sigma^p} \leq \liminf \frac{j(\sigma_n v_n)}{\sigma_n^p},$$

implying that J_1 satisfies (6.45).

Now, let $v \in V$. Then, by (6.46), applied to $w = \sigma v$ and the sequence $\{\sigma_n/\sigma\}_{n \in \mathbb{N}}$, we can find a sequence $\{w_n\} \subset V$, such that

$$w_n \rightarrow w = \sigma v$$

and

$$\frac{j\left(\frac{\sigma_n}{\sigma}(w_n)\right)}{(\sigma_n/\sigma)^p} \rightarrow J(\sigma v).$$

Letting $v_n = w_n/\sigma$, $v_n \rightarrow v$, and

$$
\begin{aligned}
\sigma^p \frac{j(\sigma_n v_n)}{\sigma_n^p} &= \frac{j\left(\frac{\sigma_n}{\sigma}(w_n)\right)}{(\sigma_n/\sigma)^p} \\
&\rightarrow J(\sigma v), \quad \text{as } n \rightarrow \infty,
\end{aligned}
$$

i.e.,

$$\frac{j(\sigma_n v_n)}{\sigma_n^p} \to \frac{J(\sigma v)}{\sigma^p} = J_1(v).$$

Thus, (6.46) holds for J_1.

Now, by the uniqueness result about J proved above, we must have $J_1 = J$, which means that

$$J(v) = \frac{J(\sigma v)}{\sigma^p}, \ \forall v \in V,$$

and (6.47) is proved.

With these settings, we can now homogenize (6.8) to obtain the following variational inequality:

$$\begin{cases} \langle \alpha(u) - f(u, \lambda), v - u \rangle + J(v) - J(u) \geq 0, \ \forall v \in V, \\ u \in V. \end{cases} \tag{6.48}$$

Before considering some properties of (6.48) and its relationships with (6.8), let us make some remarks about the situation in Chapter 3, i.e., about the variational inequality (3.1). Let K be a closed, convex subset of V, as in Section 3.1. We define j to be the indicator function of K:

$$j = I_K : V \to [0, \infty],$$

$$j(u) = \begin{cases} 0 & \text{if} \ \ u \in K, \\ \infty & \text{if} \ \ u \notin K. \end{cases} \tag{6.49}$$

Then, j is a proper, convex, lower semicontinuous functional on V, $j(0) = 0$, and $D(j) \equiv \{u \in V : j(u) \in \mathbb{R}\}$ (the effective domain of j) is K. (In fact, $K = D(j) = \ker j$.)

With j defined by (6.49), we see immediately that (3.1) can be written in the form (6.8), with $A \equiv I$.

Let K_0 be the support cone of K, defined as in Chapter 3, and let J be the indicator function of K_0:

$$J(u) = I_{K_0}(u) = \begin{cases} 0 & \text{if} \ \ u \in K_0, \\ \infty & \text{if} \ \ u \notin K_0. \end{cases} \tag{6.50}$$

Then, (6.45) and (6.46) are satisfied, with $p = 2$. In fact, in the proof of Theorem 3.2, we have already verified the following results:

$$\text{If } \sigma_n \to 0^+, v_n \rightharpoonup v, \ \text{and} \ v_n \in \sigma_n^{-1} K, \ \forall n, \ \text{then}, \ v \in K_0, \tag{6.51}$$

and, if $\sigma_n \to 0^+$ and $v \in K_0$, then, one can choose a sequence $\{v_n\} \subset V$, such that

$$v_n \in \sigma_n^{-1} K, \ \forall n, \ \text{and} \ v_n \to v \ \text{in} \ V. \tag{6.52}$$

Now, let $v_n \rightharpoonup v$ in V, $\sigma_n \to 0^+$. Then,

$$\liminf \frac{j(\sigma_n v_n)}{\sigma_n^2} \in \{0, \infty\}.$$

If $\liminf \dfrac{j(\sigma_n v_n)}{\sigma_n^2} = \infty$, then, (6.45) obviously holds. In the other case, we must have

$$\lim_{k \to \infty} \frac{j(\sigma_{n_k} v_{n_k})}{\sigma_{n_k}^2} = \liminf_{n \to \infty} \frac{j(\sigma_n v_n)}{\sigma_n^2} = 0,$$

i.e., $\sigma_{n_k} v_{n_k} \in K$ for a subsequence $\{n_k\} \subset \mathbb{N}$. Hence $v_{n_k} \in \sigma_{n_k}^{-1} K$, for all $k \in \mathbb{N}$ and $v \in K_0$, i.e., $J(v) = 0$, by (6.51). Hence, (6.45) is also satisfied in this case.

Now, suppose that $v \in V$ and $\sigma_n \to 0^+$. If $v \notin K_0$, then, there exists $\delta > 0$, such that

$$B_\delta(v) \cap \left(\bigcup_{t>0} tK \right) = \emptyset.$$

Choosing any sequence $v_n \to v, v_n \in B_\delta(v)$, $\forall n$, we see that $\sigma_n v_n \notin K$, $\forall n$, and therefore,

$$\frac{j(\sigma_n v_n)}{\sigma_n^2} = \infty, \ \forall n.$$

Hence, $\lim \dfrac{j(\sigma_n v_n)}{\sigma_n^2} = \infty = J(v)$.

If $v \in K_0$, then, by (6.52), one can choose a sequence $\{v_n\}$, such that $v_n \to v$ and $\sigma_n v_n \in K$, $\forall n$. In this case,

$$j(\sigma_n v_n) = 0, \ \forall n$$

and $J(v) = 0 = \lim \dfrac{j(\sigma_n v_n)}{\sigma_n^2}$. Thus, (6.46) holds.

We have proved that (6.45) and (6.46) are satisfied (with $p = 2$) in the particular case in Chapter 3 where j and J are defined by (6.49) and (6.50).

Now, we return to our general discussion. First, we observe that, if A satisfies (6.9), then, for $u, v \in V, t > 0$, it follows from (6.9) that

$$\langle A(tu) - A(tv), t(u - v) \rangle \ \geq \ Ct^p \| u - v \|^p,$$

i.e.,

$$\left\langle \frac{A(tu)}{t^{p-1}} - \frac{A(tv)}{t^{p-1}}, u - v \right\rangle \geq C\| u - v \|^p, \ \forall t > 0.$$

Letting $t \to 0^+$, we see from (6.40) that

$$\langle \alpha(u) - \alpha(v), u - v \rangle \geq C\| u - v \|^p, \ \forall u, v \in V. \tag{6.53}$$

Hence, α also satisfies (6.9).

In the general case, because α satisfies (A1)-(A2) and J is convex, lower semicontinuous, and nonnegative on V, we see, as before, that, for all $f \in V^*$, the variational inequality

$$\begin{cases} \langle \alpha(u) - f, v - u \rangle + J(v) - J(u) \geq 0, \; \forall v \in V, \\ u \in V \end{cases} \tag{6.54}$$

has a unique solution

$$u = u_f = P_{\alpha,J}(f).$$

For simplicity, we shall use the notation $P_0 = P_{\alpha,J}$ in the sequel. Hence, (6.48) is equivalent to

$$u = P_0[f(u, \lambda)]. \tag{6.55}$$

We may also consider (6.55) as a homogenization of (6.15).

The homogeneity of α, f, and J ((6.42) and (6.47)) implies that, if u is a solution of (6.48), then, so is tu for all $t \geq 0$. By a similar observation, we see that, if u is a solution of (6.54), then, for $t > 0$, tu is a solution of (6.54) with f replaced by $t^{p-1}f$. In other words, P_0 is positive homogeneous of degree $(p-1)^{-1}$:

$$P_0(tf) = t^{\frac{1}{p-1}} P_0(f), \; \forall t > 0, \; \forall f \in V^*. \tag{6.56}$$

These arguments lead to the following definition:

$\lambda \in \mathbb{R}$ is called an eigenvalue of (6.48) if (6.48) has a solution (u, λ) with $u \neq 0$. u is called an eigenvector of (6.48) corresponding to λ.

If u is an eigenvector of (6.48) (corresponding to λ), then, so is tu for all $t > 0$. We have the following result:

Theorem 6.4

  (I) If $(0, \lambda)$ is a bifurcation point of (6.8), then, λ is an eigenvalue of (6.48).

  (II) If a and b $(a < b)$ are not eigenvalues of (6.48) and if

$$d(I - P_0[f(\cdot, a)], B_r(0), 0) \neq d(I - P_0[f(\cdot, b)], B_r(0), 0) \tag{6.57}$$

for some $r > 0$, then, for S and C, as in Theorem 6.3, either (i) C is unbounded in $V \times \mathbb{R}$, or

  (ii) $(0, \lambda_1) \in C$ for some eigenvalue λ_1 of (6.48), $\lambda_1 \notin [a, b]$.

Proof. First, we note that, if a, b are not eigenvalues of (6.48), then, 0 is the only zero of $I - P_0[f(\cdot, a)]$ and $I - P_0[f(\cdot, b)]$ in V. The degrees in (6.57) are defined for all $r > 0$ (and are independent of r). For $\sigma \in [0,1]$, $u \in V$, and $\lambda \in \mathbb{R}$, we define

$$A_\sigma(u) = \begin{cases} \dfrac{1}{\sigma^{p-1}} A(\sigma u) & \text{if} \quad \sigma \in (0,1], \\ \alpha(u) & \text{if} \quad \sigma = 0, \end{cases} \tag{6.58}$$

$$B_\sigma(u, \lambda) = \begin{cases} \dfrac{1}{\sigma^{p-1}} B(\sigma u, \lambda) & \text{if} \quad \sigma \in (0, 1], \\ f(u, \lambda) & \text{if} \quad \sigma = 0, \end{cases} \tag{6.59}$$

$$j_\sigma(u) = \begin{cases} \dfrac{1}{\sigma^p} j(\sigma u) & \text{if} \quad \sigma \in (0, 1], \\ J(u) & \text{if} \quad \sigma = 0. \end{cases} \tag{6.60}$$

Now, let $\{\sigma_n\}$ be a sequence in $[0, 1]$, $\sigma_n \to \sigma$, as $n \to \infty$. We verify that

$$A_{\sigma_n} \to A_\sigma, \; j_{\sigma_n} \to j_\sigma$$

in the sense of (A4).

First, we consider the case $\sigma > 0$. In this case, $\sigma_n > 0$ for all n, sufficiently large. If $v_n \to v$ in V, then, by the continuity of A,

$$A_{\sigma_n}(v_n) = \frac{A(\sigma_n v_n)}{\sigma_n^{p-1}} \to \frac{A(\sigma v)}{\sigma^{p-1}} = A_\sigma(v),$$

and (6.18) follows. To prove that A_{σ_n} belongs to class (S) uniformly, we assume that $v_n \rightharpoonup v$, $w_n \to w$ in V, such that

$$\limsup \langle A_{\sigma_n}(v_n), v_n - w_n \rangle \leq 0.$$

Hence,

$$\begin{aligned} 0 \geq{} & \limsup \left\langle \frac{A(\sigma_n v_n)}{\sigma_n^{p-1}}, \frac{1}{\sigma_n}(\sigma_n v_n - \sigma_n w_n) \right\rangle \\ ={} & \frac{1}{\sigma^p} \limsup \langle A(\sigma_n v_n), \sigma_n v_n - \sigma_n w_n \rangle. \end{aligned} \tag{6.61}$$

On the other hand,

$$\sigma_n v_n \rightharpoonup \sigma v, \; \sigma_n w_n \to \sigma w \text{ in } V.$$

Because A is bounded, $\sup_{n \in \mathbb{N}} \|A(\sigma_n v_n)\| = M < \infty$. Moreover,

$$|\langle A(\sigma_n v_n), \sigma v - \sigma_n w_n \rangle| \leq M \|\sigma v - \sigma_n w_n\|.$$

Hence, $|\langle A(\sigma_n v_n), \sigma v - \sigma_n w_n \rangle| \to 0$. (6.61) implies that

$$\begin{aligned} 0 \geq{} & \limsup \langle A(\sigma_n v_n), \sigma_n v_n - \sigma_n w_n \rangle \\ ={} & \limsup[\langle A(\sigma_n v_n), \sigma_n v_n - \sigma v \rangle + \langle A(\sigma_n v_n), \sigma v - \sigma_n w_n \rangle] \\ ={} & \limsup[\langle A(\sigma_n v_n), \sigma_n v_n - \sigma v \rangle. \end{aligned}$$

Because A is in class (S), it follows that

$$\sigma_n v_n \to \sigma v,$$

i.e., $v_n \to v$ in V, proving that A_{σ_n} belongs to class (S) uniformly for the sequence $\{\sigma_n\}$.

Now, if $v_n \rightharpoonup v$, then, $\sigma_n v_n \rightharpoonup \sigma v$, and, because j is weakly lower semicontinuous,

$$
\begin{aligned}
j_\sigma(v) &= \frac{1}{\sigma^p} j(\sigma v) \\
&\leq \frac{1}{\sigma^p} \liminf j(\sigma_n v_n) \\
&= \liminf j_{\sigma_n}(v_n),
\end{aligned}
$$

and (6.17) follows. Let $v \in V$. By choosing $v_n = \dfrac{\sigma}{\sigma_n} v$, we see that $v_n \to v$ and

$$
j_{\sigma_n}(v_n) = \frac{1}{\sigma_n^p} j\left(\sigma_n v \frac{\sigma}{\sigma_n}\right) = \frac{j(\sigma v)}{\sigma_n^p}.
$$

Hence,

$$
\lim j_{\sigma_n}(v_n) = \lim \frac{j(\sigma v)}{\sigma_n^p} = \frac{j(\sigma v)}{\sigma^p} = j_\sigma(v),
$$

i.e., (6.16) holds.

In the case $\sigma = 0$, (6.15) and (6.19) follow from (6.37) and (A6) (b), and (6.16) and (6.17) are consequences of the assumptions (6.45) and (6.46).

In view of Lemma 6.1, we see that

$$
P_{A_{\sigma_n}, j_{\sigma_n}}(f_n) \to P_{A_\sigma, j_\sigma}(f) \quad \text{in } V, \tag{6.62}
$$

whenever $\sigma_n \to \sigma$ in $[0, 1]$ and $f_n \to f$ in V^*. Now we are ready to prove that the mapping

$$
(\sigma, v, \lambda) \mapsto P_{A_\sigma, j_\sigma}[B_\sigma(v, \lambda)] \tag{6.63}
$$

is completely continuous on $[0, 1] \times V \times \mathbb{R}$.

Indeed, let $v_n \rightharpoonup v, \sigma_n \to \sigma, \lambda_n \to \lambda$. Then, it follows that

$$
B_{\sigma_n}(v_n, \lambda_n) \to B_\sigma(v, \lambda) \quad \text{in } V^*. \tag{6.64}
$$

For $\sigma > 0$, this is a consequence of the complete continuity of B. For $\sigma = 0$, this follows from (6.39). Applying Lemma 6.1 with

$$
f_n = B_{\sigma_n}(v_n, \lambda_n), \quad f = B_\sigma(v, \lambda),
$$

we see from (6.62) and (6.64) that

$$
P_{A_{\sigma_n}, j_{\sigma_n}}[B_{\sigma_n}(v_n, \lambda_n)] \to P_{A_\sigma, j_\sigma}[B_\sigma(v, \lambda)] \quad \text{in } V,
$$

proving the complete continuity of the mapping in (6.63).

Now we prove (I). Suppose that $(0, \lambda)$ is a bifurcation point of (6.8) and that $\{u_n\}$ and $\{\lambda_n\}$ satisfy

$$
\|u_n\| \to 0, \ u_n \neq 0, \ \forall n, \ \text{and} \ \lambda_n \to \lambda \ (n \to \infty),
$$

and

$$\langle A(u_n) - B(u_n, \lambda_n), v - u_n \rangle + j(v) - j(u_n) \geq 0, \ \forall v \in V. \tag{6.65}$$

Letting $v_n = \|u_n\|^{-1} u_n$ and dividing both sides of (6.65) by $\|u_n\|^p$,

$$\left\langle \frac{A(u_n)}{\|u_n\|^{p-1}} - \frac{B(u_n, \lambda_n)}{\|u_n\|^{p-1}}, \frac{v}{\|u_n\|} - \frac{u_n}{\|u_n\|} \right\rangle + \frac{j(v)}{\|u_n\|^p} - \frac{j(u_n)}{\|u_n\|^p} \geq 0, \tag{6.66}$$

or, by definition of A_σ, B_σ, and j_σ,

$$\langle A_{\|u_n\|}(v_n) - B_{\|u_n\|}(v_n, \lambda_n), w - v_n \rangle + j_{\|u_n\|}(w) - j_{\|u_n\|}(v_n) \geq 0, \tag{6.67}$$

with $w = v/\|u_n\|$. Because (6.66) is true for all $v \in V$, (6.67) also holds for all $w \in V$.

Now, we see that (6.67) is equivalent to

$$v_n = P_{A_{\|u_n\|}, j_{\|u_n\|}} \left[B_{\|u_n\|}(v_n, \lambda_n) \right]. \tag{6.68}$$

Because V is reflexive, by passing to a subsequence, if necessary, we can assume that

$$v_n \rightharpoonup v \ \text{in} \ V.$$

By the complete continuity of the mapping in (6.63), we see that

$$\begin{aligned}
P_{A_{\|u_n\|}, j_{\|u_n\|}} [B_{\|u_n\|}(v_n, \lambda_n)] \quad &\rightarrow \quad P_{A_0, j_0}[B_0(v, \lambda)] \\
&= \quad P_{\alpha, J}[f(v, \lambda)] \\
&= \quad P_0[f(v, \lambda)].
\end{aligned}$$

According to (6.68), this implies that $v_n \to v$ in V, and $v = P_0[f(v, \lambda)]$. Moreover, $\|v\| = 1$ (because $\|v_n\| = 1, \ \forall n$). Hence λ is an eigenvalue of (6.48), and (I) is proved.

To prove (II), we assume that a and b are not eigenvalues of (6.48). We will prove that 0 is an isolated solution of (6.8) with $\lambda = a, b$ and, for $r > 0$, sufficiently small,

$$d(I - P[B(\cdot, a)], B_r(0), 0) = d(I - P_0[f(\cdot, a)], B_r(0), 0), \tag{6.69}$$

and

$$d(I - P[B(\cdot, b)], B_r(0), 0) = d(I - P_0[f(\cdot, b)], B_r(0), 0). \tag{6.70}$$

To prove (6.69), we need only to show that there exists $r > 0$, sufficiently small, such that, for all $\sigma \in [0, 1]$, the equation

$$u - P_{A_\sigma, j_\sigma}[B_\sigma(u, a)] = 0 \tag{6.71}$$

has no nontrivial solution in $\overline{B_r(0)}$.

Suppose that this is not the case and there exist sequences $\{u_n\} \subset V$ and $\{\sigma_n\} \subset [0,1]$, such that

$$\|u_n\| \neq 0, \ \forall n, \ \|u_n\| \to 0 \ (n \to \infty),$$

and

$$u_n = P_{A_{\sigma_n}, j_{\sigma_n}}[B_{\sigma_n}(u_n, a)], \ \forall n.$$

This equation has the following variational inequality form:

$$\langle A_{\sigma_n}(u_n) - B_{\sigma_n}(u_n, a), v - u_n \rangle + j_{\sigma_n}(v) - j_{\sigma_n}(u_n) \geq 0, \ \forall v \in V.$$

By the definitions of $A_\sigma, B_\sigma, j_\sigma$, this is equivalent to

$$\frac{1}{\sigma_n^{p-1}} \langle A(\sigma_n u_n) - B(\sigma_n u_n, a), v - u_n \rangle + \frac{1}{\sigma_n^p} j(\sigma_n v) - \frac{1}{\sigma_n^p} j(\sigma_n u_n) \geq 0,$$
$$\forall v \in V.$$

As before, by setting $v_n = u_n/\|u_n\|$ and dividing this inequality by $\|u_n\|^p$, we get the following inequality:

$$\left\langle \frac{A(\sigma_n \|u_n\| v_n)}{\sigma_n^{p-1} \|u_n\|^{p-1}} - \frac{B(\sigma_n \|u_n\| v_n, a)}{\sigma_n^{p-1} \|u_n\|^{p-1}}, \frac{v}{\|u_n\|} - v_n \right\rangle$$
$$+ \frac{1}{(\sigma_n \|u_n\|)^p} j \left(\sigma_n \|u_n\| \frac{v}{\|u_n\|} \right) - \frac{1}{(\sigma_n \|u_n\|)^p} j (\sigma_n \|u_n\| v_n) \geq 0,$$
$$\forall v \in V.$$

Letting $w = v/\|u_n\|$,

$$\langle A_{\sigma_n \|u_n\|}(v_n) - B_{\sigma_n \|u_n\|}(v_n, a), w - v_n \rangle + j_{\sigma_n \|u_n\|}(w) - j_{\sigma_n \|u_n\|}(v_n) \geq 0,$$
$$\forall w \in V.$$

This is equivalent to

$$v_n = P_{A_{\sigma_n \|u_n\|}, j_{\sigma_n \|u_n\|}} \left[B_{\sigma_n \|u_n\|}(v_n, a) \right]. \tag{6.72}$$

By assuming $v_n \rightharpoonup v$, using the complete continuity of the mapping in (6.63), and the fact that $\sigma_n \|u_n\| \to 0$, we see that

$$P_{A_{\sigma_n \|u_n\|}, j_{\sigma_n \|u_n\|}} \left[B_{\sigma_n \|u_n\|}(v_n, a) \right] \to P_0[f(v, a)] \ \text{in} \ V.$$

Hence (6.72) implies that $v_n \to v$ and $v = P_0[f(v, a)]$. Because $\|v\| = 1$, this means that a is an eigenvalue of (6.48), contradicting our assumption that a is not an eigenvalue of (6.48).

This contradiction proves that there exists $r > 0$, such that (6.71) has no solutions in $\overline{B_r(0)} \setminus \{0\}$.

Next, we observe that

$$\{I - P_{A_\sigma, j_\sigma}[B_\sigma(\cdot, a)] : \sigma \in [0,1]\}$$

is a family of compact perturbations of the identity on $\overline{B_r(0)}$. Moreover, by the above proof,

$$u - P_{A_\sigma,j_\sigma}[B_\sigma(u,a)] \neq 0,$$

for all $u \in \partial B_r(0)$, all $\sigma \in [0,1]$.

By the homotopy invariance property of the Leray–Schauder degree,

$$
\begin{aligned}
d(I - P_0[f(\cdot,a)], B_r(0),0) &= d(I - P_{A_0,j_0}[B_0(\cdot,a)], B_r(0),0) \\
&= d(I - P_{A_1,j_1}[B_1(\cdot,a)], B_r(0),0) \\
&= d(I - P[B(\cdot,a)], B_r(0),0),
\end{aligned}
$$

i.e., (6.69) holds. (6.70) follows in a similar manner. (6.69) and (6.70) imply that (6.36) is a consequence of (6.57), and, therefore, (II) is a consequence of Theorem 6.3. ∎

6.3   Elastic plates with unilateral conditions

In this section, we consider a consequence of Theorem 6.4 in the case where the homogenized variational inequality (6.48) becomes an equation. Then, we apply these considerations to derive global bifurcation results for the unilateral problems of elastic plates considered by Do ([32] and [33]).

First, we consider an abstract consequence of Theorem 6.4. Let $p = 2$, and let

$$W = D(J) = \{u \in V : J(u) < \infty\}$$

be the effective domain of J. Then, (6.48) is equivalent to

$$
\begin{cases}
\langle \alpha(u) - f(u,\lambda), v - u \rangle + J(v) - J(u) \geq 0, \ \forall v \in W, \\
u \in W.
\end{cases}
\tag{6.73}
$$

Now, we suppose that W is a closed subspace of V and that $J \equiv 0$ on W, i.e.,

$$
J(u) = I_W(u) =
\begin{cases}
0 & \text{if } u \in W(= D(J)), \\
\infty & \text{if } u \notin W.
\end{cases}
\tag{6.74}
$$

(As one can easily check, in the case where W is a subspace of V, (6.74) is satisfied if and only if J is a linear mapping on W, which is equivalent to the fact that J is additive on W.)

Under the assumption (6.74), we see that (6.73) and, then, (6.48) are equivalent to the following equation:

$$
\begin{cases}
\langle \alpha(u) - f(u,\lambda), v \rangle = 0, \ \forall v \in W, \\
u \in W.
\end{cases}
\tag{6.75}
$$

For each $g \in W^*$, we denote by $P_\alpha g = u_g$ the unique solution of the equation

$$\begin{cases} \langle \alpha(u_g) - g, v \rangle = 0, \ \forall v \in W, \\ u_g \in W. \end{cases} \tag{6.76}$$

Hence, P_α is a continuous mapping from W^* to W, and (6.75) can be written as

$$u = P_\alpha[f(u, \lambda)]. \tag{6.77}$$

(Here, because $W \subset V$, $V^* \subset W^*$, and f can be considered as a mapping from $W \times \mathbb{R}$ to W^*, which is also completely continuous.) In this particular case, Theorem 6.4 becomes

Corollary 6.5   *(I) If $(0, \lambda)$ is a bifurcation point of (6.8), then, λ is an eigenvalue of (6.75).*
   (II) If a and $b \, (a < b)$ are not eigenvalues of (6.48) and if

$$d(I - P_\alpha[f(\cdot, a)], B_r(0), 0) \neq d(I - P_\alpha[f(\cdot, b)], B_r(0), 0)$$

for some $r > 0$, then, for S and C, as in Theorem 6.3, we have the alternative in Theorem 6.4.

A particular but important situation is where α is a linear mapping and B is as in Corollary 3.3, i.e., B is of the form

$$B(u, \lambda) = \lambda \beta u + N(u, \lambda), \ u \in V, \lambda \in \mathbb{R}, \tag{6.78}$$

where $\beta \in L(V, V^*)$, $N \in C(V \times \mathbb{R}, V^*)$, β and N are completely continuous, and

$$N(u, \lambda) = o(\|u\|),$$

as $u \to 0$, uniformly for λ in bounded intervals.
   By a proof similar to that in Section 3.1, we conclude from (6.78) that

$$f(u, \lambda) = \lambda \beta u, \ \forall u \in V, \lambda \in \mathbb{R}. \tag{6.79}$$

Because α is linear, the mapping P_α defined by (6.76) is linear also. This, together with (6.79), shows that (5.11), in this case, becomes the following equation:

$$u = \lambda(P_\alpha \beta)u, \tag{6.80}$$

where $P_\alpha \beta$ is a linear, compact mapping from V to itself and from W into itself.
   In this particular case, we have the following consequence of Corollary 6.5:

Corollary 6.6 *If α is linear and B is of the form (6.78), then, (6.8) has, at most, a countable number of bifurcation points $(0, \lambda)$ with the only possible accumulation point at ∞. Moreover,*

(I) If $(0, \lambda)$ is a bifurcation point of (6.8), then, λ is an eigenvalue of $P_\alpha \beta$.

(II) If λ is an eigenvalue of $P_\alpha \beta$ of odd multiplicity, then, $(0, \lambda)$ is a bifurcation point of (6.8), and we have the alternative in Theorem 6.4.

Now, we apply the above corollary to a bifurcation problem for (thin) elastic plates (with large deflections) subject to unilateral conditions. In [32] and [33], the author studied buckling problems for a thin elastic plate with unilateral conditions on the boundary or on the plate (not necessarily over a rigid obstacle, as in Examples 4.4, 4.6), using the von Kármán model for large deflections, as follows.

Let $\Omega \subset \mathbb{R}^2$ be as in Example 4.4. We consider the following variational inequality:

$$
\begin{cases}
a(u, v - u) + \int_\Omega \sum_{i,j=1}^{2} \sigma_{ij}(u)\, \partial_i u\, \partial_j(v - u)dx \\[2mm]
\quad - \int_\Omega \lambda \sum_{i,j=1}^{2} \sigma_{ij}^0\, \partial_i u\, \partial_j(v - u)dx \ + \ j(v) \ - \ j(u) \ \geq \ 0,\ \forall v \in V, \\[2mm]
u \in V,
\end{cases}
$$

$$(6.81)$$

where a is defined by (4.46) and σ_{ij} and σ_{ij}^0 $(1 \leq i, j \leq 2)$ satisfy the assumptions stated in Example 4.4.

The space V and the functional j vary for different unilateral problems. In [32] and [33], the following cases are considered: (i) The plate satisfies unilateral conditions on the vertical displacement in Ω and classical boundary conditions on $\Gamma = \partial \Omega$, for example, the plate is clamped on the edge (see Figure 6.1).

The buckling problem for the plate has the form (6.81) with

$$V = H_0^2(\Omega) \tag{6.82}$$

and

$$j(u) = \int_\Omega \psi(u(x))dx, \tag{6.83}$$

where

$$\psi : \mathbb{R} \to [0, \infty]$$

is a proper, convex, lower semicontinuous functional, such that

$$\psi(0) = 0, \ \psi(t) > 0, \ \forall t \in \mathbb{R} \setminus \{0\} \tag{6.84}$$

and ψ is homogeneous of degree 1, i.e.,

$$\psi(t\xi) = t\psi(\xi), \ \forall t \geq 0, \ \xi \in \mathbb{R}. \tag{6.85}$$

(ii) Now, we consider the case where the plate is subjected to a unilateral condition of rotation at the edge (like a plastic hinge on edge or rotation with friction).

For a formulation of the buckling problem, we can choose the variational inequality (6.81) with (cf. [33], Problem 2.3):

$$V = H^2(\Omega) \cap H_0^1(\Omega) \tag{6.86}$$

and

$$j(u) = \int_{\partial\Omega} \psi(\partial_n u(x)) dS, \tag{6.87}$$

where ψ is as in (i) (i.e., ψ is convex, lower semicontinuous, and satisfies (6.84) and (6.85)), and $\partial_n u = \partial u / \partial n$ is the normal derivative of u on $\partial\Omega$. An example is the case where the plate is simply supported at the edge, and the moment of rotation M at boundary points satisfies a unilateral condition.

The moment $M = M(u)$ is given by

$$M(u) = -D \left[\nu \Delta u + (1 - \nu) \sum_{i,j} \partial_{ij} u \, n_i \, n_j \right],$$

where $\nu \in (0, 1/2)$ is the Poisson ratio and n is the unit outward normal vector on $\partial\Omega$.

M satisfies the following unilateral conditions on $\partial\Omega$:

$$\begin{cases} |M(u)| \leq c \text{ on } \partial\Omega, \text{ and} \\ \begin{cases} |M(u)| < c & \Rightarrow \quad \partial_n u = 0, \\ M(u) = c & \Rightarrow \quad \partial_n u \geq 0, \\ M(u) = -c & \Rightarrow \quad \partial_n u \leq 0, \end{cases} \end{cases} \tag{6.88}$$

where $c > 0$ is given.

This will lead to the functional j in (6.87) with ψ given by

$$\psi(\xi) = c|\xi|, \ \xi \in \mathbb{R}.$$

It can be easily checked that (6.84) and (6.85) are satisfied. We note that (6.88) can be used to express a friction law for rotation at the edge (cf. [33]).

(iii) As a third case, we consider that the plate is simply supported on a part $\Gamma_1 \subset \partial\Omega$ (Γ_1 is non rectilinear and $|\Gamma_1| > 0$) and is subjected to unilateral conditions of vertical displacement on the edge, whereas rotation is free (on the boundary).

In this case (cf. [33], Problem 2.4),

$$V = \{u \in H^2(\Omega) : v = 0 \text{ on } \Gamma_1\} \tag{6.89}$$

Γ_1: simply supported

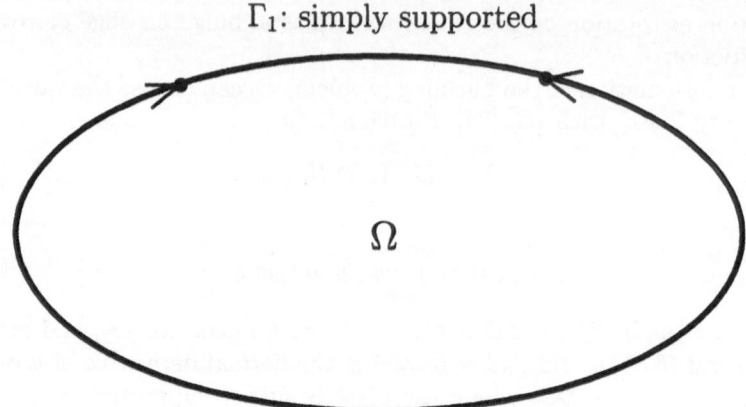

$\partial\Omega \setminus \Gamma_1$: unilateral displacement

FIGURE 6.1. Buckling of plate simply supported on Γ_1 with unilateral displacement on $\partial\Omega \setminus \Gamma_1$.

and

$$j(u) = \int_{\partial\Omega\setminus\Gamma_1} \psi(u(x))dS, \qquad (6.90)$$

where ψ again satisfies (6.84) and (6.85).

An example similar to (ii) is the following:

$$\begin{cases} u = 0 & \text{on} \quad \Gamma_1, \\ M(u) = 0 & \text{on} \quad \partial\Omega, \end{cases}$$

and the vertical displacement $F(u)$ on the boundary is given by

$$F(u) = -D\left[(1-\nu)\partial_\tau\left(\sum_{i,j}\partial_{ij}u\tau_i\tau_j\right) + \partial_n\Delta u\right],$$

where τ is the unit tangent vector on $\partial\Omega$, subject to friction on $\partial\Omega \setminus \Gamma_1$:

$$\begin{cases} |F(u)| \le c \text{ on } \partial\Omega \setminus \Gamma_1, \text{ and} \\ \begin{cases} |F(u)| < c & \Rightarrow \quad u = 0, \\ F(u) = c & \Rightarrow \quad u(x) \ge 0, \\ F(u) = -c & \Rightarrow \quad u(x) \le 0, \end{cases} \end{cases} \qquad (6.91)$$

where $c > 0$ is given. In this case, we have (6.89), (6.90) with $\psi(\xi) = c|\xi|$, $\xi \in \mathbb{R}$.

We may apply Corollary 6.6 to obtain global bifurcation results for (6.81) in these cases. We have the following result:

Corollary 6.7

(i) In the case where V and j are given by (6.82) and (6.83), the variational inequality (6.81) has no finite bifurcation points.

(ii) If V and j are given by (6.86) and (6.87), then, (6.81) has, at most, a countable number of bifurcation points. Moreover, if $(0, \lambda)$ is a bifurcation point of (6.81), then, λ is an eigenvalue of the following equation:

$$
\begin{cases}
a(u,v) - \lambda \displaystyle\int_\Omega \sum_{i,j=1}^{2} \sigma_{ij}^{0}\, \partial_i u\, \partial_j v\, dx = 0, \ \forall v \in H_0^2(\Omega), \\[2mm]
u \in H_0^2(\Omega).
\end{cases}
\tag{6.92}
$$

Conversely, if λ is an eigenvalue of (6.92) of odd multiplicity, then, $(0, \lambda)$ is a bifurcation point of (6.81) with a global bifurcation branch emanating from $(0, \lambda)$ and satisfying the alternative in Theorem 6.4.

(iii) If V and j are given by (6.89) and (6.90), then, (6.81) has, at most, a countable number of bifurcation points.

If $(0, \lambda)$ is a bifurcation point of (6.81), then, λ is an eigenvalue of the following equation:

$$
\begin{cases}
a(u,v) - \lambda \displaystyle\int_\Omega \sum_{i,j=1}^{2} \sigma_{ij}^{0}\, \partial_i u\, \partial_j v\, dx = 0, \ \forall v \in H^2(\Omega) \cap H_0^1(\Omega), \\[2mm]
u \in H^2(\Omega) \cap H_0^1(\Omega).
\end{cases}
\tag{6.93}
$$

Conversely, if λ is an eigenvalue of (6.93) of odd multiplicity, then, $(0, \lambda)$ is a bifurcation point of (6.81), and we have the same conclusion as in (ii).

Proof. We know that the spaces V given by (6.82), (6.86), and (6.89) are Hilbert spaces (with the usual inner products and norms). It is proved (see e.g., [36] or [32] and [33]) that a defined by (4.46) is coercive on these spaces. Moreover, because ψ is convex, nonnegative, and lower semicontinuous on $\mathbb{R}$, by using the fact that the mappings

$$
H^2(\Omega) \hookrightarrow C(\overline{\Omega}), \ u \mapsto u
$$

and

$$
H^2(\Omega) \to H^{1/2}(\partial\Omega), \ u \mapsto u|_{\partial\Omega}
$$

are compact ([1]), we can check directly that the functionals j defined by (6.83), (6.87), or (6.90) are proper, convex, nonnegative, and lower semicontinuous on V.

Let A, L, and $C : V \to V^*$ be defined by:

$$
\langle A(u), v \rangle = a(u, v),
\tag{6.94}
$$

$$\langle L(u), v \rangle = \int_\Omega \sum_{i,j=1}^2 \sigma_{ij}^0 \, \partial_i u \, \partial_j v \, dx, \tag{6.95}$$

and

$$\langle C(u), v \rangle = \int_\Omega \sum_{i,j=1}^2 \sigma_{ij}(u) \, \partial_i u \, \partial_j v \, dx, \tag{6.96}$$

for all $u, v \in V$ (here $\langle \cdot, \cdot \rangle$ is the inner product of V or the pairing between V^* and V). It is easily verified that A, L, and C are well defined on V (i.e., $A(u), L(u)$, and $C(u) \in V^*$, $\forall u \in V$), and (cf. [32] and [33]) that A and L are linear and continuous on V. Moreover, L is self-adjoint and compact on V, and C is a (nonlinear) completely continuous mapping that satisfies (4.47). Furthermore, by the linearity and coerciveness of a, condition (6.9) is satisfied.

Now, let $B(u, \lambda) = \lambda L u - C(u)$, $u \in V, \lambda \in \mathbb{R}$. Then, B is a completely continuous mapping from $V \times \mathbb{R}$ to V^*.

With these settings, we see that (6.81) can be written as the following variational inequality:

$$\begin{cases} \langle A(u) - B(u, \lambda), v - u \rangle + j(v) - j(u) \geq 0, \; \forall v \in V, \\ u \in V. \end{cases} \tag{6.97}$$

Now, we verify that A, B, and j satisfy the homogenization assumptions (6.37), (6.39), (6.45), and (6.46), with $p = 2$:

$$\alpha = A, \tag{6.98}$$

$$f(u, \lambda) = \lambda L u, \; u \in V, \lambda \in \mathbb{R}, \tag{6.99}$$

and

$$J = \begin{cases} I_{\{0\}} & \text{in case } (i), \\ I_{H_0^2(\Omega)} & \text{in case } (ii), \\ I_{H^2(\Omega) \cap H_0^1(\Omega)} & \text{in case } (iii) \end{cases} \tag{6.100}$$

(as usual, I_K is the indicator function of K).

In fact, because A is linear and bounded, for $v_k \to v$ in V, $\sigma_k \to 0^+$,

$$\frac{1}{\sigma_k} A(\sigma_k v_k) = A(v_k) \to A(v).$$

Hence, (6.37) and (6.98) hold.

Now (4.47) implies that $C(u) = o(\|u\|)$, as $u \to 0$. Hence, (6.39) can be verified, as in Example 4.4.

Now, we prove (6.45) and (6.46) with J given by (6.100) in case (i). One easily sees that, in all three cases, J is a proper, convex, lower semicontinuous functional from V to $[0, \infty]$ with $J(0) = 0$. Let $v_k \rightharpoonup v$ and $\sigma_k \to 0^+$.

If $v = 0$, then, (6.45) holds. Suppose $v \neq 0$, i.e., $v \not\equiv 0$ on Ω. By the compactness of the embedding

$$H^2(\Omega) \hookrightarrow C(\overline{\Omega}) \tag{6.101}$$

$v_k(x) \to v(x)$, $\forall x \in \Omega$.

Because ψ is lower semicontinuous, this implies

$$\psi(v(x)) \leq \liminf \psi(v_k(x)), \ \forall x \in \Omega.$$

By (6.84),

$$\int_\Omega \psi(v(x))dx > 0.$$

Hence, by Fatou's lemma,

$$0 < j(v) \ \leq \ \int_\Omega \liminf \psi(v_k(x))dx$$
$$\leq \ \liminf j(v_k). \tag{6.102}$$

Hence,

$$\liminf \frac{j(\sigma_k v_k)}{\sigma_k^2} \ = \ \liminf \left[\frac{1}{\sigma_k^2}\sigma_k \int_\Omega \psi(v_k(x))dx\right] \ \text{(by (6.85))}$$
$$= \ \liminf \frac{j(v_k)}{\sigma_k}.$$

However, by (6.102), $\liminf j(v_k) > 0$.
Because $\sigma_k \to 0^+$,

$$\liminf j_{\sigma_k}(v_k) \ = \ \liminf \frac{j(v_k)}{\sigma_k}$$
$$= \ \infty = \ J(v).$$

Hence, (6.45) is true for all $v \in V$.

To prove (6.46), we let $v \in V$ and choose $v_k = v$, $\forall k$. Then, clearly, $v_k \to v$. If $v = 0$, then,

$$\frac{1}{\sigma_k^2}j(\sigma_k v_k) = 0 \to 0 = J(v).$$

If $v \neq 0$, then, $j(v) > 0$, $J(v) = \infty$, and

$$\frac{1}{\sigma_k^2}j(\sigma_k v_k) = \frac{j(v)}{\sigma_k^2} \to \infty = J(v).$$

Hence, (6.46) holds for J in case (i).

Next, we establish (6.45) and (6.46) for J given by (6.100), case (ii).
Suppose that $v_k \rightharpoonup v$ in V and $\sigma_k \to 0^+$. If $v \in H_0^2(\Omega)$, then, $J(v) = 0$,

and (6.45) is satisfied. Assume $v \notin H_0^2(\Omega)$. Because $v \equiv 0$ on $\partial\Omega$, it follows that

$$\partial_n v \not\equiv 0 \text{ on } \partial\Omega.$$

Hence, $j(v) > 0$. By the compactness of the mapping

$$H^2(\Omega) \to L^2(\partial\Omega), \ u \mapsto \partial_n u \qquad (6.103)$$

([1]), without loss of generality, we can assume that

$$\partial_n v_k(x) \to \partial_n v(x) \text{ for a.e. } x \in \partial\Omega.$$

It follows from Fatou's lemma and the lower semicontinuity of ψ that

$$\begin{aligned}
0 < j(v) &\leq \int_{\partial\Omega} \liminf \psi(\partial_n v_k) dS \\
&\leq \liminf j(v_k).
\end{aligned}$$

This implies that

$$\liminf \frac{j(\sigma_k v_k)}{\sigma_k^2} = \liminf \frac{j(v_k)}{\sigma_k} = \infty = J(v),$$

and (6.45) follows.

Now let $v \in V$, and, as in the above proof, choose $v_k = v$, $\forall k$. Then $v_k \to v$, and because

$$j(v) = \begin{cases} 0 & \text{if } v \in H_0^2(\Omega) \\ > 0 & \text{if } v \notin H_0^2(\Omega), \end{cases}$$

$$\begin{aligned}
\lim \frac{j(\sigma_k v_k)}{\sigma_k^2} &= \lim \frac{j(v_k)}{\sigma_k} \\
&= \begin{cases} 0 & \text{if } v \in H_0^2(\Omega) \\ \infty & \text{if } v \notin H_0^2(\Omega) \end{cases} \\
&= J(v).
\end{aligned}$$

(6.46) is, thus, proved for J in case (ii).

The proof for (6.45) and (6.46) in case (iii) is similar, and we omit it.

Now, from (6.98), (6.99), and (6.100), we see that the homogenization (6.48) of (6.97) (or (6.81)) is of the form

$$\begin{cases} \langle A(u) - \lambda Lu, v - u \rangle + I_W(v) - I_W(u) \geq 0, \ \forall v \in V, \\ u \in V, \end{cases} \qquad (6.104)$$

with

$$
W = \begin{cases} \{0\} & \text{in case (i),} \\ H_0^2(\Omega) & \text{in case (ii),} \\ H^2(\Omega) \cap H_0^1(\Omega) & \text{in case (iii).} \end{cases} \tag{6.105}
$$

As observed in Corollaries 6.5 and 6.6, (6.104) is equivalent to

$$
\begin{cases} \langle A(u) - \lambda L u, v \rangle = 0, \ \forall v \in W, \\ u \in W. \end{cases} \tag{6.106}
$$

In case (i), because $W = \{0\}$, we see that $u = 0$ is the only solution of (6.106). Hence, in view of Corollary 6.6 (I), (6.97) has no finite bifurcation point.

In cases (ii) and (iii), (6.106) is exactly (6.92) or (6.93). Hence, the conclusions of Corollary 6.7 (ii) and (iii) follow from Corollary 5.5. ■

Remark 6.2

(a) Corollary 6.7 (i) is the content of Theorem 3.3, [33].

(b) In [32] and [33], the author studied the (local) bifurcation of (6.81) in cases (ii) and (iii), using variational methods. The main results there (Theorems 3.3, 3.7, [33], Theorem 4.3, [32]) are concerned with the existence of bifurcation points of (6.81) at the first eigenvalues of (6.92) and (6.93). Corollary 6.7 complements this work by providing results on the global behavior of the bifurcation branch of (6.81) in cases (ii) and (iii) and also provides results for higher eigenvalues of the homogenized problems.

6.4 Global behavior of bifurcation branches

In this section, we consider some consequences of Theorem 6.4 in the case where the homogenized variational inequality (6.48) is not a linear equation. Using arguments based on the Fredholm alternative for linear operators, we show how to calculate the degrees of the operators in (6.48). Thus, we derive global bifurcation results by establishing a change of degree and using Theorem 6.4.

6.4.1 Some abstract results

We shall need the following assumptions and notation. As before, we consider the variational inequality (6.8) and its homogenization (6.48). We assume throughout this section that $p = 2$ and that A satisfies (6.9) (with $p = 2$). As observed before, we know that, if (u, λ) is a solution of (6.48), then, so is (tu, λ) for all $t \geq 0$. This leads to the following definition (cf. Chapter 5).

An eigenvalue λ of (6.48) is called simple if, whenever $u_1, u_2 \neq 0$ are solutions of (6.48) (with respect to λ), then, $u_1 = Cu_2$ for some $C > 0$.

If α is a (bounded) linear operator from V to V^*, then, we denote by α^* its adjoint, i.e.,

$$\langle \alpha u, v \rangle = \langle \alpha^* v, u \rangle, \ \forall u, v \in V. \tag{6.107}$$

Similarly, for $\lambda \in \mathbb{R}$, if $f(\cdot, \lambda) : V \to V^*$ is a linear operator, then, we denote by $f^*(\cdot, \lambda)$ its adjoint,

$$\langle f(u, \lambda), v \rangle = \langle f^*(v, \lambda), u \rangle, \ \forall u, v \in V. \tag{6.108}$$

Below, we shall frequently use the following assumption about f^*: f^* can be written as a sum,

$$f^*(u, \lambda) = g(u, \lambda) + h(u), \ u \in V, \lambda \in \mathbb{R}, \tag{6.109}$$

where g is homogeneous of degree γ $(\gamma > 0)$ with respect to $\lambda \in \mathbb{R}^+$.

As noted before, 0 is always a solution of (6.48) for all λ. We assume, here, that

$$0 \text{ is not an eigenvalue of (6.48)} \tag{6.110}$$

(i.e., 0 is the unique solution of (6.48) with $\lambda = 0$), and, moreover, $\alpha - f(\cdot, 0)$ is monotone in the following sense:

$$\langle \alpha u - f(u, 0), u \rangle \geq 0, \ \forall u \in D(J). \tag{6.111}$$

We note that these assumptions are satisfied in the case where $\alpha - f(\cdot, 0)$ is a strictly monotone mapping, or more generally, if (6.111) holds and

$$\langle \alpha u - f(u, 0), u \rangle > 0, \ \forall u \in \ker J \setminus \{0\}. \tag{6.112}$$

In fact, suppose that (6.111) and (6.112) are satisfied and that u is a solution of (6.48) with $\lambda = 0$, i.e.,

$$\langle \alpha u - f(u, 0), v - u \rangle + J(v) - J(u) \geq 0, \ \forall v \in V.$$

Letting $v = 0$ in this inequality and noting that $J(0) = 0$,

$$- \langle \alpha u - f(u, 0), u \rangle \geq J(u) \geq 0. \tag{6.113}$$

From this, $J(u) \neq \infty$, i.e., $u \in D(J)$, and, hence,

$$\langle \alpha u - f(u, 0), u \rangle \geq 0,$$

by (6.111). Because $J \geq 0$, (6.113) implies that

$$J(u) = \langle \alpha u - f(u, 0), u \rangle = 0.$$

(6.112) implies that this can occur only if $u = 0$, and, thus, 0 is the unique solution of (6.48). In other words, (6.110) holds. Hence, we see that (6.110),

(6.111), and (6.112) hold in the case where $\alpha - f(\cdot, 0)$ is strictly monotone on $D(J)$, i.e.,

$$\langle \alpha u - f(u, 0), u \rangle > 0, \ \forall u \in D(J) \setminus \{0\}. \tag{6.114}$$

In several applications, f is of the form

$$f(u, \lambda) = \lambda \beta u,$$

(such as in our examples in Chapters 4 and 5), or

$$f(u, \lambda) = \lambda \beta u + h(u), \tag{6.115}$$

where β and h are bounded linear mappings from V to V^*. In the first case, $f(\cdot, 0) \equiv 0$, and (6.114) is obviously satisfied by (6.53). In the second case, (6.114) is satisfied if $\alpha - h$ is strictly monotone, i.e.,

$$\langle (\alpha - h)u, u \rangle > 0, \ \forall u \in D(J) \setminus \{0\}. \tag{6.116}$$

In what follows, we shall usually assume that (6.110) and (6.111) (or (6.112)) are satisfied.

With these settings, we can now prove the following theorem:

Theorem 6.8 *Suppose that λ_0 is an eigenvalue of (6.48). Let*

$$\begin{aligned}
K(\lambda_0) &= \ker[\alpha^* - f^*(\cdot, \lambda_0)] \cap \ker J \\
&= \{u \in V : \alpha^* u - f^*(u, \lambda_0) = 0, J(u) = 0\}.
\end{aligned}$$

Assume that either one of the following conditions is satisfied:

(a) λ_0 is a simple eigenvalue of (6.48) with an eigenvector u_1. Moreover, there exists $u_0 \in K(\lambda_0)$, such that

$$\langle \alpha^* u_0 - h(u_0), u_1 \rangle > 0. \tag{6.117}$$

(b) $K(\lambda_0)$ is not symmetric, and for each eigenvector u_1 of (6.48) corresponding to λ_0, there exists $u_0 = u_0(u_1) \in K(\lambda_0)$, such that (6.117) holds.

Then, there exists a global bifurcation branch of nontrivial solutions of (6.8), which emanates from $[0, \lambda_0]$ and satisfies the alternative in Theorem 6.4.

Proof. (a) It follows from (6.110) that 0 is not an eigenvalue of (6.48). We show that

$$d(I - P_0[f(\cdot, 0)], B_r(0), 0) = 1 \tag{6.118}$$

for all $r > 0$. In fact, because

$$u - P_0[f(u, 0)] \neq 0, \ \forall u \neq 0,$$

the degree in (6.118) is well defined and does not depend on r. Consider the following family of compact perturbations of the identity:

$$H(u,t) = u - P_0[tf(u,0)], \ 0 \le t \le 1.$$

We prove that

$$H(u,t) \ne 0, \ \forall u \in \partial B_r(0), t \in [0,1]. \tag{6.119}$$

In fact, suppose this is not the case, i.e.,

$$u = P_0[tf(u,0)]$$

for some u, $\|u\| = r > 0$, and $t \in [0,1]$. By (6.110), $t < 1$, and, thus, u is a solution of

$$\langle \alpha u - tf(u,0), v - u \rangle + J(v) - J(u) \ge 0, \ \forall v \in V.$$

Letting $v = 0$,

$$-\langle \alpha u - tf(u,0), u \rangle - J(u) \ge 0,$$

or

$$(1-t)\langle \alpha u, u \rangle + t\langle \alpha u - f(u,0), u \rangle + J(u) \le 0.$$

Because $u \in D(J)$, (6.111) implies that the terms in the left-hand side of this inequality are all nonnegative. Hence,

$$(1-t)\langle \alpha u, u \rangle = t\langle \alpha u - f(u,0), u \rangle = J(u) = 0.$$

However, $(1-t)\langle \alpha u, u \rangle > 0$ by (6.53) because $t < 1$ and $u \ne 0$. This contradiction proves (6.119).

Now, by the homotopy invariance property of the degree,

$$
\begin{aligned}
\mathrm{d}(I - P_0[f(\cdot,0)], B_r(0), 0) &= \mathrm{d}(H(\cdot,1), B_r(0), 0) \\
&= \mathrm{d}(H(\cdot,0), B_r(0), 0) \\
&= \mathrm{d}(I, B_r(0), 0) \\
&= 1,
\end{aligned}
$$

proving (6.118). Now, to apply Theorem 6.4, we need only to show that

$$\mathrm{d}(I - P_0[f(\cdot,\lambda)], B_r(0), 0) \ne 1$$

for $\lambda > \lambda_0$, close to λ_0. In fact, we will show that there exists $\lambda_1 > \lambda_0$, such that

$$\mathrm{d}(I - P_0[f(\cdot,\lambda)], B_r(0), 0) = 0 \tag{6.120}$$

for $\lambda_0 < \lambda < \lambda_1, r > 0$ sufficiently small.

First, we remark that, because V is reflexive, we can, by the Lindenstrauss-Asplund-Trojanski theorem, find a norm $\| \cdot \|_0$ on V equivalent to $\| \cdot \|$, i.e.,

$$C_0\|u\| \leq \|u\|_0 \leq C_0^{-1}\|u\|, \ \forall u \in V, \tag{6.121}$$

for some $C_0 > 0$ fixed, such that $(V, \| \cdot \|_0)$ and $(V^*, \| \cdot \|_0^*)$ are locally uniformly convex. Let ϕ be the duality mapping in $(V, \| \cdot \|_0)$ corresponding to the gauge function

$$\Phi(r) = r, \ r \geq 0.$$

Then (see, e.g., [17]) ϕ is a strictly monotone and bicontinuous mapping from V to V^* such that

$$\begin{cases} \|\phi(x)\|_0^* &= \|x\|_0, \\ \langle\phi(x), x\rangle &= \|x\|_0^2, \\ \phi(\lambda x) &= \lambda\phi(x), \ \forall x \in V, \lambda \geq 0. \end{cases} \tag{6.122}$$

Consider the following family of compact perturbations of the identity:

$$H(t, u, \lambda) = u - P_0[(1 - t)f(u, \lambda) + tf(u, \lambda_0) + t\phi(u_0)] \tag{6.123}$$

with $t \in [0, 1], u \in V, \lambda \in \mathbb{R}$. We shall prove that there exist $R_0 > 0, \lambda_1 > \lambda_0$, such that

$$H(t, u, \lambda) \neq 0, \ \forall \lambda \in (\lambda_0, \lambda_1), \ \forall u \in V, \|u\| \geq R_0, \tag{6.124}$$

$$H(0, u, \lambda) \neq 0, \ \forall u \neq 0, \ \forall \lambda \in (\lambda_0, \lambda_1), \tag{6.125}$$

$$H(1, u, \lambda_0) \neq 0, \ \forall u \in V. \tag{6.126}$$

Suppose that $u_n \in V, t_n \in [0, 1], \lambda_n \in \mathbb{R}, n = 1, 2, \ldots$ satisfy

$$H(t_n, u_n, \lambda_n) = 0,$$

i.e.,

$$u_n = P_0[(1 - t_n)f(u_n, \lambda_n) + t_nf(u_n, \lambda_0) + t_n\phi(u_0)]. \tag{6.127}$$

Then, $u_n \in D(J)$, and

$$\langle\alpha u_n - (1-t_n)f(u_n, \lambda_n) - t_nf(u_n, \lambda_0) - t_n\phi(u_0), v - u_n\rangle + J(v) - J(u_n) \geq 0 \tag{6.128}$$

for all $v \in V$.

Letting $v = u_n + u_0$ in this inequality,

$$\langle\alpha u_n - (1-t_n)f(u_n, \lambda_n) - t_nf(u_n, \lambda_0) - t_n\phi(u_0), u_0\rangle + J(u_n + u_0) - J(u_n) \geq 0. \tag{6.129}$$

Now, we remark that because $J(u_0) = 0$, we must have

$$J(u + u_0) \le J(u), \quad \forall u \in V. \tag{6.130}$$

In fact, for $\xi \in (0, 1)$, by the convexity and homogeneity of J,

$$
\begin{aligned}
J(u + u_0) &= J\left[\xi \frac{u}{\xi} + (1 - \xi)\frac{u_0}{1 - \xi}\right] \\
&\le \xi J\left(\frac{u}{\xi}\right) + (1 - \xi)J\left(\frac{u_0}{1 - \xi}\right) \\
&= \xi \frac{1}{\xi^2} J(u) + (1 - \xi)\frac{1}{(1 - \xi)^2} J(u_0) \quad \text{(by (6.47))} \\
&= \frac{J(u)}{\xi}.
\end{aligned}
$$

Because this holds for all $\xi \in (0, 1)$, we obtain (6.130) by letting $\xi \to 1^-$. Now from (6.129) and (6.130),

$$
\begin{aligned}
0 &\le \langle \alpha u_n - (1 - t_n)f(u_n, \lambda_n) - t_n f(u_n, \lambda_0) - t_n \phi(u_0), u_0 \rangle \\
&= \langle \alpha^* u_0 - (1 - t_n)f^*(u_0, \lambda_n) - t_n f^*(u_0, \lambda_0), u_n \rangle - t_n \langle \phi(u_0), u_0 \rangle \\
&= \langle \alpha^* u_0 - f^*(u_0, \lambda_0), u_n \rangle + (1 - t_n)\langle f^*(u_0, \lambda_0) - f^*(u_0, \lambda_n), u_n \rangle \\
&\quad - t_n \langle \phi(u_0), u_0 \rangle.
\end{aligned}
$$

Because $u_0 \in K(u_0)$,

$$
\begin{aligned}
0 &= \alpha^* u_0 - f^*(u_0, \lambda_0) \\
&= \alpha^* u_0 - g(u_0, \lambda_0) - h(u_0) \quad \text{by (6.109)}.
\end{aligned}
$$

Hence, the above inequality becomes

$$
\begin{aligned}
0 &\le (1 - t_n)\langle g(u_0, \lambda_0) + h(u_0) - g(u_0, \lambda_n) - h(u_0), u_n \rangle \\
&\quad - t_n \langle \phi(u_0), u_0 \rangle \\
&= (1 - t_n)\left\langle g(u_0, \lambda_0) - g\left(u_0, \left(\frac{\lambda_n}{\lambda_0}\right)\lambda_0\right), u_n \right\rangle - t_n \langle \phi(u_0), u_0 \rangle \\
&= (1 - t_n)\left[1 - \left(\frac{\lambda_n}{\lambda_0}\right)^\gamma\right]\langle g(u_0, \lambda_0), u_n \rangle - t_n \langle \phi(u_0), u_0 \rangle, \tag{6.131}
\end{aligned}
$$

and, then,

$$(1 - t_n)\left[1 - \left(\frac{\lambda_n}{\lambda_0}\right)^\gamma\right]\langle \alpha^* u_0 - h(u_0), u_n \rangle - t_n \langle \phi(u_0), u_0 \rangle \ge 0. \tag{6.132}$$

Next, we prove (6.124), (6.125), and (6.126).

If $t_n = 1$, then, (6.128) becomes $\langle \phi(u_0), u_0 \rangle \le 0$. However, from (6.121) and (6.122),

$$\langle \phi(u_0), u_0 \rangle = \|u_0\|_0^2 \ge C_0^2 \|u_0\|^2 > 0. \tag{6.133}$$

This contradiction proves that

$$H(1, u, \lambda_0) \neq 0$$

for all $u \in V$, and, thus, (6.126) holds. Now suppose that (6.125) does not hold, i.e., there exist sequences $\{u_n\}$, $\{\lambda_n\}$, $u_n \neq 0$, $\forall n$, and $\lambda_n \to \lambda_0^+$ such that

$$H(0, u_n, \lambda_n) = 0, \ \forall n \in N. \tag{6.134}$$

Letting $t_n = 0$ in (6.132),

$$\left[1 - \left(\frac{\lambda_n}{\lambda_0} \right)^\gamma \right] \langle \alpha^* u_0 - h(u_0), u_n \rangle \geq 0.$$

Because $\lambda_n / \lambda_0 > 1$, $\forall n$,

$$\langle \alpha^* u_0 - h(u_0), u_n \rangle \leq 0. \tag{6.135}$$

Setting $v_n = u_n / \|u_n\|$, from this inequality,

$$\langle \alpha^* u_0 - h(u_0), v_n \rangle \leq 0. \tag{6.136}$$

On the other hand, (6.134) is the same as (cf. (6.127))

$$u_n = P_0[f(u_n, \lambda_n)].$$

Dividing both sides of this equation by $\|u_n\|$ and using the homogeneity of P_0 and $f(\cdot, \lambda_n)$ ((6.45) and (6.46)),

$$\begin{aligned} v_n &= \frac{1}{\|u_n\|} P_0[f(u_n, \lambda_n)] \\ &= P_0[f(v_n, \lambda_n)]. \end{aligned} \tag{6.137}$$

Without loss of generality, we can assume that $v_n \rightharpoonup v$ in V, as $n \to \infty$. By the complete continuity of $P_0 \circ f$, we conclude from (6.137) that, in fact, $v_n \to v$ in V, and

$$v = P_0[f(v, \lambda_0)]. \tag{6.138}$$

Because $\|v\| = 1$, v is an eigenvector of (6.48) with respect to $\lambda = \lambda_0$, and, because λ_0 is a simple eigenvalue of (6.48), we must have

$$v = C u_1, \tag{6.139}$$

for some $C > 0$.

Now, letting $n \to \infty$ in (6.136),

$$\langle \alpha^* u_0 - h(u_0), v \rangle \leq 0. \tag{6.140}$$

In view of (6.139), this implies that

$$\langle \alpha^* u_0 - h(u_0), u_1 \rangle \leq 0,$$

contradicting (6.117). This contradiction proves (6.125).

To prove (6.124), again, we proceed indirectly, i.e., we assume there exist sequences $\{u_n\} \subset V, \{\lambda_n\} \subset (\lambda_0, \infty)$, and $\{t_n\}$, such that $\|u_n\| \to \infty, \lambda_n \to \lambda_0^+, t_n \in [0,1], \forall n$, and

$$H(t_n, u_n, \lambda_n) = 0, \ \forall n.$$

As above, we have (6.127) and (6.132). From (6.133) and (6.132), it follows that $t_n < 1$ for all n.

Hence, again by (6.133) and (6.132),

$$(1 - t_n) \left[1 - \left(\frac{\lambda_n}{\lambda_0} \right)^\gamma \right] \langle \alpha^* u_0 - h(u_0), u_n \rangle \geq 0,$$

and

$$\left[1 - \left(\frac{\lambda_n}{\lambda_0} \right)^\gamma \right] \langle \alpha^* u_0 - h(u_0), u_n \rangle \geq 0, \ \ \forall n. \tag{6.141}$$

Because $\lambda_n > \lambda_0, \forall n$, we have (6.135). Letting $v_n = u_n / \|u_n\|$, as above, again, we have (6.136).

Now, dividing both sides of (6.127) by $\|u_n\|$ and using (6.45) and (6.56),

$$v_n = P_0 \left[(1 - t_n) f(v_n, \lambda_n) + t_n f(v_n, \lambda_0) + \frac{t_n \phi(u_0)}{\|u_n\|} \right]. \tag{6.142}$$

We may assume that $v_n \rightharpoonup v$ in V. Because $\|u_n\| \to \infty$,

$$\frac{\phi(u_0)}{\|u_n\|} \to 0, \ (n \to \infty).$$

Because $f(v_n, \lambda_n) \to f(v, \lambda_0)$ in V^*, by the complete continuity of f, we see that the right-hand side of (6.132) tends to $P_0[f(v, \lambda_0)]$ as $n \to \infty$.

This shows that $v_n \to v$ in V and

$$\begin{cases} v = P_0[f(v, \lambda_0)] \\ \|v\| = 1. \end{cases}$$

Again, v is an eigenvector of (6.48) with $\lambda = \lambda_0$, and, then, (6.139) holds. Also, by letting $n \to \infty$ in (6.136), we obtain (6.140), which contradicts the assumption (6.117) and completes the proof of (6.124). Now, from (6.126), we see that the equation

$$u - P_0[f(u, \lambda_0) + \phi(u_0)] = 0$$

has no solution in V. Hence,

$$d(I - P_0[f(\cdot, \lambda_0) + \phi(u_0)], B_R(0), 0) = 0,$$

for all $R > 0$.

On the other hand, (6.124) and the homotopy invariance property of the Leray–Schauder degree, show that, for $R > R_0$, $\lambda \in (\lambda_0, \lambda_1)$,

$$
\begin{aligned}
d(I - P_0[f(\cdot, \lambda)], B_R(0), 0) &= d(I - H(0, \cdot, \lambda), B_R(0), 0) \\
&= d(I - H(1, \cdot, \lambda), B_R(0), 0) \\
&= d(I - P_0[f(\cdot, \lambda_0) + \phi(u_0)], B_R(0), 0) \\
&= 0.
\end{aligned}
$$

(6.143)

Let $R > R_0 > r > 0$. From (6.125), we know that the equation

$$H(0, u, \lambda) \equiv u - P_0[f(u, \lambda)] = 0$$

has no solution in $\overline{B_R(0)} \setminus B_r(0)$. Hence, by the excision property,

$$
\begin{aligned}
d(I - P_0[f(\cdot, \lambda)], B_r(0), 0) &= d(I - P_0[f(\cdot, \lambda)], B_R(0), 0) \\
&= 0 \ \ (\text{by } (6.143)),
\end{aligned}
$$

proving(6.120).

Our result now follows from (6.118) and (6.120) by applying Theorem 6.4, with $a = 0, b = \lambda_0 + \epsilon$, and $\epsilon > 0$, sufficiently small. We have proved (a).

(b) First, we note that $\ker J$ is a closed convex cone in V. In fact, let $u, v \in \ker J, w = \theta u + (1 - \theta)v, 0 \leq \theta \leq 1$. Then,

$$0 \leq J(w) \leq \theta J(u) + (1 - \theta)J(v) = 0.$$

Hence, $J(w) = 0$, and, then, J is convex. Let $\{v_n\} \subset \ker J, v_n \to v$. By the lower semicontinuity of J,

$$0 \leq J(v) \leq \liminf J(v_n) = 0,$$

proving that $v \in \ker J$ and, then, the closedness of J. The homogeneity of J ((6.47)) implies that $\ker J$ is a cone. Now, because $\ker[\alpha^* - f^*(\cdot, \lambda_0)]$ is a closed subspace of V, $K(\lambda_0)$ is a closed convex cone in V.

Let $W = \overline{K(\lambda_0) - K(\lambda_0)}$ be the closed subspace of V spanned by $K(\lambda_0)$. W is a Banach subspace of V, and, because $K(\lambda_0)$ is not symmetric,

$$W \neq K(\lambda_0).$$

Let $x_0 \in W \setminus K(\lambda_0)$. By a consequence of the Hahn-Banach Theorem (theorem on separation of convex sets, [14]), there exists $\psi \in V^*$, such that

$$\begin{cases} \langle \psi, u \rangle \geq 0, \ \forall u \in K(\lambda_0) \\ \langle \psi, x_0 \rangle < 0. \end{cases} \tag{6.144}$$

Because $W \neq K(\lambda_0)$, this implies the existence of $\bar{u} \in K(\lambda_0)$, such that

$$\langle \psi, \bar{u} \rangle > 0. \tag{6.145}$$

Now, we already proved (6.118) in (a) and shall show that (6.120) also holds in the present case. We define the following family of compact perturbations of the identity similar to that in (6.123):

$$H(t, u, \lambda) = u - P_0[(1 - t)f(u, \lambda) + tf(u, \lambda_0) + t\psi], \tag{6.146}$$

$t \in [0, 1], u \in V, \lambda \in \mathbb{R}$, ($\psi$ defined in (6.144)), and prove that (6.124), (6.125), and (6.126) are also true in this case.

Let $(u_n, t_n, \lambda_n) \in V \times [0, 1] \times \mathbb{R}$, $n = 1, 2, \ldots$ be such that

$$H(u_n, t_n, \lambda_n) = 0,$$

or, equivalently,

$$u_n = P_0[(1 - t_n)f(u_n, \lambda_n) + t_n f(u_n, \lambda_0) + t_n \psi]. \tag{6.147}$$

Let $u_0 \in K(\lambda_0)$. Arguing as in (a) (equations (6.131) and (6.132)), we can conclude from this equation that

$$0 \leq (1 - t_n) \left[1 - \left(\frac{\lambda_n}{\lambda_0} \right)^\gamma \right] \langle g(u_0, \lambda_0), u_n \rangle - t_n \langle \psi, u_0 \rangle \tag{6.148}$$

and, therefore,

$$0 \leq (1 - t_n) \left[1 - \left(\frac{\lambda_n}{\lambda_0} \right)^\gamma \right] \langle \alpha^* u_0 - h(u_0), u_n \rangle - t_n \langle \psi, u_0 \rangle, \tag{6.149}$$

for all $u_0 \in K(\lambda_0)$.

If $t_n = 1$, then, from (6.149),

$$\langle \psi, u_0 \rangle \leq 0,$$

for all $u_0 \in K(\lambda_0)$. This contradicts (6.145). Hence, $H(1, u, \lambda_0) \neq 0$, $\forall u \in V$. We have (6.126).

Similarly, letting $t_n = 0$ in (6.149), it follows that (6.137) and (6.138) are satisfied, i.e., v is an eigenvector of (6.48) corresponding to $\lambda = \lambda_0$, and, moreover, (6.140) also holds, i.e.,

$$\langle \alpha^* u_0 - h(u_0), v \rangle \leq 0.$$

Because this is satisfied for all $u_0 \in K(\lambda_0)$, we have a contradiction to (6.117), proving (6.125).

Now, to prove (6.124), we need only to notice that (6.149) and (6.144) still imply (6.141) and, therefore, (6.135). However, equation (6.132), then, becomes

$$v_n = P_0 \left[(1 - t_n)f(v_n, \lambda_n) + t_n f(v_n, \lambda_0) + t_n \frac{\psi}{\|u_n\|} \right].$$

By letting $n \to \infty$ in this equation, again, we obtain (6.138) (because $\psi/\|u_n\| \to 0, n \to \infty$).

The remaining part of the proof of (6.124) can be carried out in the same manner as in (a). Now, from (6.124), (6.125), and (6.126), the excision property and the homotopy invariance property of the Leray–Schauder degree, we obtain (6.120), as was already done in (a). Theorem 6.4 is, then, applied to yield the desired bifurcation conclusion.   ∎

Remark 6.3 (a) In the case where f is symmetric and homogeneous of order $\gamma > 0$ and α is the identity mapping (or, more generally, when α is a self-adjoint operator), then, Theorem 6.8 (a) has Theorem 4.4, Chapter 5 as a consequence. In fact, in this case, $j = I_K$ is the indicator function of the support cone K_0 of K. Then,

$$K(\lambda_0) = \{ u \in K_0 : u - f(u, \lambda_0) = 0 \}.$$

If λ_0 is a simple eigenvalue of (3.6) (or (6.48)), whose eigenvector u_1 is also an eigenvector of (3.6), then, $u_1 \in K(\lambda_0)$. Consequently, (6.117) in Theorem 6.8 is satisfied by choosing $u_0 = u_1$ ($\langle u_0, u_1 \rangle = \|u_1\|^2 > 0$).

Hence, Theorem 4.4 is a particular case of Theorem 6.8 (a).

(b) In Theorem 6.8 (b), we do not need to assume the simplicity of λ_0 and may replace this assumption by the condition that $K(\lambda_0)$ is not symmetric.

(c) It can be directly verified that Theorem 6.8 (a) contains the main bifurcation results (Theorem 4.2, Lemma 4.3) of [45]. In this paper, the authors considered the following variational inequality of von Kármán's type (similar to those considered in Examples 4.4 and 4.6):

$$\begin{cases} \langle A_1 u + A_2 u - \lambda L u + T(u), v - u \rangle \geq 0, \ \forall v \in K, \\ u \in K, \end{cases} \tag{6.150}$$

where K is a (closed, convex) cone in a Hilbert space $(X, \langle \cdot, \cdot \rangle)$. A_1, A_2, and L are bounded linear mappings from X to X, A_1 is coercive, and L and A_2 are compact. T is a completely continuous mapping from X to X, T is strictly positive (i.e., $T(0) = 0$, and $\langle T(u), u \rangle > 0$, $\forall u \in K \setminus \{0\}$) and homogeneous of degree $p > 1$.

Theorem 4.2 of [45] states that under those above assumptions on A_1, A_2, L, and K (and some technical assumptions), if we let ρ be given by

$$\rho^{-1} = \sup_{x \in K \setminus \{0\}} \frac{\langle Lx, x \rangle}{\langle (A_1 + A_2)x, x \rangle},$$

and assume that ρ is an isolated eigenvalue of the inequality

$$u \in K : \langle (A_1 + A_2)u, v - u \rangle \geq \langle \lambda Lu, v - u \rangle, \ \forall v \in K$$

and of the operator $(A_1 + A_2) - \lambda L$, such that $\dim \ker(A_1 + A_2 - \rho L) = 1$ and there exist $u_\rho \in \ker(A_1 + A_2 - \rho L) \cap \overset{\circ}{K}$, $\overline{u}_\rho \in \ker(A_1^* + A_2^* - \rho L^*) \cap \overset{\circ}{K}$, such that $\langle (A_1 + A_2)u_\rho, \overline{u}_\rho \rangle > 0$, then,

$$\Gamma_\rho := \overline{\{(\lambda, u) \in \mathbb{R} \times K \setminus \{0\} : u \text{ solution of (6.150)}\}}$$

contains a subcontinuum Γ_0, such that $(\rho, 0) \in \Gamma_0$, which either (i) is unbounded, or (ii) $\Gamma_0 \cap \{\{0\} \times \mathbb{R} \neq \{(\rho, 0)\}$.

(6.150) can be written as a particular case of (6.8) with $j = I_K, A = A_1$, and $B(u, \lambda) = \lambda Lu - A_2 u - T(u)$. Because K is a cone, K coincides with its support cone, and, then, the homogenized functional in this case is the same functional $j = I_K$. The homogenization in this situation is straightforward, and $f(u, \lambda) = \lambda Lu - A_2 u$. It can be easily verified that Theorem 6.8, when restricted to the special case of the variational inequality (6.150), contains Lemma 4.3, (i) and (iii) and Theorem 4.2 of [45]. If the conditions in the quoted theorems are satisfied (i.e., there exist $u_\rho \in \ker(A_1 + A_2 - \rho L) \cap \overset{\circ}{K}$, $\overline{u}_\rho \in \ker(A_1^* + A_2^* - \rho L^*) \cap \overset{\circ}{K}$, such that $\langle (A_1 + A_2)u_\rho, \overline{u}_\rho \rangle > 0$, and ρ is the first eigenvalue of the corresponding variational inequality and is assumed to be simple), then, the assumptions in Theorem 6.8 (a) are also satisfied. Moreover, Theorem 6.8 (a) is stronger than the results in [45].

As seen in the proof of Theorem 6.8 (a), to obtain the global bifurcation results, it is not necessary to assume that T is homogeneous or strictly positive. The results are for the general case of Banach spaces, with general convex functionals j, including the problems on convex sets (not necessarily convex cones) as particular cases. Moreover, the problem is concerned with nonlinear principal operators A (which allow us to study quasilinear or nonlinear variational inequalities).

Restricted to the case of variational inequalities of the form (6.150), Theorem 6.8 (a) contains the results in [45] mentioned above, e.g. in those results, the eigenvectors u_ρ and $\overline{u}_\rho$ are assumed in the interior part of K, whereas Corollary 4.6, Chapter 4, assumes only that the eigenvectors are demi-interior points of K. Theorem 6.8, needs only that the eigenvectors u_0 and u_1 belong to $\ker J$ (or to K, in particular). There are also cases (as shown in the following example) where Theorem 6.8 is applicable while the assumptions in the results of [45] are not satisfied.

Example 6.1 Let V be $\mathbb{R}^2$ with the usual Euclidean inner product. Consider the following variational inequality:

$$\begin{cases} \langle u - \lambda\beta u, v - u \rangle \geq 0, \ \forall v \in K, \\ u \in K, \end{cases} \tag{6.151}$$

where

$$K = \{(x_1, x_2) \in \mathbb{R}^2 : x_1 \geq 0, x_2 \geq 0\}$$

is the (closed) first quadrant, and $\beta : \mathbb{R}^2 \to \mathbb{R}^2$ is given by

$$\beta(x) = \begin{pmatrix} -1 & 0 \\ 0 & 1 \end{pmatrix} \begin{pmatrix} x_1 \\ x_2 \end{pmatrix} = \begin{pmatrix} -x_1 \\ x_2 \end{pmatrix}.$$

K is a convex cone. Hence, $K = K_0$. Moreover, $\beta(K) \not\subset K$. It follows that

$$u - \lambda\beta u = 0 \tag{6.152}$$

if and only if

$$u_1 = \lambda(-u_1), \ u_2 = \lambda u_2,$$

or

$$\begin{cases} u_1(1 + \lambda) & = & 0, \\ u_2(1 - \lambda) & = & 0. \end{cases}$$

There are two eigenvalues $\lambda_1 = 1$ and $\lambda_2 = -1$, and the corresponding eigenvectors are

$$X_1 = (0, 1)^T \text{ and } X_2 = (1, 0)^T.$$

First, consider $\lambda_1 > 0$. We see that λ_1 is a simple eigenvalue of the equation (6.152). Moreover, it is a simple eigenvalue of the variational inequality (6.151).

In fact, suppose that $u = (u_1, u_2)^T$ is a solution of (6.151) with $\lambda = \lambda_1 = 1$. Because K is a cone, we see that (6.151) (with $\lambda = \lambda_1 = 1$) is equivalent to the following system

$$\begin{cases} \langle u - \beta u, u \rangle = 0, \\ \langle u - \beta u, v \rangle \geq 0, \ \forall v \in K. \end{cases}$$

Hence,

$$\begin{aligned} 0 & = & \langle (u_1, u_2)^T - (-u_1, u_2)^T, (u_1, u_2)^T \rangle \\ & = & \langle (2u_1, 0)^T, (u_1, u_2)^T \rangle \\ & = & 2u_1^2. \end{aligned}$$

Therefore, $u_1 = 0$. Because $u \in K$, $u_2 \geq 0$, and, then,

$$(u_1, u_2) = (0, u_2) = u_2(0, 1),$$

with $u_2 \geq 0$. This shows that $\lambda_1 = 1$ is a simple eigenvalue of (6.151).

In this case, we see that β is self-adjoint and $j = J = I_K$. Hence, $D(J) = \ker K = K$, and

$$
\begin{aligned}
K(\lambda_1) &= \ker(I - \beta^*) \cap \ker J \\
&= \{u \in K : u - \beta u = 0\} \\
&= \{t(0, 1)^T : t \geq 0\}.
\end{aligned}
$$

In this case, we can choose $u_0 = u_1 = (0, 1)^T$, and (6.117) is satisfied (because, in this example, $h \equiv 0$ and $\langle \alpha^* u_0 - h(u_0), u_1 \rangle = \|u_0\|^2 = 1$). On the other hand, because $\lambda_1 = 1$ is simple and the associated eigenvector $X_1 = (0, 1)^T$ is in ∂K, $\ker(I - \lambda_1 \beta) \cap \overset{\circ}{K} = \emptyset$, and we can not find u_ρ and $\bar{u}_\rho$, as in the theorems of [45].

We also note that Theorem 6.8 is valid for bifurcation from higher eigenvalues. As an example, we consider Example 4.3 and Corollary 4.7 in Chapter 4.

Example 6.2 We consider Example 4.3 (c) in the particular case where $a = \pi$ and $I_1 = \{1\}$, $I_2 = \{3\}$, (this happens when we choose, for example,

$$A = B = 1, \ C = D = 3,$$

i.e., the obstacles are concentrated at the points 1 and 3 (see Figure 6.2)). Because $1 < \pi/2 < 2$, we have proved in Example 4.3 (c) (Corollary 4.7) that global bifurcation occurs on the interval $[0, 4]$, 4 being the second eigenvalue of (4.42), which is the linear equation associated with the homogenized variational inequality (4.43). We know that (4.42) has eigenvalues $\lambda_k = k^2$ and eigenvectors

$$u_k(x) = \sin kx, \ x \in [0, \pi], k = 1, 2, 3, \ldots.$$

We now check that $\lambda_2 = 4$ is in fact greater than the first eigenvalue ρ of (4.43). Indeed, in this case,

$$K = \{u \in H^2(0, \pi) \cap H^1_0(0, \pi) : u(1) \geq 0 \geq u(3)\}.$$

K is a closed, convex cone, and, thus, $K = K_0$. On the other hand,

$$\langle Au, v \rangle = \int_0^\pi u'' v'', \ \langle Lu, v \rangle = \int_0^\pi u' v'$$

FIGURE 6.2. Example 6.2.

for $u, v \in H^2(0, \pi) \cap H_0^1(0, \pi)$. Hence, ρ is given by

$$\rho = \left[\sup_{u \in K \setminus \{0\}} \frac{\displaystyle\int_0^\pi (u')^2}{\displaystyle\int_0^\pi (u'')^2} \right]^{-1}.$$

We consider the function

$$u(x) = x(x-3)^2 \chi_{[0,3]} = \begin{cases} x(x-3)^2 & \text{if } 0 \le x \le 3, \\ 0 & \text{if } 3 \le x \le \pi. \end{cases}$$

Because $u(0) = u(\pi) = 0, u(3) = u'(3) = 0,\ u \in C^1[0,\pi], u'' \in L^2(0,\pi)$, and, then, $u \in H^2(0,\pi) \cap H_0^1(0,\pi)$. Moreover, $u(1) = 4 > 0 = u(3)$, i.e., $u \in K \setminus \{0\}$. By direct calculations,

$$\frac{\displaystyle\int_0^\pi (u')^2}{\displaystyle\int_0^\pi (u'')^2} = \frac{\displaystyle\int_0^3 [(x-3)^2 + 2x(x-3)]^2 dx}{\displaystyle\int_0^3 (6x-12)^2 dx} = \frac{3}{10}.$$

Hence,

$$\rho \le \left[\frac{\displaystyle\int_0^\pi (u')^2}{\displaystyle\int_0^\pi (u'')^2} \right]^{-1} = \frac{10}{3} < 4 = \lambda_2.$$

Another simple example is provided by the following bifurcation problem of an elastic beam on $[0, \pi]$ with two obstacles at $\pi/4$ and $3\pi/4$:

$$\begin{cases} \displaystyle\int_0^\pi u'(v-u)' - \lambda \int_0^\pi u(v-u) \ge 0, \ \forall v \in K, \\ u \in K, \end{cases} \tag{6.153}$$

with

$$K = \{u \in H_0^1(0, \pi) : u(\pi/4) \geq 0 \geq u(3\pi/4)\}.$$

(K is a closed convex cone in $H_0^1(0, \pi)$.) The linear equation associated with (6.153) is the following:

$$\begin{cases} \int_0^\pi u'v' - \lambda \int_0^\pi uv = 0, \ \forall v \in H_0^1(0, \pi), \\ u \in H_0^1(0, \pi), \end{cases}$$

or, equivalently,

$$\begin{cases} u'' + \lambda u = 0 \ \text{ in } \ (0, \pi), \\ u(0) = u(\pi) = 0. \end{cases} \tag{6.154}$$

This equation has eigenvalues $\lambda_k = k^2$, and corresponding eigenfunctions

$$u_k(x) = \sin kx, \ x \in [0, \pi], k = 1, 2, 3, \ldots.$$

Arguing as in Example 4.3 (c), we see that $u_1(x) = \sin x$ does not belong to K. However, $u_2 \in \overset{o}{K}$ (because $u_2(\pi/4) > 0 > u_2(3\pi/4)$). Moreover, because $\lambda_2 = 4$ is simple and $B = f$ defined by

$$B(u, \lambda) = f(u, \lambda) = \lambda \int_0^\pi uv, \ u, v \in H_0^1(0, \pi)$$

is symmetric with respect to u, we can apply Corollary 4.6 in this case. However, it is easily seen that $\lambda_2 = 4$ is greater than the first eigenvalue of (6.153):

$$\rho = \left[\sup_{u \in K \setminus \{0\}} \frac{\int_0^\pi u^2}{\int_0^\pi (u')^2} \right]^{-1}.$$

In fact, consider

$$u(x) = \begin{cases} \sin(4x/3), & 0 \leq x \leq 3\pi/4 \\ 0, & 3\pi/4 \leq x \leq \pi. \end{cases}$$

Then, $u \in C[0, \pi]$, and $u' \in L^2(0, \pi)$, and, therefore, $u \in H_0^1(0, \pi)$. Moreover, $u \in K \setminus \{0\}$ (because $u(\pi/4) = \sin(\pi/3) > 0 = u(3\pi/4)$). On the other hand,

$$\int_0^\pi u^2 = \int_0^{3\pi/4} \sin^2\left(\frac{4x}{3}\right) dx = \frac{3\pi}{8},$$

and

$$\int_0^\pi (u')^2 = \frac{16}{9} \int_0^{3\pi/4} \cos^2\left(\frac{4x}{3}\right) dx = \frac{2\pi}{3}.$$

Hence,

$$\rho \leq \left(\frac{3\pi/8}{2\pi/3}\right)^{-1} = \frac{16}{9} < 4 = \lambda_2.$$

As above, if ρ has an associated eigenvector that is also an eigenvector of (6.154), then, ρ must be λ_1, and $u_1 \in K$ or $-u_1 \in K$.

6.4.2   Unbounded bifurcation branches

Next, we consider cases where the bifurcating solution continua are in fact unbounded. The degree calculations in this situation are motivated by a result of Szulkin (Theorem 3, [119]).

For $u_0 \in V, \lambda_0 \in \mathbb{R}^+$, we define

$$D(J, u_0, \lambda_0) = \{u \in D(J) : \max\{\langle \phi(u_0), u \rangle, \langle g(u, \lambda_0), u \rangle, \langle h(u), u \rangle\} \geq 0\}.$$

($\phi : V \to V^*$ is the duality mapping considered in the proof of Theorem 6.8, and g, h are given as in (6.109).) In the case where $h \equiv 0$, i.e., $f(u, \lambda)$ is γ-homogeneous with respect to λ, we define

$$D(J, u_0, \lambda_0) = \{u \in D(J) : \max\{\langle \phi(u_0), u \rangle, \langle f(u, \lambda_0), u \rangle\} \geq 0\}.$$

We have the following theorem:

Theorem 6.9

(a) *Assume that (6.109) is satisfied, and that, for $\lambda_0 > 0$, there exists $u_0 \in K(\lambda_0)$, such that*

$$\langle g(u_0, \lambda_0), u \rangle \geq 0, \ \forall u \in D(J, u_0, \lambda_0). \tag{6.155}$$

Then, if $\lambda > \lambda_0$ is not an eigenvalue of (6.48),

$$d(I - P_0[f(\cdot, \lambda)], B_r(0), 0) = 0$$

for all $r > 0$, and there exists a continuum of nontrivial solutions of (6.8), bifurcating from $[0, \lambda]$ and satisfying the alternative in Theorem 6.4.

(b) *If the operator $A - B(\cdot, 0)$ is strictly monotone on V and if (6.155) is replaced by the following stronger condition:*

$$\langle g(u_0, \lambda_0), u \rangle > 0, \ \forall u \in D(J, u_0, \lambda_0) \setminus \{0\}, \tag{6.156}$$

then, every $\lambda > \lambda_0$ is not an eigenvalue of (6.48), and there exists an unbounded branch of solutions of (6.8) bifurcating from $\{0\} \times [0, \lambda_0]$.

Proof. (a) Using the proof of Theorem 6.8,

$$d(I - P_0[f(\cdot, 0)], B_r(0), 0) = 1 \tag{6.157}$$

for all $r > 0$.

Assuming $\lambda > \lambda_0$ is not an eigenvalue of (6.48) and that (6.155) is satisfied, we establish that

$$d(I - P_0[f(\cdot, \lambda)], B_r(0), 0) = 0. \tag{6.158}$$

First, we note that, because of

$$u - P_0[f(u, \lambda)] \neq 0, \; \forall u \neq 0,$$

this degree exists and does not depend on $r > 0$. Consider the following family

$$H(u, t) = u - P_0[f(u, \lambda) + t\phi(u_0)],$$

$(0 \leq t \leq 1, u \in V)$ of completely continuous perturbations of the identity. We prove that there exists $R_0 > 0$, such that

$$H(u, t) \neq 0, \; \forall t \in [0, 1], \; \forall u \in V \text{ such that } \|u\| \geq R_0. \tag{6.159}$$

Suppose this is not the case. Then there exist sequences $\{u_n\} \subset V$ and $\{t_n\} \subset [0, 1]$ such that $\|u_n\| \to \infty \; (n \to \infty)$ and

$$H(u_n, t_n) = 0, \; \forall n,$$

i.e.,

$$u_n = P_0[f(u_n, \lambda) + t_n\phi(u_0)], \; \forall n.$$

Setting $v_n = u_n/\|u_n\|$ and dividing this equation by $\|u_n\|$, (by using the homogeneity of P_0 and $f(\cdot, \lambda)$),

$$v_n = P_0\left[f(v_n, \lambda) + \frac{t_n\phi(u_0)}{\|u_n\|}\right].$$

Now letting $n \to \infty$ and assuming (without loss of generality) that $v_n \to v$ in V, one concludes, from the complete continuity of f and the fact that

$$\frac{t_n\phi(u_0)}{\|u_n\|} \to 0, \; n \to \infty,$$

the following:

$$v_n \to v \text{ in } V \text{ (hence, } \|v\| = 1),$$

and

$$v = P_0[f(v, \lambda)].$$

This contradicts the assumption that λ is not an eigenvalue of (6.48), proving (6.159).

Applying the homotopy invariance property, from (6.159), for $R \geq R_0$ and $r > 0$, the following holds:

$$
\begin{aligned}
d(I - P_0[f(\cdot, \lambda)], B_r(0), 0) &= d(I - P_0[f(\cdot, \lambda)], B_R(0), 0) \\
&= d(H(\cdot, 0), B_R(0), 0) \\
&= d(H(\cdot, 1), B_R(0), 0) \\
&= d(I - P_0[f(\cdot, \lambda) + \phi(u_0)], B_R(0), 0).
\end{aligned}
\tag{6.160}
$$

Now, we prove that the equation

$$
u - P_0[f(u, \lambda) + \phi(u_0)] = 0
\tag{6.161}
$$

has no solution in V. In fact, assume that u is a solution of (6.161), i.e.,

$$
\langle \alpha u - [f(u, \lambda) + \phi(u_0)], v - u \rangle + J(v) - J(u) \geq 0, \ \forall v \in V.
\tag{6.162}
$$

Letting $v = u + u_0$ in (6.162),

$$
\langle \alpha u - f(u, \lambda) - \phi(u_0), u_0 \rangle + J(u + u_0) - J(u) \geq 0.
$$

However, because $J(u_0) = 0$, by (6.130), $J(u + u_0) - J(u) \leq 0$. Hence,

$$
\begin{aligned}
0 &\leq \langle \alpha u - f(u, \lambda) - \phi(u_0), u_0 \rangle \\
&= \langle \alpha^* u_0 - f^*(u_0, \lambda), u \rangle - \langle \phi(u_0), u_0 \rangle \\
&= \langle \alpha^* u_0 - f^*(u_0, \lambda_0), u \rangle + \langle f^*(u_0, \lambda_0) - f^*(u_0, \lambda), u \rangle - \langle \phi(u_0), u_0 \rangle \\
&= \langle g(u_0, \lambda_0) - g(u_0, \lambda), u \rangle - \langle \phi(u_0), u_0 \rangle \\
&= \left[1 - \left(\frac{\lambda}{\lambda_0} \right)^\gamma \right] \langle g(u_0, \lambda_0), u \rangle - \langle \phi(u_0), u_0 \rangle.
\end{aligned}
\tag{6.163}
$$

On the other hand, by letting $v = 0$ in (6.162),

$$
-\langle \alpha u - f(u, \lambda) - \phi(u_0), u \rangle - J(u) \geq 0.
$$

Therefore,

$$
\langle f(u, \lambda) + \phi(u_0), u \rangle \geq \langle \alpha u, u \rangle + J(u) \geq 0,
$$

and, thus,

$$
\langle \phi(u_0), u \rangle \geq 0 \ \text{or} \ \langle f(u, \lambda), u \rangle \geq 0.
$$

In the latter case, because

$$
\begin{aligned}
\langle f(u, \lambda), u \rangle &= \langle f^*(u, \lambda), u \rangle \\
&= \langle g(u, \lambda), u \rangle + \langle h(u), u \rangle \\
&= \left(\frac{\lambda}{\lambda_0} \right)^\gamma \langle g(u, \lambda_0), u \rangle + \langle h(u), u \rangle,
\end{aligned}
$$

we must have

$$\langle g(u, \lambda_0), u \rangle \geq 0 \text{ or } \langle h(u), u \rangle \geq 0.$$

This means that at least one of the three terms $\langle \phi(u_0), u \rangle$, $\langle g(u, \lambda_0), u \rangle$, or $\langle h(u), u \rangle$ must be nonnegative. By definition, $u \in D(J, u_0, \lambda_0)$.

In the case where f is homogeneous of degree γ with respect to λ, $h \equiv 0$, and, then, $f = g$. It follows that

$$\langle f(u, \lambda), u \rangle = \left(\frac{\lambda}{\lambda_0} \right)^{\gamma} \langle f(u, \lambda_0), u \rangle$$

and, therefore, $\langle f(u, \lambda), u \rangle \geq 0$ if and only if $\langle f(u, \lambda_0), u \rangle \geq 0$. Also, $u \in D(J, u_0, \lambda_0)$.

Now by (6.155),

$$\langle g(u_0, \lambda_0), u \rangle \geq 0.$$

This implies, together with (6.163), that

$$
\begin{aligned}
0 &\leq \left[1 - \left(\frac{\lambda}{\lambda_0} \right)^{\gamma} \right] \langle g(u_0, \lambda_0), u \rangle - \langle \phi(u_0), u_0 \rangle \\
&\leq -\langle \phi(u_0), u_0 \rangle \\
&= -\|u_0\|_0^2 \\
&< 0.
\end{aligned}
$$

This contradiction proves that (6.161) has no solution in V, which implies that

$$d(I - P_0[f(\cdot, \lambda) + \phi(u_0)], B_R(0), 0) = 0$$

for all $R > 0$, and, thus,

$$d(I - P_0[f(\cdot, \lambda)], B_r(0), 0) = 0$$

for all $r > 0$.

Using also (6.157) and Theorem 6.4, we complete the proof of (a).

(b) Now, suppose that (6.156) is satisfied. We prove that, for all $\lambda > \lambda_0$, λ is not an eigenvalue of (6.48).

Suppose, otherwise, that there exist λ and $u \neq 0$, such that

$$\langle \alpha u - f(u, \lambda), v - u \rangle + J(v) - J(u) \geq 0, \ \forall v \in V.$$

Letting $v = u + u_0$ in this inequality,

$$\langle \alpha u - f(u, \lambda), u_0 \rangle + J(u + u_0) - J(u) \geq 0.$$

Arguing as in (6.163), it follows that

$$\langle \alpha u - f(u, \lambda), u_0 \rangle \geq 0$$

and, therefore,

$$0 \leq \left[1 - \left(\frac{\lambda}{\lambda_0}\right)^{\gamma}\right] \langle g(u_0, \lambda_0), u \rangle.$$

Then,

$$\langle g(u_0, \lambda_0), u \rangle \leq 0. \tag{6.164}$$

Also, letting $v = 0$ in (6.48), again,

$$\langle f(u, \lambda), u \rangle \ \geq \ 0.$$

Hence, also, $u \in D(J, u_0, \lambda_0)$. However, because $u \neq 0$, this and (6.156) contradict (6.164).

Now, let $\mathcal{C}$ be the continuum of nontrivial solutions of (6.8) emanating form $[0, \lambda_0]$ ($\mathcal{C}$ is defined as in Theorem 6.3).

By (a), we know that $\mathcal{C}$ must be either unbounded, or there exists

$$(0, \lambda_1) \in \mathcal{C} \cap (V \times (\mathbb{R} \setminus [0, \lambda_0])), \tag{6.165}$$

where λ_1 is an eigenvalue of (6.48) (cf. Theorem 6.4). If $\mathcal{C}$ is bounded, then, (6.165) holds. Hence, λ_1 can not lie in (λ_0, ∞). Because $\lambda_1 \notin [0, \lambda_0]$, we must have $\lambda_1 < 0$. Therefore,

$$\mathcal{C} \cap (V \times [0, \lambda_0]) \neq \emptyset, \ \mathcal{C} \cap (V \times (-\infty, \lambda_1]) \neq \emptyset,$$

with $\lambda_1 < 0$. Because $\mathcal{C}$ is connected, there exists a sequence $\{(x_n, \lambda_n)\} \subset \mathcal{C}$, such that $\lambda_n < 0$, $\forall n$ (then, $x_n \neq 0$), and $\lambda_n \to 0^-$, as $n \to \infty$.

Because $\mathcal{C} \subset \mathcal{S}$, (x_n, λ_n) satisfies (6.8) for all n, i.e.,

$$x_n = P[B(x_n, \lambda_n)], \ \forall n. \tag{6.166}$$

Because $\mathcal{C}$ is bounded, $\{x_n\}$ is also bounded in V. Hence, by passing to a subsequence, if necessary, we can assume, without loss of generality, that $x_n \rightharpoonup x_0$ in V. By the complete continuity of B,

$$B(x_n, \lambda_n) \to B(x_0, 0) \ \text{in} \ V^*,$$

and, therefore, from (6.166),

$$x_0 = P[B(x_0, 0)],$$

or, equivalently,

$$\langle A(x_0) - B(x_0, 0), v - x_0 \rangle + j(v) - j(x_0) \geq 0, \ \forall v \in V. \tag{6.167}$$

Because $A - B(\cdot, 0)$ is strictly monotone, this variational inequality has, at most, one solution (cf. Chapter 2, [65], Chapter 2). However, we already know that 0 is a solution of (6.167). Hence $x_0 = 0$, which implies that

$$(x_n, \lambda_n) \to (0, 0), \ \text{as} \ n \to \infty.$$

Because $x_n \neq 0$, $\forall n$, $(0,0)$ is a bifurcation point of (6.8). By Theorem 6.4 (I), 0 is an eigenvalue of (6.48), in contradiction to (6.110). Hence, $\mathcal{C}$ is unbounded in $V \times \mathbb{R}$.   ∎

As applications for this result, we consider bifurcation problems for a quasilinear variational inequality and one concerning the Stokes equation.

Example 6.3 [Bifurcation for quasilinear variational inequalities] In this example we consider a bifurcation problem for a variational inequality containing a quasilinear second-order elliptic operator.

Consider the following variational inequality:

$$
\begin{cases}
\displaystyle \int_\Omega \left[\sum_{i=1}^N a_i(x, u(x), \nabla u(x))\, \partial_i(v-u)(x) \right. \\
\qquad \left. + a_0(x, u(x), \nabla u(x))(v-u)(x) \right] dx \\
\qquad + \displaystyle \int_\Omega b(x, u(x), \lambda)(v-u)(x)dx + j(v) - j(u) \geq 0, \ \forall v \in H_0^1(\Omega), \\
u \in H_0^1(\Omega).
\end{cases}
$$

$$(6.168)$$

Here, Ω is a bounded domain in $\mathbb{R}^N$ ($N \geq 1$) with a smooth boundary. For $0 \leq i \leq N$,

$$a_i : \Omega \times \mathbb{R}^{N+1} \to \mathbb{R}$$

are Carathéodory functions satisfying

$$a_i(x, 0) = 0, \quad \text{for a.e. } x \in \Omega, \tag{6.169}$$

and the following growth condition

$$|a_i(x, u, \xi)| \leq M(|u| + |\xi|) \tag{6.170}$$

for a.e. $x \in \Omega$, all $u \in \mathbb{R}$, $\xi \in \mathbb{R}^N$, where M is a given positive constant.

We assume that, a_i ($0 \leq i \leq N$) satisfy the usual uniform monotonicity condition:

$$\sum_{i=1}^N [a_i(x, u, \xi) - a_i(x, v, \eta)](\xi_i - \eta_i) + [a_0(x, u, \xi) - a_0(x, v, \eta)](u - v)$$

$$\geq C|\xi - \eta|^2 \tag{6.171}$$

for a.e. $x \in \Omega$, all $u, v \in \mathbb{R}^N$, $\xi, \eta \in \mathbb{R}^N$, where $C > 0$ is a constant.

We also assume that for almost all $x \in \Omega$, the functions $a_i(x, \cdot, \cdot)$ are differentiable at $(0,0)$, and

$$
\begin{aligned}
a_{i0}(x) &= \partial_u a_i(x, 0, 0), \\
a_{ij}(x) &= \partial_{\xi_j} a_i(x, 0, 0), \ 1 \leq j \leq N, \ x \in \Omega,
\end{aligned}
\tag{6.172}
$$

are bounded functions in $x \in \Omega$. From these assumptions, it follows that

$$\lim_{\sigma \to 0} \frac{1}{\sigma} a_i(x, \sigma u, \sigma \xi) = a_{i0}(x)u + \sum_{j=1}^{N} a_{ij}(x)\xi_j, \qquad (6.173)$$

for a.e. $x \in \Omega$, all $u \in \mathbb{R}$, $\xi \in \mathbb{R}^N$. Also, (6.170) implies that $a_{ij} \in L^\infty(\Omega)$ with

$$|a_{ij}| \le M, \ 0 \le i, j \le N. \qquad (6.174)$$

According to (6.170) and the differentiability of $a_i(x, \cdot, \cdot)$, we see that, for sequences $\{\sigma_n\}, \{u_n\} \subset \mathbb{R}$, and $\{\xi_n\} \subset \mathbb{R}^N$ satisfying

$$\sigma_n \to 0, u_n \to u, \xi_n \to \xi,$$

we always have

$$\lim_{n \to \infty} \frac{1}{\sigma_n} a_i(x, \sigma_n u_n, \sigma_n \xi_n) = a_{i0}(x)u + \sum_{j=1}^{N} a_{ij}(x)\xi_j. \qquad (6.175)$$

Now, letting $v = 0$ and $\eta = 0$ in (6.171) and replacing u and ξ by σu and $\sigma\xi$, respectively ($\sigma > 0$), we obtain

$$\left| \sum_{i=1}^{N} a_i(x, \sigma u, \sigma \xi)\xi_i + a_0(x, \sigma u, \sigma \xi)u \right| \ge C\sigma|\xi|^2.$$

Dividing this inequality by $\sigma > 0$ and letting $\sigma \to 0^+$, it follows from (6.173) that

$$\sum_{i,j=1}^{N} a_{ij}(x)\xi_i\xi_j + u \sum_{i=1}^{N}(a_{i0} + a_{0i})\xi_i + a_{00}u^2 \ge C|\xi|^2, \qquad (6.176)$$

for a.e. $x \in \Omega$, all $u \in \mathbb{R}$, $\xi \in \mathbb{R}^N$. Hence, we have the usual ellipticity condition for linear operators. Now, we assume that

$$b : \Omega \times \mathbb{R}^2 \to \mathbb{R}$$

is a Carathéodory function, such that

$$b(x, 0, \lambda) = 0 \ \text{ for all } \ \lambda \in \mathbb{R}, \ \text{ a.e. } x \in \Omega,$$

and b satisfies the following growth condition

$$|b(x, u, \lambda)| \le C(\lambda)[K(x)|u| + L(x)|u|^{q-1}], \qquad (6.177)$$

where $C(\lambda)$ is bounded for λ in bounded intervals of $\mathbb{R}$, $L \in L^\infty(\Omega)$, $K \in L^{\frac{q}{q-2}}(\Omega)$ with

$$2 \le q < 2^* = \begin{cases} \dfrac{2N}{N-2} & \text{if } N \ge 3 \\[2mm] \infty & \text{if } N = 1, 2. \end{cases}$$

We assume, furthermore, that $b(x, \cdot, \lambda)$ has a partial derivative at 0, which is continuous with respect to λ in the sense that, for almost all $x \in \Omega$, $\partial_u b(x, 0, \cdot)$ is continuous in $\mathbb{R}$, and, if $u_n \to u$, $\sigma_n \to 0^+$ and $\lambda_n \to \lambda$ in $\mathbb{R}$, then, for a.e. $x \in \Omega$,

$$\frac{1}{\sigma_n} b(x, \sigma_n u_n, \lambda_n) \to \partial_u b(x, 0, \lambda) u \text{ in } \mathbb{R}. \tag{6.178}$$

Note that this condition is satisfied if B has a partial derivative with respect to $u \in \mathbb{R}$, and, for almost all $x \in \Omega$, $\partial_u b(x, u, \lambda)$ is continuous with respect to $(u, \lambda) \in \mathbb{R}^2$. We assume that j is a convex, lower semicontinuous functional from $H_0^1(\Omega)$ to $[0, \infty]$ such that $j(0) = 0$.

We rewrite (6.168) in operator form. From (6.170), it follows that $a_i(\cdot, u, \nabla u) \in L^2(\Omega)$, whenever $u \in H^1(\Omega)$. Hence, the integral

$$\langle Au, v \rangle = \int_\Omega \left[\sum_{i=1}^N a_i(x, u(x), \nabla u(x))\, \partial_i v(x) + a_0(x, u(x), \nabla u(x)) v(x) \right] dx \tag{6.179}$$

is well defined for all $u, v \in V = H_0^1(\Omega)$. Moreover, $Au \in V^*$, $\forall u \in V$. In fact, $\langle Au, \cdot \rangle$ is a linear mapping and, by (6.170) and Poincaré's inequality,

$$
\begin{aligned}
|\langle Au, v \rangle| &\leq \int_\Omega \left[\sum_{i=1}^N |a_i(x, u, \nabla u)|\, |\partial_i v| + |a_0(x, u, \nabla u)|\, |v| \right] \\
&\leq M \int_\Omega (|u| + |\nabla u|)(|\partial_i v| + |v|) \\
&\leq C_0 M \left[\left(\int_\Omega |u|^2 \right)^{1/2} + \left(\int_\Omega |\nabla u|^2 \right)^{1/2} \right] \int_\Omega |\nabla v|^2 \\
&\leq C_1 M \|u\|\, \|v\|
\end{aligned}
$$

($C_1 = C_1(\Omega) > 0$ is a fixed constant). Hence, A is a mapping from V to V^* with

$$\|Au\| \leq C_1 M \|u\|, \ \forall u \in V.$$

In particular, A is bounded on V. Moreover, one can check that A is continuous on V.

It immediately follows from (6.179) that $A(0) = 0$, proving condition (A1). The monotonicity condition (6.171) implies that

$$
\begin{aligned}
\langle Au - Av, u - v \rangle &= \int_\Omega \{ \sum_{i=1}^N [a_i(x, u, \nabla u) - a_i(x, v, \nabla v)] \partial_i(v - u) \\
&\quad + [a_0(x, u, \nabla u) - a_0(x, v, \nabla v)](v - u) \} dx \\
&\geq \int_\Omega C |\nabla u - \nabla v|^2 \\
&= C \|u - v\|^2,
\end{aligned}
$$

i.e., A satisfies (6.9), and, therefore, (A2) is satisfied. Now, for $u, v \in V$, we define

$$\langle B(u, \lambda), v \rangle = \int_\Omega b(x, u(x), \lambda) v(x) dx. \tag{6.180}$$

Let $u, v \in V$. Because the embedding $H_0^1(\Omega) \hookrightarrow L^q(\Omega)$, $q < 2^*$ (2^* is the Sobolev conjugate of 2) is continuous (and moreover, compact), from (6.177) and Hölder's inequality,

$$\int_\Omega |b(x, u(x), \lambda)| \, |v(x)| dx$$

$$\leq \int_\Omega [K(x)|u| + L(x)|u|^{q-1}] \, |v|$$

$$\leq C(\lambda) \left[\|K\|_{L^{\frac{q}{q-2}}(\Omega)} \|u\|_{L^q(\Omega)} \|v\|_{L^q(\Omega)} + \|L\|_{L^\infty(\Omega)} \|u\|_{L^q(\Omega)}^{q-1} \|v\|_{L^q(\Omega)} \right]$$

$$\leq C_3 C(\lambda) \left[\|K\|_{L^{\frac{q}{q-2}}(\Omega)} \|u\| + \|L\|_{L^\infty(\Omega)} \|u\|^{q-1} \right] \|v\|.$$

This shows that B, given by (6.180), is well defined and is a mapping from $V \times \mathbb{R}$ to V. Next, one easily verifies that B is completely continuous.

With this setting, we can write (6.168) in the equivalent operator form (6.8) with $V = H_0^1(\Omega)$ and A, B given by (6.179) and (6.180).

Next, we consider the homogenization of (6.168).

Let α be given by

$$\langle \alpha u, v \rangle = \int_\Omega \left\{ \sum_{i,j=1}^{N} a_{ij}(x) \, \partial_j u(x) \, \partial_i v(x) \right.$$

$$+ \sum_{i=1}^{N} [a_{i0}(x) \, u(x) \, \partial_i v(x) + a_{0i}(x) \, \partial_i u(x) \, v(x)]$$

$$\left. + d_0(x) u(x) v(x) \right\} dx, \tag{6.181}$$

for $u, v \in H_0^1(\Omega)$, where a_{ij}, $0 \leq i, j \leq N$, are given by (6.172) and $d_0 = a_{00}$.

As remarked before, $\alpha : V \to V$ is a linear, second-order elliptic operator in weak form. We prove that A_σ tends to α as $\sigma \to 0^+$ in the sense of (6.37) with $p = 2$. In fact, let $\{u_n\} \subset V, \{\sigma_n\} \subset \mathbb{R}_*^+$ be such that $u_n \to u$ in V, $\sigma_n \to 0^+$.

For all $v \in V$,

$$\left| \left\langle \frac{1}{\sigma_n} A(\sigma_n u_n) - \alpha u, v \right\rangle \right|$$

$$\leq \sum_{i=1}^{N} \int_{\Omega} \left| \frac{1}{\sigma_n} a_i(\cdot, \sigma_n u_n, \sigma_n \nabla u_n) - \left[\sum_{j=1}^{N} a_{ij} \, \partial_j u + a_{i0} u \right] \right| |\partial_i v|$$

$$+ \int_{\Omega} \left| \frac{1}{\sigma_n} a_0(\cdot, \sigma_n u_n, \sigma_n \nabla u_n) - \left[\sum_{j=1}^{N} a_{0j} \, \partial_j u + d_0 u \right] \right| |v|$$

$$\leq \sum_{i=1}^{N} \left(\int_{\Omega} \left| \frac{1}{\sigma_n} a_i(\cdot, \sigma_n u_n, \sigma_n \nabla u_n) - \left[\sum_{j=1}^{N} a_{ij} \, \partial_j u + a_{i0} u \right] \right|^2 \right)^{\frac{1}{2}} \|\partial_i v\|_{L^2(\Omega)}$$

$$+ \left(\int_{\Omega} \left| \frac{1}{\sigma_n} a_0(\cdot, \sigma_n u_n, \sigma_n \nabla u_n) - \left[\sum_{j=1}^{N} a_{0j} \, \partial_j u + d_0 u \right] \right|^2 \right)^{\frac{1}{2}} \|v\|_{L^2(\Omega)}$$

$$\leq C \left\{ \sum_{i=0}^{N} \left(\int_{\Omega} \left| \frac{1}{\sigma_n} a_i(\cdot, \sigma_n u_n, \sigma_n \nabla u_n) \right. \right. \right.$$

$$\left. \left. \left. - \left[\sum_{j=1}^{N} a_{ij} \, \partial_j u + a_{i0} \, u \right] \right|^2 \right)^{\frac{1}{2}} \right\} \|v\|,$$

with $C = C(\Omega) > 0$. Hence,

$$\left\| \frac{1}{\sigma_n} A(\sigma_n u_n) - \alpha u \right\|$$

$$\leq C \left\{ \sum_{i=0}^{N} \left(\int_{\Omega} \left| \frac{1}{\sigma_n} a_i(\cdot, \sigma_n u_n, \sigma_n \nabla u_n) - \left[\sum_{j=1}^{N} a_{ij} \, \partial_j u + a_{i0} \, u \right] \right|^2 \right)^{\frac{1}{2}} \right\}$$

$$\tag{6.182}$$

for all $n \in \mathbb{N}$. From (6.175), it follows that

$$\lim_{n \to \infty} \frac{1}{\sigma_n} a_i(x, \sigma_n u_n(x), \sigma_n \nabla u_n(x)) = a_{i0}(x) u(x) + \sum_{j=1}^{N} a_{ij}(x) \, \partial_j u(x),$$

for a.e. $x \in \Omega$.

On the other hand, using (6.170) and (6.174), it follows that, for $0 \leq i \leq N$,

$$\left| \frac{1}{\sigma_n} a_i(x, \sigma_n u_n(x), \sigma_n \nabla u_n(x)) - \left[a_{i0}(x) u(x) + \sum_{j=1}^{N} a_{ij}(x) \partial_j u(x) \right] \right|$$

$$\leq \frac{M}{\sigma_n}\left[|\sigma_n u_n| + |\sigma_n \nabla u_n|\right] + M\left(|u| + \sum_{j=1}^{N}|\partial_j u|\right)$$

$$\leq CM(|g| + |u| + |\nabla u|)$$

where C is a positive constant. Because $|g| + |u| + |\nabla u| \in L^2(\Omega)$,

$$\lim_{n\to\infty}\int_\Omega \left|\frac{1}{\sigma_n}a_i(x,\sigma_n u_n(x),\sigma_n \nabla u_n(x))\right.$$

$$\left. - \left[a_{i0}(x)u(x) + \sum_{j=1}^{N}a_{ij}(x)\partial_j u(x)\right]\right|^2 dx = 0$$

for $i = 0, 1, \ldots, N$. This and (6.182) give (6.37). Similarly, we define

$$\langle f(u,\lambda), v\rangle = \int_\Omega \partial_u b(x,0,\lambda)\, u(x)\, v(x)\, dx, \ u, v \in V. \tag{6.183}$$

(6.177) and (6.178) imply that, for $x \in \Omega, u \in \mathbb{R}$,

$$\begin{aligned}
|\partial_u b(x,0,\lambda)u| &= \lim_{\sigma\to 0^+}\left|\frac{1}{\sigma}b(x,\sigma u,\lambda)\right| \\
&\leq \lim_{\sigma\to 0^+}\left\{C(\lambda)\left[K(x)\frac{|\sigma u|}{\sigma} + L(x)\frac{|\sigma u|^{q-1}}{\sigma}\right]\right\} \\
&= \lim_{\sigma\to 0^+}\left\{C(\lambda)\left[K(x)|u| + L(x)|u|^{q-1}\sigma^{q-2}\right]\right\} \\
&\leq C(\lambda)[K(x)|u| + L(x)|u|^{q-1}] \text{ (because } q \geq 2\text{)}.
\end{aligned}$$

For the particular case where $u = 1$,

$$|\partial_u b(x,0,\lambda)| \leq C(\lambda)[K(x) + L(x)], \ x \in \Omega. \tag{6.184}$$

Hence, $\partial_u b(\cdot,0,\lambda) \in L^{\frac{q}{q-2}}(\Omega)$, showing that the integral in (6.183) is well defined and, therefore, f maps $V \times \mathbb{R}$ into V with

$$\begin{aligned}
\|f(u,\lambda)\| &\leq \|\partial_u b(\cdot,0,\lambda)\|_{L^{\frac{q}{q-2}}(\Omega)}\|u\|_{L^q(\Omega)} \\
&\leq C\|\partial_u b(\cdot,0,\lambda)\|_{L^{\frac{q}{q-2}}(\Omega)}\|u\|, \ \forall u \in V, \lambda \in \mathbb{R}.
\end{aligned} \tag{6.185}$$

Thus, f is a bounded, linear mapping with respect to u. If

$$u_n \rightharpoonup u \text{ in } V, \ \lambda_n \to \lambda \text{ in } \mathbb{R},$$

then, it follows that

$$\lim_{n\to\infty}\int_\Omega \left|\frac{1}{\sigma_n}b(x,\sigma_n u_n(x),\lambda_n) - \partial_u b(x,0,\lambda)u(x)\right|^{\frac{q}{q-1}} dx = 0. \tag{6.186}$$

By a straightforward calculation, we see that

$$\left\|\frac{1}{\sigma_n}B(\sigma_n u_n, \lambda_n) - f(u,\lambda)\right\|$$

$$\leq C\left(\int_\Omega \left|\frac{1}{\sigma_n}b(x,\sigma_n u_n(x),\lambda_n) - \partial_u b(x,0,\lambda)u(x)\right|^{\frac{q}{q-1}} dx\right)^{\frac{q-1}{q}}$$

$(C = C(\Omega)$ is a fixed positive constant). This and (6.186) imply that

$$\lim_{n\to\infty}\left\|\frac{1}{\sigma_n}B(\sigma_n u_n, \lambda_n) - f(u,\lambda)\right\| = 0,$$

and (6.39) holds.

Let $J : V \to [0,\infty]$ be the homogenization of j with $p = 2$, i.e., J is a convex, lower semicontinuous functional that satisfies (6.45)-(6.46). With these settings, we see that the homogenized variational inequality corresponding to (6.168) is of the form,

$$\begin{cases} \langle \alpha u - f(u,\lambda), v - u\rangle + J(v) - J(u) \geq 0, \ \forall v \in V, \\ u \in V \end{cases} \tag{6.187}$$

with α and f given by (6.181) and (6.183).

We may now use Theorem 6.4 to obtain bifurcation properties of (6.168) by studying the homogenized variational inequality (6.187). Depending on j and J, we can apply the results in Section 6.3 or 6.4 to compute the degrees of the operators associated with (6.187). For instance, we consider the case where j is given by (6.238), as in Example 6.6 with $1 < \gamma < 2$ (the other cases can be handled similarly). We assume that $b(x,u,0) = 0$ for a.e. $x \in \Omega$, all $u \in \mathbb{R}$, and that $\partial_u b(x,0,\lambda)$ is of the form

$$\partial_u b(x,0,\lambda) = \lambda p(x), \ x \in \Omega, \ \lambda \in \mathbb{R},$$

where $p \in L^{\frac{q}{q-2}}(\Omega)$ and $p(x) > 0$ for a.e. $x \in \Omega$. We have the following result.

Corollary 6.10 *Let $\lambda_0 \in \mathbb{R}$ be an eigenvalue of the linear equation*

$$\begin{cases} \int_\Omega \left[\sum_{i,j=1}^N a_{ij}\,\partial_j u\,\partial_i v + \sum_{i=1}^N (a_{i0}\,u\,\partial_i v + a_{0i}\,\partial_i u\,v) + d_0 uv\right] dx \\ \quad -\alpha \int_\Omega puv dx = 0, \ \forall v \in H_0^1(\Omega), \\ u \in H_0^1(\Omega). \end{cases} \tag{6.188}$$

Suppose that there exists an eigenvector u_0 of (6.188) corresponding to $\lambda = \lambda_0$, such that $u_0(x) > 0$ for a.e. $x \in \Omega$. Then, $(0,\lambda_0)$ is a bifurcation point of (6.168), and, for all $\lambda > \lambda_0$, $(0, \lambda)$ is not a bifurcation point of (6.168). Moreover, there exists an unbounded branch of solutions of (6.168) bifurcating from $\{0\} \times [0, \lambda_0]$.

Proof. Note that, with α given by (6.181), for $u, v \in V$,

$$\langle \alpha^* u, v \rangle = \langle \alpha v, u \rangle = \int_\Omega \left\{ \sum_{i,j=1}^N a_{ij}(x)\, \partial_i u(x)\, \partial_j v(x) + \sum_{i=1}^N [a_{i0}(x)\, v(x)\, \partial_i u(x) \right.$$

$$\left. + a_{0i}(x)\, \partial_i v(x)\, u(x)] + d_0(x) u(x) v(x) \right\} dx.$$

Because $f(u, \lambda)$ is symmetric with respect to u and homogeneous with respect to λ, $f = f^* = g$. Hence, (6.188) is the same as $\alpha^* u - f^*(u, \lambda) = 0$. As in Example 6.6, we know that

$$D(J) = \ker J = K_1 = \{u \in V : u \geq 0 \text{ a.e. on } \Omega\}.$$

Hence, $u_0 \in K(\lambda_0)$. Moreover, for $u \in D(J) \setminus \{0\}$,

$$\langle g(u_0, \lambda_0), u \rangle = \lambda_0 \int_\Omega p u_0 u > 0,$$

i.e., (6.156) holds. Because $B(\cdot, 0) = 0$, we see that $A - B(\cdot, 0) = A$ is strictly monotone on V. Our conclusion follows from Theorem 6.9 (b). ∎

Note that we have similar results if j represents an obstacle in the domain Ω or on the boundary $\partial\Omega$ as in Example 6.5. The above analysis can also be generalized in a straightforward way to variational inequalities containing elliptic operators of order $2m$ ($m \geq 1$).

Example 6.4 In this example, we reconsider variational inequalities associated with the Stokes problem. Now, we assume that, for all points of a subdomain $\Omega_0 \subset \Omega$, the flow is only in the positive directions of all three axes x, y, and z (see Figure 6.3). The set K of admissible velocity fields, therefore is given by

$$K = \{w \in V : w_i(x) \geq 0 \text{ for a.e. } x \in \Omega_0, i = 1, 2, 3\}. \tag{6.189}$$

Because K is a cone, it follows that $K_0 = K$. We also assume (4.63), where k is a positive matrix on Ω_0 in the sense that, for all $i, j = 1, 2, 3$,

$$k_{ij}(x) > 0 \text{ for a.e. } x \in \Omega_0. \tag{6.190}$$

A direct application of Theorem 6.9 yields the following result:

Corollary 6.11 *Assume that $\lambda_0 > 0$ is an eigenvalue of the equation (4.64) with an eigenvector u, such that*

$$u_i(x) > 0 \text{ for a.e. } x \in \Omega_0. \tag{6.191}$$

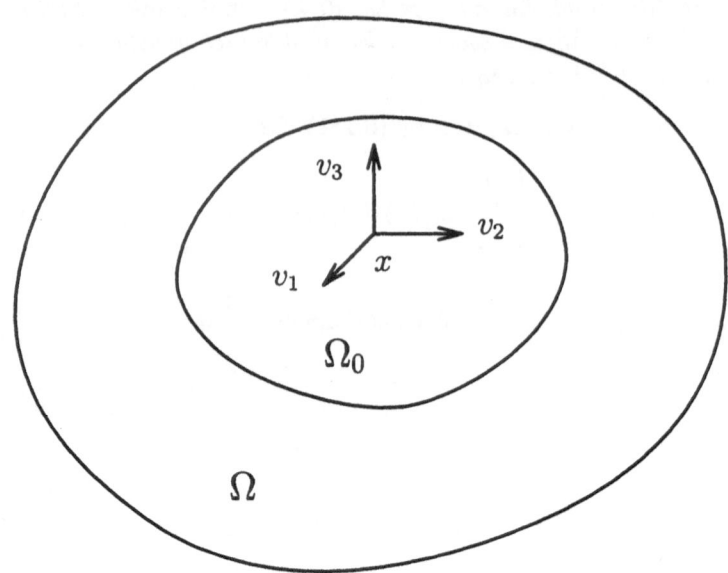

FIGURE 6.3. Example 6.4.

*Then, there exists an unbounded branch of nontrivial solutions of (4.61)
(with K given by (6.189)) that bifurcates from $\{0\} \times [0, \lambda_0]$.*

To prove this result, we observe that (4.64) is the same as $\alpha^*(u) -
f^*(u, \lambda) = 0$. Also, because $f^* = g$, it follows from (6.191) that

$$
\begin{aligned}
\langle g(u, \lambda_0), v \rangle &= \lambda_0 \int_\Omega v^T k^T u \\
&= \int_\Omega \sum_{i,j} k_{ij} u_j v_i > 0,
\end{aligned}
$$

for all $v \in K \setminus \{0\}$.

6.4.3 Some corollaries

Next, we consider some further consequences of Theorem 6.8. First, we
need a lemma.

Lemma 6.12
   (a) If

$$
\ker[\alpha^* - f^*(\cdot, \lambda_0)] \cap (\ker J)^I \neq \emptyset \tag{6.192}
$$

*($(\ker J)^I$ is the demi-interior part of $\ker J$, i.e., the set of all demi-interior
points of $\ker J$), then, $u \in V$ is a solution of the homogenized variational*

inequality (6.48) (with $\lambda = \lambda_0$) if and only if $u \in \ker J$ and u is a solution of the corresponding equation

$$\langle \alpha(u) - f(u, \lambda_0), v \rangle = 0, \ \forall v \in V. \tag{6.193}$$

(b) Suppose the following: λ_0 is an eigenvalue of (6.48) with an eigenvector

$$u_1 \in (\ker J)^I \tag{6.194}$$

(respectively, $u_1 \in D(J)^I$ which is also a solution of (6.193), and $D(J)$ is closed in V),

$$J \not\equiv 0 \quad (respectively, \ D(J) \neq V), \tag{6.195}$$

$$\lambda_0 \ \text{is a simple eigenvalue of (6.193)}, \tag{6.196}$$

(respectively, $D(J) \neq V$), and (6.192) is satisfied. Then, λ_0 is a simple eigenvalue of (6.48).

(c) Assume that α and $f(\cdot, \lambda_0)$ are self-adjoint operators (i.e., $\alpha = \alpha^*$ and $f(\cdot, \lambda_0) = f^*(\cdot, \lambda_0)$). Suppose that $J \not\equiv 0$ on V and that λ_0 is a simple eigenvalue of (6.193) with an eigenvector $u_0 \in (\ker J)^I$. Then, λ_0 is a simple eigenvalue of (6.48).

Proof. (a) Suppose $u_0 \in \ker [\alpha^* - f^*(\cdot, \lambda_0)] \cap (\ker J)^I$. Then,

$$\langle \alpha^* u_0 - f^*(u_0, \lambda_0), v \rangle = 0, \ \forall v \in V, \tag{6.197}$$

and there exists a subset D, dense in V, such that, for each $v \in D$, there exists $\epsilon > 0$, such that

$$J(u_0 + \epsilon v) = 0. \tag{6.198}$$

Now let u_1 be a solution of (6.48) with $\lambda = \lambda_0$, i.e.,

$$\langle \alpha u_1 - f(u_1, \lambda_0), v - u_1 \rangle + J(v) - J(u_1) \geq 0, \ \forall v \in V. \tag{6.199}$$

Let $v \in D$, and choose $\epsilon > 0$, such that (6.198) is satisfied. Letting $v = u_1 + u_0 + \epsilon v$ in (6.199),

$$\langle \alpha u_1 - f(u_1, \lambda_0), u_0 + \epsilon v \rangle + J(u_1 + u_0 + \epsilon v) - J(u_1) \geq 0.$$

It follows from (6.198), as in the proof of Theorem 6.8 (cf. (6.130)), that $J(u_1 + u_0 + \epsilon v) - J(u_1) \leq 0$. Hence,

$$\langle \alpha u_1 - f(u_1, \lambda_0), u_0 + \epsilon v \rangle \geq 0.$$

Using (6.197),

$$\begin{aligned} 0 \ &\leq \ \langle \alpha u_1 - f(u_1, \lambda_0), u_0 \rangle + \epsilon \langle \alpha u_1 - f(u_1, \lambda_0), v \rangle \\ &= \ \langle \alpha^* u_0 - f^*(u_0, \lambda_0), u_1 \rangle + \epsilon \langle \alpha u_1 - f(u_1, \lambda_0), v \rangle \\ &= \ \epsilon \langle \alpha u_1 - f(u_1, \lambda_0), v \rangle. \end{aligned}$$

Hence, because $\overline{D} = V$,

$$\langle \alpha u_1 - f(u_1, \lambda_0), v \rangle \geq 0, \ \forall v \in V.$$

Replacing v by $-v$,

$$\langle \alpha u_1 - f(u_1, \lambda_0), v \rangle = 0, \ \forall v \in V,$$

i.e., u_1 is a solution of (6.193). Now, (6.193) implies that

$$\langle \alpha u_1 - f(u_1, \lambda_0), v - u_1 \rangle = 0, \ \forall v \in K. \tag{6.200}$$

Substituting this in (6.199), $J(v) - J(u_1) \geq 0, \ \forall v \in K$. Because $J \geq 0$ and $J(0) = 0$, $J(u_1) = 0$, i.e., $u_1 \in \ker J$.

The converse is clear.

(b) Suppose that λ_0 satisfies (6.194), (6.195), (6.196), and (6.192). We prove that λ_0 is a simple eigenvalue of (6.48). Let u_2 be another eigenvector of (6.48) associated with λ_0.

Consider the case where $u_1 \in (\ker J)^I$. Then, u_1 is also a solution of (6.193). In fact, let $w \in D$, and choose $\epsilon > 0$, such that $u_1 + \epsilon w \in \ker J$. Hence,

$$\langle \alpha u_1 - f(u_1, \lambda_0), v - u_1 \rangle + J(v) - J(u_1) \geq 0, \ \forall v \in V.$$

Because $J(u_1 + \epsilon w) = J(u_1) = 0$, letting $v = u_1 + \epsilon w$,

$$\epsilon \langle \alpha u_1 - f(u_1, \lambda_0), w \rangle \geq 0, \ \forall w \in D.$$

Hence $\langle \alpha u_1 - f(u_1, \lambda_0), w \rangle \geq 0, \ \forall w \in D$. Because $\overline{D} = V$, u_1 is a solution of (6.193).

By (a), we see that (6.192) implies that u_2 is also a solution of (6.193). By (6.196),

$$u_2 = t u_1 \text{ for some } t \in \mathbb{R} \setminus \{0\}. \tag{6.201}$$

Suppose $t < 0$. By the homogeneity of α and $f(\cdot, \lambda_0)$ (of degree 1) and of J (of degree 2), we see that $|t|^{-1} u_2 = -u_1$ is also a solution of both (6.48) and (6.193). Hence, for all $v \in V$, by (6.193) for $-u_1$,

$$\begin{aligned} 0 \ &\leq \ \langle \alpha(-u_1) - f(-u_1, \lambda_0), v + u_1 \rangle + J(v) - J(-u_1) \\ &= \ J(v) - J(-u_1) \ (\text{because } \alpha(-u_1) - f(-u_1, \lambda_0) = 0). \end{aligned}$$

Letting $v = 0$, $J(-u_1) \leq 0$. Hence, $J(-u_1) = 0$, and $-u_1 \in \ker J$. For $w \in D$ and $\epsilon > 0$, such that $u_1 + \epsilon w \in \ker J$, by the convexity of $\ker J$,

$$\frac{1}{2}(-u_1) + \frac{1}{2}(u_1 + \epsilon w) \in \ker J.$$

Hence $\frac{\epsilon}{2} w \in \ker J$, and, therefore, $w \in \ker J$, $\forall w \in D$.

Because $D \subset \ker J$, $V = \ker J$, i.e., $J \equiv 0$ on V, contradicting the assumption that $J \not\equiv 0$ on V.

This contradiction proves that $t > 0$. Therefore λ_0 is a simple eigenvalue of (6.48).

Now, consider the case where $u_1 \in [D(J)]^I$ is also a solution of (6.193). Using similar arguments, we see that (6.192) implies that u_2 is a solution of (6.193). Hence, we also have (6.201).

Moreover, because u_2 is a solution of (6.48), $u_2 \in D(J)$. Hence, $\xi u_2 \in D(J)$, $\forall \xi \geq 0$. If $t < 0$, then, $-u_1 = |t|^{-1} u_2 \in D(J)$.

Let $w \in D$ and $\epsilon > 0$, such that $u_1 + \epsilon w \in D(J)$. Because $D(J)$ is a convex cone, $\frac{\epsilon}{2} w = -\frac{1}{2} u_1 + \frac{1}{2}(u_1 + \epsilon w) \in D(J)$, and $w \in D(J)$.

Hence, $D \subset D(J)$, and, again, $V = D(J)$, which contradicts the assumption that $D(J) \neq V$. Hence, $t > 0$, and λ_0 is simple.

(c) Now, assume that α and $f(\cdot, \lambda_0)$ are self-adjoint operators and that λ_0 is an eigenvalue of (6.193) with an eigenvector $u_0 \in (\ker J)^I$. Because $\alpha = \alpha^*$, $f(\cdot, \lambda_0) = f^*(\cdot, \lambda_0)$, (6.192) is satisfied, and $u_0 \in \ker [\alpha^* - f^*(\cdot, \lambda_0)] \cap (\ker J)^I$. Moreover, (6.194), (6.195), and (6.196) also hold. Hence, by (b), λ_0 is a simple eigenvalue of (6.48). ∎

From this lemma, we have the following corollary of Theorem 6.8.

Corollary 6.13

(a) Suppose that λ_0 is an eigenvalue of (6.48) that satisfies (6.194), (6.195), (6.196), (6.192), and (6.117). Then, there exists a continuum of nontrivial solutions of (6.8), emanating from $\{0\} \times [0, \lambda_0]$ and satisfying the alternative in Theorem 6.4.

(b) Assume that α and $f(\cdot, \lambda_0)$ are self-adjoint, $h \equiv 0$, and that λ_0 is a simple eigenvalue of (6.48) with an eigenvector $u_1 \in \ker J$, which is also a solution of (6.193). Then, we arrive at the conclusion in (a).

(c) Assume that α and $f(\cdot, \lambda_0)$ are self-adjoint and that $h \equiv 0$. If λ_0 is a simple eigenvalue of (6.193) with an eigenvector $u_1 \in (\ker J)^I$, then, we arrive at the conclusion in (a).

We specialize to the case where V is a Hilbert space, $\alpha = I$ is the identity mapping, $j = I_K$ (resp. $J = I_{K_0}$) is the indicator function of a closed convex set K (respectively the support cone K_0 of K), and $f(u, \lambda)$ is homogeneous and symmetric with respect to u and homogeneous of degree γ with respect to λ. In this case, $h \equiv 0$, and $D(J) = \ker J = K_0$. Corollary 6.13 (c), then, reduces to Corollary 4.6.

From Lemma 6.12 and Theorem 6.8 (b), we have the following result.

Corollary 6.14

(a) Suppose that

$$[(\ker J)^I \setminus (-\ker J)] \cap \ker [\alpha^* - f^*(\cdot, \lambda_0)] \neq \emptyset, \tag{6.202}$$

and for each $u_1 \in \ker J \setminus \{0\}$ satisfying the linear equation (6.193), there exists $u_0 \in K(\lambda_0)$, such that (6.117) holds. Then, we have the conclusion of Theorem 6.8.

(b) Suppose that α and $f(\cdot, \lambda_0)$ are self-adjoint and $h \equiv 0$. If (6.202) is satisfied, then, we have the conclusion of Theorem 6.8.

Proof. (a) Using (6.202), we see that

$$\ker [\alpha^* - f^*(\cdot, \lambda_0)] \cap (\ker J)^I \neq \emptyset.$$

Hence, (6.192) holds. Moreover, there exists

$$\bar{u} \in \ker [\alpha^* - f^*(\cdot, \lambda_0)] \cap (\ker J)^I,$$

such that $-\bar{u} \notin \ker J$. Hence, $\bar{u} \in K(\lambda_0)$, and $-\bar{u} \notin K(\lambda_0)$, i.e., $K(\lambda_0)$ is not symmetric. Let u_1 be an eigenvector of (6.48) corresponding to λ_0. By Lemma 6.12, $u_1 \in \ker J \setminus \{0\}$, and u_1 satisfies (6.193). By hypothesis, there exists $u_0 \in K(\lambda_0)$, such that (6.117) is satisfied. Our conclusion now follows from Theorem 6.8 (b).

(b) Now, suppose that

$$\alpha = \alpha^*, \; f(\cdot, \lambda_0) = f^*(\cdot, \lambda_0), \; h = 0,$$

and (6.202) is satisfied. Note that, in this case,

$$K(\lambda_0) = \ker [\alpha - f(\cdot, \lambda_0)] \cap \ker J.$$

It follows that, if $u_1 \in \ker J \setminus \{0\}$ satisfies (6.193), then, $u_1 \in K(\lambda_0)$. Therefore, we can choose $u_0 = u_1$, for which

$$\langle \alpha u_0, u_1 \rangle = \langle \alpha u_1, u_1 \rangle > 0$$

by the coerciveness of α. We have (6.117) and our conclusion follows from (a).   ∎

We note that (6.202) is equivalent to the following condition:

$$\begin{cases} \ker [\alpha^* - f^*(\cdot, \lambda_0)] \cap (\ker J)^I \neq \emptyset, \\ \ker J \neq V. \end{cases} \tag{6.203}$$

In fact, if (6.202) is satisfied, then, it follows immediately that

$$\ker [\alpha^* - f^*(\cdot, \lambda_0)] \cap (\ker J)^I$$

is not empty because it contains the set in (6.202). If $\ker J = V$, i.e., $J \equiv 0$, then, $V = \ker J = (\ker J)^I = -\ker J$. Therefore,

$$[(\ker J)^I \setminus (-\ker J)] \cap \ker [\alpha^* - f^*(\cdot, \lambda_0)] \subset (\ker J)^I \setminus (-\ker J)$$

$$= \emptyset,$$

contradicting (6.202). Hence, ker $J \neq V$, and we have (6.203).

Conversely, suppose (6.203) holds, and let

$$u \in \ker [\alpha^* - f^*(\cdot, \lambda_0)] \cap (\ker J)^I.$$

Then, $-u \notin \ker J$, because, otherwise, $J(-u) = 0$. Because $u \in (\ker J)^I$, there exists $D \subset V$, $\overline{D} = V$, such that, for each $v \in V$, we can choose $\epsilon = \epsilon(v) > 0$, such that

$$u + \epsilon v \in \ker J,$$

i.e., $J(u + \epsilon v) = 0$. By the homogeneity and the convexity of J,

$$
\begin{aligned}
0 &\leq \left(\frac{\epsilon}{2}\right)^p J(v) = J\left(\frac{\epsilon}{2}v\right) \\
&= J\left[\frac{1}{2}(-u) + \frac{1}{2}(u + \epsilon v)\right] \\
&\leq \frac{1}{2}J(-u) + \frac{1}{2}J(u + \epsilon v) \\
&= 0.
\end{aligned}
$$

Hence, $J(v) = 0$, $\forall v \in D$. Then, $D \subset \ker J$, and $V = \overline{D} = \ker J$, contradicting the second condition in (6.203). Therefore, $u \notin (-\ker J)$, and

$$u \in [(\ker J)^I \setminus (-\ker J)] \cap \ker [\alpha^* - f^*(\cdot, \lambda_0)],$$

proving (6.202) and establishing the equivalence between (6.202) and (6.203).

In view of (6.203), we note that, whenever ker $J \neq V$ and α, $f(\cdot, \lambda_0)$ are self-adjoint, then, Corollary 6.14 (b) is stronger than Corollary 6.13 (c). In the case where ker $J = V$, (6.48) is equivalent to (6.193), and the global bifurcation property of (6.8) is given by the theorems in Section 6.3 (Corollaries 6.5 and 6.6). Corollaries 6.13 and 6.14 give us the corresponding results for the case ker $J \neq V$.

In the following section, we consider some further applications of the above abstract results to various kinds of variational inequalities.

6.5   More applications and examples

Example 6.5   In this example, we consider a bifurcation problem for a second-order, elliptic, boundary value problem with a lower dimensional obstacle on the boundary.

Let Ω be a bounded domain in $\mathbb{R}^N$ ($N \geq 1$) with a smooth boundary. Let $a_{ij}, a_i, a_0 \in L^\infty(\Omega)$, $1 \leq i, j \leq N$ such that

$$
\begin{aligned}
&\sum_{i,j=1}^N a_{ij}(x)\xi_i\xi_j \geq C_0|\xi|^2, \\
&a_0(x) \geq C_1, \ \forall \xi \in \mathbb{R}^N, \ \text{a.e.} \ x \in \mathbb{R},
\end{aligned}
\tag{6.204}
$$

C_0 and C_1 are given positive constants. Let $g : \Omega \times \mathbb{R} \to \mathbb{R}$ be a bounded Carathéodory function. We consider the following nonlinear eigenvalue problem:

$$-\sum_{i,j=1}^{N} \partial_j(a_{ij}\, \partial_i u) + \sum_{i=1}^{N} a_i\, \partial_i u + a_0 u = \lambda g(x,u)u \text{ in } \Omega \qquad (6.205)$$

with the following unilateral boundary condition

$$\begin{cases} u - \psi \geq 0, \\ \partial_L u \geq 0, \\ (u - \psi)\partial_L u = 0 \text{ on } \partial\Omega, \end{cases} \qquad (6.206)$$

where $\psi \in C(\partial\Omega)$ is a given function and

$$\partial_L u = -\sum_{i,j=1}^{N} a_{ij}\, \partial_i u\, n_j$$

is the normal derivative associated with the operator

$$L = -\sum_{i,j=1}^{N} \partial_j(a_{ij}\partial_i(\cdot)) + \sum_{i=1}^{N} a_i\partial_i(\cdot) + a_0(\cdot),$$

(n is the outward normal vector on $\partial\Omega$). A particular but interesting case of this problem is where $\psi = 0$.

As usual, the weak formulation of (6.205)-(6.206) is the following variational inequality:

$$\begin{cases} \displaystyle\int_{\Omega}\left[\sum_{i,j=1}^{N} a_{ij}\, \partial_i u\, \partial_j(v - u) + \sum_{i=1}^{N} a_i\, \partial_i u\,(v - u) + a_0 u(v - u)\right] \\ \qquad \geq \lambda \displaystyle\int_{\Omega} g(\cdot, u)u(v - u), \; \forall v \in K, \\ u \in K, \end{cases}$$
$$(6.207)$$

or in the operator form,

$$\begin{cases} \langle Au, v - u\rangle - \langle B(u, \lambda), v - u\rangle \geq 0, \; \forall v \in K, \\ u \in K, \end{cases} \qquad (6.208)$$

where
- $V = H^1(\Omega)$ with the usual norm $\|\cdot\|$ and inner product $\langle\cdot,\cdot\rangle$,

$$K = \{u \in V : u \geq \psi \text{ on } \partial\Omega\}, \qquad (6.209)$$

- $A : V \to V$,

$$\langle Au, v \rangle = \int_{\Omega} \left[\sum_{i,j=1}^{N} a_{ij}(x) \, \partial_i u(x) \, \partial_j v(x) + \sum_{i=1}^{N} a_i(x) \, \partial_i u(x) \, v(x) \right.$$

$$\left. + a_0 u(x) v(x) \right] dx \qquad (6.210)$$

for all $u, v \in V$,
- $B : V \times \mathbb{R} \to V$,

$$\langle B(u, \lambda), v \rangle = \int_{\Omega} \lambda g(x, u(x)) u(x) v(x) dx, \; u, v \in V, \lambda \in \mathbb{R}, \qquad (6.211)$$

- $\psi \le 0$ on $\partial \Omega$ (so that 0 is always a (trivial) solution of (6.208)).

Now, we consider some properties of these mappings.

It is clear that A is bounded and linear from V to V, and B is a completely continuous mapping from $V \times \mathbb{R}$ to V. We assume that A is coercive on V in the sense that we have the following inequality

$$\langle Au, u \rangle \ge C \|u\|^2, \; \forall u \in V, \qquad (6.212)$$

for some constant $C > 0$. Then,

$$\int_{\Omega} \left(\sum_{i,j} a_{ij} \, \partial_i u \, \partial_j u + a_0 u^2 \right) \ge C_0 \int_{\Omega} |\nabla u|^2 + C_1 \int_{\Omega} |u|^2,$$

and

$$\sum_i \left| \int_{\Omega} a_i \partial_i u u \right| \le \left(\sum_i \|a_i\|_{L^\infty(\Omega)} \right) \left(\int_{\Omega} |\nabla u|^2 \right)^{\frac{1}{2}} \left(\int_{\Omega} |u|^2 \right)^{\frac{1}{2}}.$$

Hence, (6.212) is satisfied, whenever

$$C_0 C_1 > \frac{1}{4} \left(\sum_i \|a_i\|_{L^\infty(\Omega)} \right)^2. \qquad (6.213)$$

Furthermore, suppose that

$$\lim_{u \to \infty} g(x, u) = b(x) \; \text{ for a.e. } \; x \in \Omega, \qquad (6.214)$$

where $b \in L^\infty(\Omega)$ is a given function. From this, we see that, if $u_n \rightharpoonup u$ in $H^1(\Omega)$, $\sigma_n \to 0^+$, and $\lambda_n \to \lambda$ in $\mathbb{R}$, then,

$$\frac{1}{\sigma_n} B(\sigma_n u_n, \lambda_n) \to f(u, \lambda) \; \text{ in } \; H^1(\Omega), \qquad (6.215)$$

with

$$\langle f(u, \lambda), v \rangle = \lambda \int_{\Omega} buv, \ \forall u, v \in H^1(\Omega). \tag{6.216}$$

Now, we show that the support cone of K is given by

$$K_0 = \{ u \in H^1(\Omega) : u(x) \geq 0 \text{ a.e. on } S \}, \tag{6.217}$$

with $S = S_\psi = \{ x \in \partial\Omega : \psi(x) = 0 \}$. In fact, if $u \in K$, then, $u(x) \geq \psi(x) = 0$ a.e. on S. Hence $tu \geq 0$ a.e. on S, i.e., $tu \in K_0$, $\forall t \geq 0$, which means that

$$\bigcup_{t \geq 0} tK \subset K_0. \tag{6.218}$$

Because K_0 is closed, $\overline{\bigcup_{t \geq 0} tK} \subset K_0$. To prove the converse, let $u \in K_0$. We choose $\phi \in C^1(\overline{\Omega})$, such that $\phi \geq 1$ on S. Now, let $\epsilon > 0$. Because $u \geq 0$ a.e. on S, one can choose $\overline{u} \in C^1(\overline{\Omega})$, such that $\overline{u} \geq 0$ on S, and $\|u - \overline{u}\| \leq \epsilon$. Let $u_\epsilon = \overline{u} + \epsilon\phi$. Then, $u_\epsilon \in C^1(\overline{\Omega})$, and because $\overline{u} \geq 0$ on S,

$$u_\epsilon \geq \epsilon\phi \geq \epsilon \text{ on } S.$$

Hence, there exists $\delta > 0$, such that

$$u_\epsilon(x) > \frac{\epsilon}{2}, \ \forall x \in B_{\delta, \partial\Omega}(S) (= \{ x \in \partial\Omega : \text{dist}(x, S) < \delta \}). \tag{6.219}$$

Because $\psi(x) < 0$ for x in the compact set $\partial\Omega \setminus B_{\delta, \partial\Omega}(S)$ and ψ is continuous,

$$\psi(x) \leq -\eta, \ x \in \partial\Omega \setminus B_{\delta, \partial\Omega}(S) \tag{6.220}$$

for some $\eta > 0$. From (6.219), $u_\epsilon(x) > \frac{\epsilon}{2} > 0$, and, then,

$$\frac{\eta}{\|u_\epsilon\|_\infty} u_\epsilon(x) > 0 \geq \psi(x) \text{ for } x \in B_{\delta, \partial\Omega}(S).$$

On the other hand,

$$\frac{\eta}{\|u_\epsilon\|_\infty} u_\epsilon(x) \geq - \left| \frac{\eta}{\|u_\epsilon\|_\infty} u_\epsilon(x) \right| \geq -\eta \geq \psi(x)$$

for $x \in \partial\Omega \setminus B_{\delta, \partial\Omega}(S)$. Hence,

$$\frac{\eta}{\|u_\epsilon\|_\infty} u_\epsilon \geq \psi \text{ on } \partial\Omega,$$

i.e., $\frac{\eta}{\|u_\epsilon\|_\infty} u_\epsilon \in K$. Therefore,

$$u_\epsilon \in \frac{\|u_\epsilon\|_\infty}{\eta} K \subset \bigcup_{t \geq 0} tK.$$

Because $\|u - u_\epsilon\| \leq \|u - \bar{u}\| + \|\bar{u} - u_\epsilon\| \leq \epsilon(1 + \|\phi\|)$, we see that $u_\epsilon \to u$ in $H^1(\Omega)$ as $\epsilon \to 0$. Hence,

$$u \in \overline{\bigcup_{t \geq 0} tK}.$$

$K_0 \subset \overline{\bigcup_{t \geq 0} tK}$, which, together with (6.218), proves that K_0, given by (6.217), is the support cone of K.

Because A is linear and bounded, $A = \alpha$. From (6.215), (6.217), we see that the homogenized variational inequality corresponding to (6.208) is the following:

$$\begin{cases} \langle Au, v - u \rangle - \langle f(u, \lambda), v - u \rangle \geq 0, \ \forall v \in K_0 \\ u \in K_0, \end{cases} \tag{6.221}$$

with f and K_0 given by (6.216) and (6.217). On the other hand, $f(u, \lambda)$ is symmetric with respect to u, and, from (6.210), we see that the adjoint A^* of A is given by

$$\langle A^* u, v \rangle = \langle Av, u \rangle = \int_\Omega \left[\sum_{i,j} a_{ji} \, \partial_i u \, \partial_j v + \sum_i a_i \, u \, \partial_i v + a_0 uv \right]. \tag{6.222}$$

Hence, the linear equation associated with (6.221) is

$$\begin{cases} \langle Au, v \rangle - \lambda \int_\Omega buv = 0, \ \forall v \in H^1(\Omega), \\ u \in H^1(\Omega), \end{cases} \tag{6.223}$$

and the corresponding adjoint equation is

$$\begin{cases} \langle A^* u, v \rangle - \lambda \int_\Omega buv = 0, \ \forall v \in H^1(\Omega), \\ u \in H^1(\Omega), \end{cases} \tag{6.224}$$

where A^* is the second-order elliptic operator given by (6.222)((We see that 6.223) and (6.224) are usual second-order, linear, elliptic equations with Neumann boundary conditions.)

Applying Theorem 6.8 to this particular problem, we get the following result.

Corollary 6.15

(a) *Suppose that $\lambda_0 > 0$ is a simple eigenvalue of (6.221) with an associated eigenvector u_1. Moreover, assume that there exists a solution u_0 of (6.224) (with $\lambda = \lambda_0$), such that*

$$u_0 \geq 0 \quad on \ S_\psi \tag{6.225}$$

and

$$\int_\Omega bu_0u_1 > 0. \tag{6.226}$$

Then, there exists a global branch of nontrivial solutions of (6.207), which bifurcates from $\{0\} \times [0, \lambda_0]$ and satisfies the alternative in Theorem 6.4.

(b) Suppose that λ_0 is an eigenvalue of (6.221) and that there exists a solution $\bar{u}$ of (6.224) (with $\lambda = \lambda_0$) satisfying (6.225), but

$$\bar{u} \not\equiv 0 \quad on \ S_\psi. \tag{6.227}$$

Furthermore, assume that, for each eigenvector u_1 of (6.221) corresponding to λ_0, there exists a solution u_0 of (6.224) (with $\lambda = \lambda_0$), such that (6.225) and (6.226) are satisfied. Then, we also arrive at the conclusion of (a).

Proof. This is a direct corollary of Theorem 6.8. We just note that, in this case,

$$K(\lambda_0) = \{u \in K_0 : u \text{ satisfies } (6.48)\}.$$

For the proof of (b), we note that (6.227) implies that $K(\lambda_0)$ is not symmetric. ∎

Now we apply Corollary 6.13 to obtain some more specific conditions to guarantee the existence of global bifurcation branches for the variational inequality (6.207). We have the following result.

Corollary 6.16 *Suppose that λ_0 is a simple eigenvalue of the linear equation (6.223), which corresponds to an eigenvector u_1 satisfying (6.225).*
Furthermore, assume that there exists a solution u_0 of (6.224), such that

$$u_0 \geq \gamma > 0 \quad a.e. \ on \ S_\psi, \tag{6.228}$$

for some $\gamma > 0$, and that (6.226) holds.
Then we arrive at the conclusion of Corollary 6.15.

Proof. First, we verify that, if u_0 satisfies (6.228), then, $u_0 \in K_0^I$. In fact, we choose $D = C^1(\overline{\Omega})$, which is dense in $H^1(\Omega)$. For $w \in D$, there exists $M > 0$, such that

$$|w(x)| \leq M, \ \forall x \in \overline{\Omega}.$$

Hence, with $\epsilon = \frac{\gamma}{2M} > 0$,

$$u_0 + \epsilon w \geq \gamma - \frac{\gamma}{2M}M = \frac{\gamma}{2} > 0 \text{ for a.e. } x \in S.$$

Hence, $u_0 + \epsilon w \in K_0$.

This shows that u_0 is a demi-interior point of K_0. Corollary 6.16 now follows directly from Corollary 6.13 (a); we need only to notice that, in our problem, $J = I_{K_0}$. Hence,

$$D(J) = \ker J = K_0. \tag*{∎}$$

We remark that, for u_0 satisfying (6.228), in general, u_0 is not in the interior of K_0. Now we consider the case where A is symmetric. We see that A given by (6.210) is symmetric if and only if

$$a_{ij} = a_{ji}, \ a_i = 0 \ \text{on} \ \Omega, \ \forall i, j = 1, \ldots, N.$$

We also observe that, in this case, (6.213) is immediately satisfied. Hence, A is coercive. Note that $\ker J = K_0 \neq V$. Applying Corollaries 6.13 (b) and 6.14 (b), we obtain the following result.

Corollary 6.17 *Suppose that A is a self-adjoint operator.*
(a) If λ_0 is a simple eigenvalue of (6.221) with an eigenvector u_1 which is also a solution of the equation (6.223), then we arrive at the conclusion of Corollary 6.15.
(b) If λ_0 is an eigenvalue of (6.223) with an eigenvector u_1 satisfying (6.228), then, we arrive at the conclusion of Corollary 6.15.

We illustrate this result by the following simple semilinear ODE: Consider the equation,

$$- u'' + u = \lambda u \cos u, \ \text{on} \ (0, 1), \tag{6.229}$$

with the unilateral boundary condition,

$$\begin{cases} u(0), u(1) \geq 0, \\ u'(0) \leq 0 \leq u'(1), \\ u(0)u'(0) = u(1)u'(1) = 0. \end{cases} \tag{6.230}$$

(6.229)-(6.230) has the following weak formulation as a variational inequality:

$$\begin{cases} \displaystyle\int_0^1 [u'(v - u)' + u(v - u)] \geq \lambda \int_0^1 u \cos u \, (v - u), \ \forall u \in K, \\ u \in K, \end{cases} \tag{6.231}$$

with

$$K = \{u \in H^1(0, 1) : u(0), u(1) \geq 0\}.$$

(6.231) is of the form (6.207) with

$$\langle Au, v \rangle = \int_0^1 (u'v' + uv),$$

$$g(u) = \cos u, \ \langle B(u, \lambda), v \rangle = \lambda \int_0^1 (\cos u)uv, \ u, v \in H^1(0, 1),$$

and $\psi \equiv 0$. Hence, (6.212) is satisfied, and A is a self-adjoint operator on $H^1(0,1)$.

We have (6.214) with $b \equiv 1$ on $(0,1)$, and $K = K_0$ is a convex cone in $H^1(0,1)$. Therefore, the homogenized variational inequality (6.221) of (6.231) becomes

$$\begin{cases} \int_0^1 [u'(v-u)' + u(v-u)] \geq \lambda \int_0^1 u(v-u), \ \forall u \in K, \\ u \in K. \end{cases} \tag{6.232}$$

The linear equation associated with (6.232) is the following:

$$\begin{cases} \int_0^1 (u'v' + uv) = \lambda \int_0^1 uv, \ \forall u \in H^1(0,1), \\ u \in H^1(0,1), \end{cases}$$

or, equivalently,

$$\begin{cases} -u'' + (1-\lambda)u = 0 \ \text{on} \ (0,1), \\ u'(0) = u'(1) = 0. \end{cases} \tag{6.233}$$

It is easy to see that $\lambda_0 = 1$ is the first eigenvalue of (6.233), which is simple, and that the corresponding eigenspace is $\mathbb{R}$. This implies that $u_1 \equiv 1$ is in K and is an eigenvector of (6.232) with $\lambda_0 = 1$. Moreover, $u_1 \in \overset{\circ}{K}$. Therefore, we may apply Corollary 6.17 to obtain the following result:

$\lambda_0 = 1$ *is a bifurcation point of (6.231) (and, then, of (6.229)-(6.230)) and the corresponding global bifurcation branch satisfies the alternative in Theorem 6.4.*

Example 6.6 [Bifurcation for a second-order variational inequality containing a convex functional] We consider the bifurcation problem for the following semilinear equation containing a nonsmooth nonlinear term:

$$\begin{cases} \Delta u + \lambda u + q(u^-)^{\gamma-1} = 0 \ \text{in} \ \Omega, \\ u = 0 \ \text{on} \ \partial\Omega. \end{cases} \tag{6.234}$$

Here, Ω is a bounded domain in $\mathbb{R}^N$ ($N \geq 1$) with a smooth boundary, $\gamma > 1$, q is a given function on Ω, and u^- denotes the negative part of u,

$$u^- = \begin{cases} 0 & \text{if} \ u \geq 0, \\ -u & \text{if} \ u < 0. \end{cases}$$

We assume that

$$q \in L^\infty(\Omega) \ \text{and ess inf}_\Omega \ q > 0. \tag{6.235}$$

In the case where $N = 2$, this boundary value problem describes the bifurcation of a flat membrane occupying the domain $\Omega \subset \mathbb{R}^2$ with a fixed edge and resting (without adhesion) on an elastic foundation (see Figure 6.4). The membrane is subject to a variable loading, whose magnitude is represented by the parameter λ, and the restoring force $F = F(u)$ is given by a nonlinear law $F(u) = q(u^-)^{\gamma-1}$, where q and γ depend on the membrane and the foundation. $F(u) \neq 0$ only if $u < 0$, i.e., when a portion of the membrane enters the foundation.

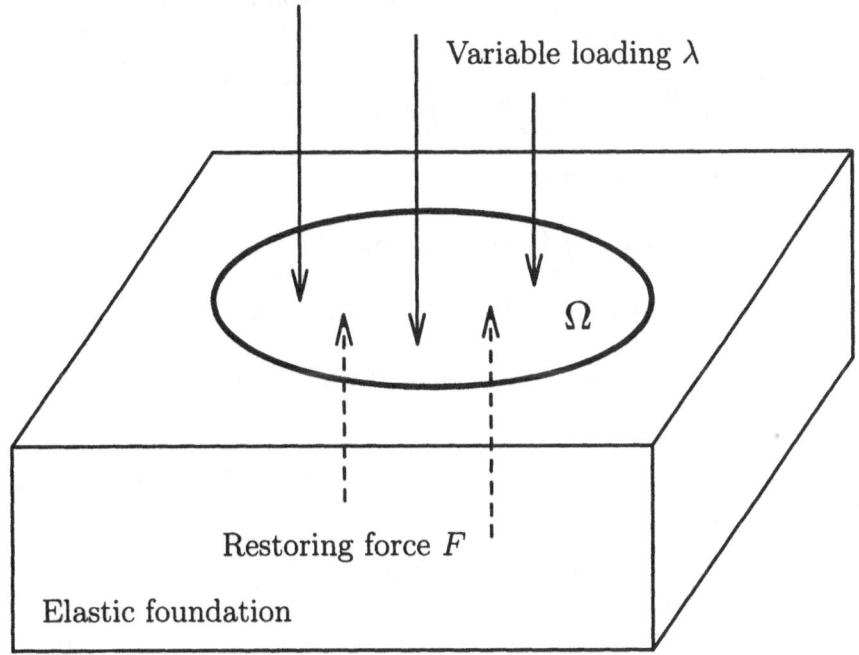

FIGURE 6.4. Example 6.6.

The weak formulation of (6.234) is the variational inequality:

$$\begin{cases} \int_\Omega \nabla u \nabla(v-u) - \lambda \int_\Omega u(v-u) + \int_\Omega k(v^-)^\gamma - \int_\Omega k(u^-)^\gamma \\ \quad \geq 0, \quad \forall v \in H_0^1(\Omega), \\ u \in H_0^1(\Omega), \end{cases} \tag{6.236}$$

where $k = q/\gamma$. In fact, if u satisfies (6.234), then, for $v = 0$ on $\partial\Omega$, multiplying (6.234) by $v - u$ and integrating over Ω,

$$\int_\Omega \nabla u \nabla(v-u) - \lambda \int_\Omega u(v-u) - \int_\Omega q(u^-)^{\gamma-1}(v-u) = 0. \tag{6.237}$$

Because $\gamma > 1$, the function $h(\xi) = (\xi^-)^\gamma$ is of class C^1 and $h'(\xi) = -\gamma(\xi^-)^{\gamma-1}$ is increasing on $\mathbb{R}$. Hence, h is convex, and

$$(\eta^-)^\gamma - (\xi^-)^\gamma \geq -\gamma(\xi^-)^{\gamma-1}(\eta - \xi), \ \forall \eta, \xi \in \mathbb{R}.$$

Because $q > 0$,

$$\frac{1}{\gamma} \int_\Omega q(v^-)^\gamma - \frac{1}{\gamma} \int_\Omega q(u^-)^\gamma \geq - \int_\Omega q(u^-)^{\gamma-1}(v - u).$$

Substituting this in (6.237), we obtain (6.236). Conversely, suppose that u satisfies (6.236). For $w = 0$ on $\partial\Omega$, letting $v = u + tw, (t > 0)$ in (6.236),

$$0 \leq \int_\Omega \nabla u \nabla w - \lambda \int_\Omega uw + \int_\Omega k \left\{ \frac{[(u + tw)^-]^\gamma - (u^-)^\gamma}{t} \right\}.$$

Letting $t \to 0^+$, we see that

$$0 \leq \int_\Omega [-\Delta u - \lambda u - \gamma k(u^-)^{\gamma-1}]w.$$

Because this holds for all w with $w = 0$ on $\partial\Omega$, replacing w by $-w$,

$$\int_\Omega [\Delta u + \lambda u + q(u^-)^{\gamma-1}]w = 0, \ \forall w, \ w = 0 \ \text{on} \ \partial\Omega,$$

which implies (6.234).

Now, (6.236) is of the form (6.8) with $V = H_0^1(\Omega)$, and $A = V \to V^*$, $B : V \times \mathbb{R} \to V$, $j : V \to \mathbb{R} \cup \{\infty\}$, given by

$$\begin{cases} \langle Au, v \rangle &= \displaystyle\int_\Omega \nabla u \nabla v, \\[2mm] \langle B(u, \lambda), v \rangle &= \lambda \displaystyle\int_\Omega uv, \\[2mm] j(u) &= \displaystyle\int_\Omega k(u^-)^\gamma, \forall u, v \in V, \end{cases} \qquad (6.238)$$

($\langle \cdot, \cdot \rangle$ and $\| \cdot \|$ denote the usual inner product and norm on $H_0^1(\Omega)$).

Using standard arguments, we can show that A is a linear, bounded, and self-adjoint operator on V that satisfies the coerciveness condition (6.9), B is completely continuous, and $B(u, \lambda)$ is linear with respect to u and λ. Hence, the assumptions in Section 6.1 are satisfied.

By the linearity and continuity of A, we have (6.37) with $\alpha = A$. Similarly, from the complete continuity and linearity of B (with respect to u), (6.39) follows with

$$f(u, \lambda) = B(u, \lambda), \ u \in V, \lambda \in \mathbb{R}.$$

We see that f is symmetric with respect to u and homogeneous (of degree 1) with respect to λ.

Next, let us consider the conditions on j. First, because h is convex and $k \in L^\infty(\Omega)$,

$$k \geq k_0 > 0 \text{ a.e. on } \Omega. \qquad (6.239)$$

Hence, (6.238) implies that j is a convex functional from V to $[0, \infty]$, and $j(0) = 0$. Let $\{v_n\} \subset V, v_n \to v$ in V. We choose a subsequence $\{v_{n_k}\} \subset \{v_n\}$, such that

$$\lim_{k \to \infty} j(v_{n_k}) = \liminf j(v_n).$$

Because $v_n \to v$ in $L^2(\Omega)$, by passing once more to a subsequence, if necessary, we can assume that

$$v_{n_k}(x) \to v(x) \text{ for a.e. } x \in \Omega.$$

Hence, $v_{n_k}^- \to v^-$ a.e. on Ω, and

$$k(v_{n_k}^-)^\gamma \to k(v^-)^\gamma \text{ a.e. on } \Omega.$$

By Fatou's lemma,

$$\begin{aligned} j(v) = \int_\Omega k(v^-)^\gamma &\leq \liminf \int_\Omega k(v_{n_k}^-)^\gamma \\ &= \liminf j(v_{n_k}) \\ &= \lim j(v_{n_k}) \\ &= \liminf j(v_n). \end{aligned}$$

This shows that j is lower semicontinuous on V.

Now, we consider the convex set K_1 given by

$$K_1 = \{u \in H_0^1(\Omega) : u \geq 0 \text{ a.e. on } \Omega\},$$

and define the functional J from V to $[0, \infty]$ by

$$J(u) = \begin{cases} I_{K_1}(u) & \text{if } 1 < \gamma < 2, \\ j(u) & \text{if } \gamma = 2, \\ 0 & \text{if } \gamma > 2. \end{cases} \qquad (6.240)$$

Let j_σ be defined by (6.44). We have the following lemma:

Lemma 6.18 *Under the assumptions above,*

$$j_\sigma \to J \ (as \ \sigma \to 0^+)$$

in the sense of (6.45)-(6.46).

Proof. We consider the cases $1 < \gamma < 2$, $\gamma = 2$, and $\gamma > 2$ separately.

(a) $1 < \gamma < 2$. Let $v_n \to v$ and $\sigma_n \to 0^+$. If $v \geq 0$ on Ω, i.e., $v \in K_1$, then, $J(v) = 0$. Next, we show that

$$J(v) \leq \liminf j_{\sigma_n}(v_n). \tag{6.241}$$

Suppose, otherwise, that $v < 0$ on a subset Ω_0 of Ω of positive measure. Then, $v^- > 0$ on this subset, and, by (6.239),

$$\int_\Omega k(v^-)^\gamma \geq k_0 \int_\Omega (v^-)^\gamma > 0.$$

Therefore, $v_n \to v$ in $L^2(\Omega)$, and, by passing to a subsequence, if necessary, we can also assume that

$$v_n \to v \text{ a.e. in } \Omega.$$

Hence, $(v_n^-)^\gamma \to (v^-)^\gamma$ a.e. in Ω, and, in view of Fatou's lemma,

$$0 < \int_\Omega k(v^-)^\gamma \leq \liminf \int_\Omega k(v_n^-)^\gamma.$$

It follows that

$$\liminf j_{\sigma_n}(v_n) = \liminf \frac{1}{\sigma_n^2}\left[\int_\Omega k(v_n^-)^\gamma\right] = \infty,$$

(because $\frac{1}{\sigma_n^2} \to \infty$), proving (6.241). Now, let $v \in H^1(\Omega)$, and choose $v_n = v$, $\forall n$.

If $v \geq 0$ a.e. on Ω, then,

$$\sigma_n v_n = \sigma_n v \geq 0 \text{ a.e. on } \Omega,$$

and, hence, $j(\sigma_n v_n) = \int_\Omega k(\sigma_n v_n)^- = 0$, $\forall n$, implying that

$$J(v) = 0 = \lim \frac{j(\sigma_n v_n)}{\sigma_n^2}.$$

Hence, (6.46) holds. In the other case, it follows, as above, that $\int_\Omega k(v_n^-)^\gamma = \int_\Omega k(v^-)^\gamma > 0$. Hence,

$$j_{\sigma_n}(v_n) = \frac{1}{\sigma_n^2} \int_\Omega k(v^-)^\gamma \to \infty \text{ as } n \to \infty.$$

Because $J(v) = I_{K_1}(v) = \infty$, (6.46) also holds.

(b) $\gamma = 2$. In this case $j(u) = \int_\Omega k(u^-)^2$, $u \in V$, and, therefore, j is homogeneous of degree 2. It remains to be shown that $j = J$.

It follows from the homogeneity of j that

$$j_\sigma = j, \forall \sigma > 0. \tag{6.242}$$

Letting $v_n \rightharpoonup v, \sigma_n \to 0^+$, it follows, from the weak lower semicontinuity of j, that

$$j(v) \leq \liminf j(v_n) = \liminf j_{\sigma_n}(v_n).$$

Thus, (6.45) follows. For $v \in V$, let $v_n = v$, $\forall n$. By (6.242),

$$J(v) = j(v) = \lim j(v_n) = \lim j_{\sigma_n}(v_n),$$

proving (6.46).

(c) $\gamma > 2$. Because $J \equiv 0 \leq j$, we, clearly, have (6.45). For $v \in V$, again choosing $v_n = v$, $\forall n$, it follows that

$$j_{\sigma_n}(v_n) = \frac{1}{\sigma_n^2} \int_\Omega k(\sigma_n v^-)^\gamma = \sigma_n^{\gamma-2} \int_\Omega k(v^-)^\gamma \to 0,$$

because $\sigma_n \to 0$ and $\gamma - 2 > 0$. Hence, $\lim j_{\sigma_n}(v_n) = 0 = J(v)$, and (6.46) holds. ∎

In view of Lemma 6.18, the homogenized variational inequality corresponding to (6.236) is given by

$$\begin{cases} \int_\Omega \nabla u \nabla(v-u) - \lambda \int_\Omega u(v-u) + J(v) - J(u) \geq 0, \ \forall v \in H_0^1(\Omega), \\ u \in H_0^1(\Omega), \end{cases}$$

$$(6.243)$$

and the associated linear equation is expressed by

$$\begin{cases} \int_\Omega \nabla u \nabla v - \lambda \int_\Omega uv = 0, \ \forall v \in H_0^1(\Omega), \\ u \in H_0^1(\Omega), \end{cases} \qquad (6.244)$$

which, in turn, is equivalent to

$$\begin{cases} \Delta u + \lambda u = 0 \ \text{ in } \ \Omega, \\ u = 0 \ \text{ on } \ \partial\Omega. \end{cases} \qquad (6.245)$$

We also note that, in the case where $1 < \gamma < 2$, (6.243) can be written as

$$\begin{cases} \int_\Omega \nabla u \nabla(v-u) - \lambda \int_\Omega u(v-u) \geq 0, \ \forall v \in K_1, \\ u \in K_1. \end{cases} \qquad (6.246)$$

Now, we can apply Corollaries 6.6 and 6.13 to our problem and obtain the following result.

Corollary 6.19 *Assume that $\gamma > 2$. Then, (6.236) has, at most, a count-able number of bifurcation points, and, if $(0, \lambda)$ is a bifurcation point of (6.236), then, λ is an eigenvalue of (6.244). Conversely, if λ is an eigen-value of (6.244) of odd multiplicity, then, $(0, \lambda)$ is a bifurcation point of (6.236) with an associated global bifurcation branch satisfying the alterna-tive in Theorem 6.4.*

Proof. In this case, $J \equiv 0$ on V. Hence, (6.243) becomes

$$
\begin{cases}
\displaystyle \int_\Omega \nabla u \nabla (v - u) - \lambda \int_\Omega u(v - u) \geq 0, \ \forall v \in H_0^1(\Omega), \\
u \in H_0^1(\Omega).
\end{cases}
\tag{6.247}
$$

Letting $v = u \pm w \in H_0^1(\Omega)$ for $w \in H_0^1(\Omega)$, we see that (6.247), in this situation, is equivalent to (6.244).

Hence, with the notations of Section 6.3, $W = V = H_0^1(\Omega)$, and (6.244) is the same as (6.75) and (6.80). Corollary 6.19 follows directly from Corollary 6.6. ∎

Corollary 6.20 *Assume that $1 < \gamma \leq 2$. Let λ_0 be the first eigenvalue of (6.244). Then, λ_0 is a simple eigenvalue of (6.243), and $(0, \lambda_0)$ is a bifurcation point of (6.236), which an associated global bifurcation branch satisfying the alternative in Theorem 6.4.*

Proof. Let $1 < \gamma < 2$. Note that λ_0 is a simple eigenvalue of (6.244) with an associated eigenvector u_0, which can be chosen, such that

$$
u_0(x) > 0, \ \forall x \in \Omega.
\tag{6.248}
$$

As observed before,

$$
u_0 \in K_1^I.
\tag{6.249}
$$

In fact, let $w \in C_0^\infty(\Omega)$, such that $\text{supp}\, w \subset \Omega_0$, $\overline{\Omega}_0 \subset \Omega$. By standard regularity results for second-order elliptic operators, we know that $u_0 \in C(\overline{\Omega})$. Because $u_0(x) > 0$, $\forall x \in \overline{\Omega}_0$,

$$
\inf_{\overline{\Omega}_0} u_0 \geq m > 0.
$$

Let $\epsilon = m(2\|w\|_{L^\infty(\Omega)})^{-1} > 0$. Then,

$$
(u_0 + \epsilon w)(x) = u_0(x) > 0, \ \forall x \in \Omega \setminus \overline{\Omega}_0.
$$

For $x \in \overline{\Omega}_0$,

$$
\begin{aligned}
(u_0 + \epsilon w)(x) &\geq \inf_{\overline{\Omega}_0} u_0 - \frac{m}{2\|w\|_{L^\infty(\Omega)}} w(x) \\
&\geq \frac{m}{2} > 0.
\end{aligned}
$$

Hence, $u_0 + \epsilon w > 0$ on Ω, i.e., $u_0 + \epsilon w \in K_1$. Because $C_0^\infty(\Omega)$ is dense in $H_0^1(\Omega)$, we have (6.249). In the case where $\gamma = 2$, from (6.240), $J(u) = 0$ if and only if $j(u) = \int_\Omega k(v^-)^\gamma = 0$. This happens if and only if $v^- = 0$ a.e. on Ω, i.e., $v \geq 0$ a.e. on Ω.

In other words, $\ker J = K_1$. Because $B = f$, given by (6.238), is symmetric and homogeneous of degree 1 with respect to λ, $h \equiv 0$.

Hence, the result follows from Corollary 6.13 (c). ∎

In the case where $1 < \gamma < 2$, by applying Theorem 6.9, we can conclude more about the global bifurcation branch of (6.236) with the following result:

Corollary 6.21 *Let $1 < \gamma < 2$ and let λ_0 be as in Corollary 6.20. Then, there exist no bifurcation points $(0, \lambda)$ of (6.236) such that $\lambda > \lambda_0$, and the bifurcation branch of (6.236) emanating from $[0, \lambda_0]$ is, in fact, unbounded in $V \times \mathbb{R}$.*

Proof. From (6.238), $A - B(\cdot, 0) = A$ is strictly monotone on V. Because $f(u, \lambda) = B(u, \lambda)$ is symmetric with respect to u and is homogeneous with respect to λ, $g(u, \lambda) = f(u, \lambda)$. For $1 < \gamma < 2$,

$$D(J) = D(I_{K_1}) = K_1.$$

If $u \in D(J) \supset D(J, u_0, \lambda_0)$, then, $u \geq 0$ on Ω, and, hence,

$$\langle g(u_0, \lambda_0), u \rangle = \lambda_0 \int_\Omega u_0 u \geq 0,$$

as follows from (6.248). Moreover, $\langle g(u_0, \lambda_0), u \rangle = 0$ only if $u = 0$. Therefore, (6.156) is satisfied, and our conclusion follows from Theorem 6.9 (b). ∎

We note that the above arguments can be applied to study the bifurcation problem for variational inequalities of the form (6.234)-(6.236) that contain general second-order (not necessarily symmetric) elliptic operators instead of the Laplacian.

Example 6.7 [Buckling problems for beams subject to unilateral constraints represented by convex functionals] In this example, we consider a bifurcation problem for a beam (as considered in Example 4.3) resting between two foundations (one above and one below, with partial contact along its length) (see Figure 6.5) with nonlinear elastic laws similar to that in Example 6.6. This problem can be modeled by the following variational inequality:

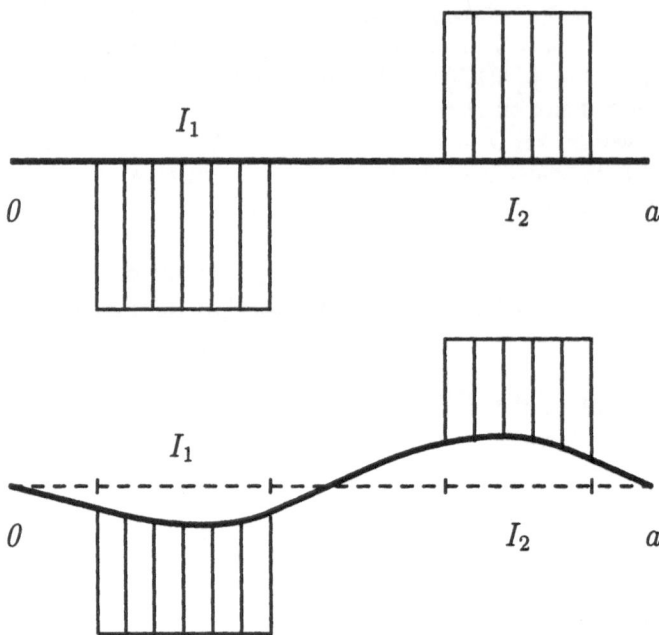

FIGURE 6.5. Example 6.7 - Beam with elastic obstacles.

$$
\begin{cases}
\displaystyle \int_0^a u''(v-u)'' dx - \lambda \int_0^a \frac{u'}{\sqrt{1+u'^2}}(v-u)' dx. \\[2mm]
\displaystyle + \left[\int_{I_1} k_1(v^-)^\gamma dx + \int_{I_2} k_2(v^+)^\beta dx \right] \\[2mm]
\displaystyle - \left[\int_{I_1} k_1(u^-)^\gamma dx + \int_{I_2} k_2(u^+)^\beta dx \right] \geq 0, \ \forall v \in V, \\[2mm]
u \in V.
\end{cases}
\tag{6.250}
$$

Here, $[0, a]$ $(a > 0)$ is the interval occupied by the beam, and $V = H_0^2(0, a)$, or $V = H^2(0, a) \cap H_0^1(0, a)$ depending on whether the beam is clamped or is simply supported at the ends 0 and a. $I_1, I_2 \subset (0, a), |I_1|, |I_2| > 0$ are closed sets representing the domain of possible contact between the beam and the foundations.

The variational inequality (6.250) can also be seen as the formulation for the equilibrium problem of a beam resting between rigid foundations in an interface model.

In Signorini models (as considered in Examples 4.2, 4.3, and 4.4) the contact is represented by a nonpenetrating condition. The key ingredients of the interface model, initiated by Oden and Martins ([86], [96], and [97]), are the introduction of the "normal response" at points where contact may occur, and the substitution of nonpenetrating conditions by those related to normal responses. The idealized contact surfaces are different from real contact surfaces by small asperities. Hence removing the Signorini nonpenetrating assumptions does not mean penetration of the two bodies in contact into each other, but merely allows the average surfaces to get closer to each other. Detailed discussions of the physical and experimental motivations of the interface model were presented in [86] and [96] and the references therein.

As in Examples 4.2, 4.3, and 4.4, the Signorini nonpenetrating constraints are represented by various closed convex sets. In interface models, we replace these constraints by the normal responses $F = F(u)$ caused by the (normal) displacement u:

$$F(u) = \tilde{\phi}(u),$$

where $\tilde{\phi} = \tilde{\phi}(u)$ is the Nemitskii operator associated with a Carathéodory function,

$$\phi = \phi(x, u),$$

representing the interface condition.

Experimental evidence (cf. [86]) shows that for small a displacement u, $\phi(x, u)$ has a power-like behavior, i.e.,

$$\phi(x, u) = c(x)(u^+)^\gamma, \quad \text{or} \quad \phi(x, u) = c(x)(u^-)^\gamma.$$

Hence, $F(u) = \tilde{\phi}(u) = c(u^\pm)^\gamma$. Using arguments similar to those in Example 6.6, we see that, under the interface model, the equilibrium problem for a beam resting (without adhesion) between two rigid obstacles ω_1 and ω_2, respectively, on I_1 and I_2 can be formulated as (6.250). k_1, γ and k_2, β depend on the interface conditions of ω_1 and ω_2, and the integrals

$$\int_{I_1} k_1(u^-)^\gamma \quad \text{and} \quad \int_{I_2} k_2(u^+)^\beta$$

represent the normal responses caused by the contact between the beam and ω_1 and ω_2.

Therefore,

$$j(u) = \int_{I_1} k_1(u^-)^\gamma + \int_{I_2} k_2(u^+)^\beta, \ u \in V, \tag{6.251}$$

denotes the effect of the foundations ω_1 and ω_2 on the beam, where $\gamma, \beta > 1$ and $k_1, k_2 \in L^\infty(I)$ are characterized by the foundations and the beam. Note that, if $k_2 \equiv 0$, we have a convex functional similar to that of the previous example, and the analysis will be the same as above. Now, we suppose that

$$k_1, k_2 \geq k_0 > 0 \ \text{a.e. in } I. \tag{6.252}$$

Because $u \mapsto u^+, u^-, \ u \in \mathbb{R}$ are nonnegative and convex, we see that j is well defined, with values in $[0, \infty]$. Moreover, j is convex and nonnegative, and $j(0) = 0$. Using Fatou's lemma, as in Example 6.6, we find that j is lower semicontinuous on V.

Therefore, (6.250) is of the form (6.8) with

$$\langle Au, v \rangle = \int_0^a u''v'',$$

$$\langle B(u,\lambda), v \rangle = \lambda \int_0^a \frac{u'v'}{\sqrt{1+u'^2}}, \ \forall u, v \in V, \ \lambda \in \mathbb{R},$$

and j given by (6.251).

A is linear and continuous and satisfies (6.9) with $p = 2$. Hence (6.18), (6.19), and (6.20) are immediately satisfied with

$$\alpha = A. \tag{6.253}$$

In Example 4.3, we have shown that B is a completely continuous mapping from $V \times \mathbb{R}$ to V, and, furthermore, B satisfies (6.39) with

$$\langle f(u,\lambda), v \rangle = \lambda \int_0^a u'v', \ u, v \in V, \ \lambda \in \mathbb{R}. \tag{6.254}$$

Let $J_1, J_2 : V \to [0, \infty]$ be given by

$$J_1(u) = \begin{cases} I_{K_1}(u) & \text{if } 1 < \gamma < 2, \\ \int_{I_1} k_1(u^-)^2 & \text{if } \gamma = 2, \\ 0 & \text{if } \gamma > 2, \end{cases} \tag{6.255}$$

$$J_2(u) = \begin{cases} I_{K_2}(u) & \text{if } 1 < \beta < 2, \\ \int_{I_2} k_2(u^+)^2 & \text{if } \beta = 2, \\ 0 & \text{if } \beta > 2, \end{cases} \tag{6.256}$$

where

$$K_1 = \{u \in V : u \geq 0 \text{ a.e. on } I_1\},$$
$$K_2 = \{u \in V : u \leq 0 \text{ a.e. on } I_2\}. \tag{6.257}$$

Let $J : V \to [0, \infty]$,

$$J(u) = J_1(u) + J_2(u)$$

$$= \begin{cases} I_{K_1}(u) + I_{K_2}(u) = I_{K_1 \cap K_2}(u) & \text{if } 1 < \gamma, \beta < 2, \\[2mm] I_{K_1}(u) + \displaystyle\int_{I_2} k_2(u^+)^2 & \text{if } 1 < \gamma < 2 = \beta, \\[2mm] I_{K_1}(u) & \text{if } 1 < \gamma < 2 < \beta, \\[2mm] \displaystyle\int_{I_1} k_1(u^-)^2 + I_{K_2}(u) & \text{if } 1 < \beta < 2 = \gamma, \\[2mm] j(u) = \displaystyle\int_{I_1} k_1(u^-)^2 + \int_{I_2} k_2(u^+)^2 & \text{if } \gamma = \beta = 2, \\[2mm] \displaystyle\int_{I_1} k_1(u^-)^2 & \text{if } \gamma = 2 < \beta, \\[2mm] I_{K_2}(u) & \text{if } 1 < \beta < 2 < \gamma, \\[2mm] \displaystyle\int_{I_2} k_2(u^+)^2 & \text{if } \beta = 2 < \gamma, \\[2mm] 0 & \text{if } \gamma, \beta > 2, \end{cases} \tag{6.258}$$

for $u \in V$. Then, K_1 and K_2 are closed, convex sets in V, and, hence, J_1 and J_2 are convex and lower semicontinuous on V.

Therefore,

$$J = J_1 + J_2$$

also has this property.

Moreover, by using the arguments in the proof of Lemma 6.18, we can prove that

$$j_\sigma \text{ tends to } J \text{ in the sense of (6.45) and (6.46).} \tag{6.259}$$

From (6.253), (6.254), and (6.259), it follows that the homogenized variational inequality corresponding to (6.250), in this case, is expressed by

$$\begin{cases} \displaystyle\int_0^a u''(v-u)''dx - \lambda \int_0^a u'(v-u)'dx + J(v) - J(u) \geq 0, \ \forall v \in V, \\[2mm] u \in V, \end{cases} \tag{6.260}$$

where J is given by (6.258).

The linear equation associated with (6.260) is therefore

$$\begin{cases} \displaystyle\int_0^a u''v''dx - \lambda \int_0^a u'v'dx = 0, \ \forall v \in V, \\[2mm] u \in V, \end{cases} \tag{6.261}$$

which is equivalent to (4.42) if $V = H^2(0, a) \cap H_0^1(0, a)$, or

$$\begin{cases} u^{(4)} + \lambda u'' = 0 \quad \text{on} \quad [0, a], \\ u(0) = u(a) = u'(0) = u'(a) = 0, \end{cases} \qquad (6.262)$$

if $V = H_0^2(0, a)$. We have the following global bifurcation result for (6.250).

Corollary 6.22 *(a) Assume that $\gamma, \beta > 2$. Then, $(0, \lambda)$ is a bifurcation point of (6.32) if and only if λ is an eigenvalue of (6.261). Moreover, at each eigenvalue λ of (6.261), the global bifurcation branch emanating from $(0, \lambda)$ satisfies the alternative in Theorem 6.4. (b) Assume that $1 < \gamma \leq 2 < \beta$. Let $\lambda_0 > 0$ be an eigenvalue of (6.261) with a corresponding eigenvector u_0 satisfying*

$$u_0(x) > 0, \ \forall x \in I_1. \qquad (6.263)$$

Then, there exists a branch of nontrivial solutions (6.250) that bifurcates from $\{0\} \times [0, \lambda_0]$ and satisfies the conclusion of Theorem 6.4.

(c) Assume that $1 < \beta \leq 2 < \gamma$. Let $\lambda_0 > 0$ be an eigenvalue of (6.261) with an associated eigenvector u_0 satisfying

$$u_0(x) < 0, \ \forall x \in I_2. \qquad (6.264)$$

Then, we arrive at the conclusion of (b).

(d) Assume that $\beta, \gamma \leq 2$. Let $\lambda_0 > 0$ be an eigenvalue of (6.261) with a corresponding eigenvector u_0 satisfying both (6.263) and (6.264). Then we arrive at the conclusion of (b).

Proof. First, we note that, with the choice of spaces $V = H^2(0, a) \cap H_0^1(0, a)$ or $V = H_0^2(0, a)$, all the eigenvalues of (6.261) are simple. In fact, (6.261) is equivalent to (6.262) if $V = H_0^2(0, a)$, and to (4.42) in the case where $V = H^2(0, a) \cap H_0^1(0, a)$. By direct calculations, we know that (6.262) has eigenvalues

$$\lambda_k^{(1)} = \left(\frac{2k\pi}{a} \right)^2, \ k = 1, 2, \ldots,$$

and

$$\lambda_k^{(2)} = \left(\frac{2\mu_k}{a} \right)^2, \ k = 1, 2, \ldots,$$

where μ_k $(k \in \mathbb{N})$ are positive solutions of the equation $\tan \mu = \mu$, and the corresponding eigenvectors are

$$u_k^{(1)}(x) = \cos \left(\frac{2k\pi}{a} x \right) - 1$$

and

$$u_k^{(2)}(x) = \mu_k \left[1 - \cos \left(\frac{2\mu_k}{a} x \right) \right] + \left[\sin \left(\frac{2\mu_k}{a} x \right) - \frac{2\mu_k}{a} x \right], \ 0 \leq x \leq a.$$

Hence, the $\lambda = \lambda_k^{(1)}, \lambda_k^{(2)}$ are simple. In the second case, as in Example 4.3, we know that (4.42) has eigenvalues

$$\lambda_k = \left(\frac{k\pi}{a}\right)^2, \ k = 1, 2, \ldots,$$

with the corresponding eigenvectors

$$u_k(x) = \sin\left(\frac{k\pi}{a}\right), \ k = 1, 2, \ldots, \ 0 \le x \le a,$$

and the λ_k, again, are simple.

(a) From (6.258), we see that

$$J \equiv 0$$

in the case where $\gamma, \beta > 2$. (6.260) is therefore equivalent to (6.261).

As in Section 6.3, $W = V$. By the above remark, we see that all eigenvalues of (6.261) are simple, hence, in particular, are of odd multiplicity.

Our conclusion therefore follows directly from Corollary 6.6.

(b) Assume that $1 < \gamma < 2 < \beta$. Then, by (6.258),

$$J = I_{K_1}.$$

Hence ker $J = D(J) = K_1$. Arguing as in the proof of Corollary 6.20, we see that (6.263) implies that

$$u_0 \in K_1^I = (\ker J)^I.$$

If $\gamma = 2 < \beta$, then, $J(u) = \int_0^a k_1(u^-)^2$, and ker $J = K_1$.

Again, if u_0 satisfies (6.263), then, $u_0 \in (\ker J)^I$. Also, $\alpha = A$ and $f(\cdot, \lambda)$ are symmetric, and $h = 0$.

Moreover, because λ_0 is simple, we can apply Corollary 6.13 (c) to obtain the desired conclusion.

The proof of (c) is similar to that of (b).

(d) If $\gamma = \beta = 2$, then,

$$J(u) = j(u) = \int_{I_1} k_1(u^-)^2 + \int_{I_2} k_2(u^+)^2, \ u \in V.$$

It follows that

$$J(u) = 0 \quad \Longleftrightarrow \quad \int_{I_1} k_1(u^-)^2 = \int_{I_2} k_2(u^+)^2 = 0$$

$$\Longleftrightarrow \quad u \ge 0 \text{ on } I_1 \text{ and } u \le 0 \text{ on } I_2.$$

Hence, in this case,

$$\ker J = K_1 \cap K_2. \tag{6.265}$$

If $\gamma, \beta < 2$, then, $J = I_{K_1 \cap K_2}$, and (6.265) obviously holds. If $\gamma < 2 = \beta$, then,

$$J(u) = I_{K_1}(u) + \int_{I_2} k_2(u^+)^2.$$

Again,

$$J(u) = 0 \quad \Longleftrightarrow \quad u \in K_1 \cap K_2.$$

Similarly, for the case $\gamma = 2 > \beta$, we see that (6.265) holds whenever $1 < \gamma, \beta \le 2$.

Now, suppose that u_0 satisfies both (6.263) and (6.264), i.e.,

$$u_0 > 0 \text{ on } I_1 \quad \text{and} \quad u_0 < 0 \text{ on } I_2. \tag{6.266}$$

We show that

$$u_0 \in (K_1 \cap K_2)^I = (\ker J)^I. \tag{6.267}$$

In fact, from (6.266), it follows that $I_1 \cap I_2 = \emptyset$, and, because I_1, I_2 are compact, there exists $\delta > 0$, such that

$$u_0(x) \ge \delta, \ \forall x \in I_1 \quad \text{and} \quad u_0(x) \le -\delta, \ \forall x \in I_2.$$

Let $w \in C_0^\infty(0, a)$, and set $\epsilon = \delta / \|w\|_{L^\infty(0,a)} > 0$. Then,

$$\begin{cases} u_0 + \epsilon w \ge \delta - \epsilon \|w\|_{L^\infty(0,a)} = 0 & \text{on} \quad I_1, \text{ and} \\ u_0 + \epsilon w \le -\delta + \epsilon \|w\|_{L^\infty(0,a)} = 0 & \text{on} \quad I_2. \end{cases}$$

Hence $u_0 + \epsilon w \in K_1 \cap K_2$. Because $D = C_0^\infty(0, a)$ is dense in $H_0^1(0, a)$, we have (6.267). Because λ_0 is a simple eigenvalue of (6.261), our claim follows from Corollary 6.13 (c). ■

We now illustrate this result by a simple example.

Example 6.8 Let $a = 1$, and suppose that the beam is simply supported at the ends, i.e., $V = H^2(0, 1) \cap H_0^1(0, 1)$. We consider (6.250) with

$$\gamma = \beta = 2, \ k_1 = k_2 \equiv 1,$$

and

$$I_1 = \left[\frac{1}{6}, \frac{1}{4}\right] \cup \left[\frac{3}{4}, \frac{11}{12}\right], \quad I_2 = \left[\frac{5}{12}, \frac{1}{2}\right]$$

(see Figure 6.6). As in Example 4.3, we know that (6.261), which is equivalent to (4.42), has eigenvalues

$$\lambda = \lambda_k = k^2 \pi^2$$

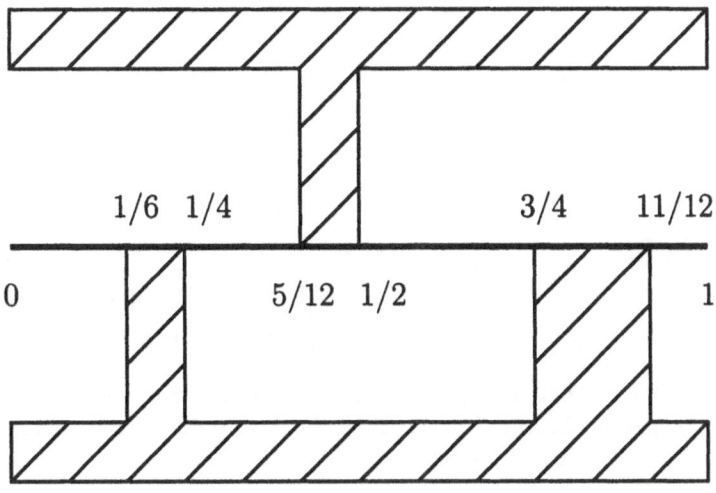

FIGURE 6.6. Example 6.8.

and eigenvectors $u_k = \sin(k\pi x)$, $k \in \mathbb{N}$. All eigenvalues are simple, and although

$$\mathbb{R}u_1 \cap (K_1 \cap K_2) = \mathbb{R}u_2 \cap (K_1 \cap K_2) = \emptyset,$$

$$u_3(x) > 0, \ \forall x \in I_1, \ u_3(x) < 0, \ \forall x \in I_2,$$

implying that $u_3 \in (K_1 \cap K_2)^I$.

We can, therefore, apply Corollary 6.22 (d) to get a global bifurcation branch of (6.250) emanating from $\{0\} \times [0, 9\pi^2]$.

In the above example (and in Example 6.6), we note that the arguments also apply for the case where the bases of the foundations do not coincide with the horizontal line (or plane), but, instead, lie above and below this line. In these cases, the integrals $\int_{I_1} k_1(v^-)^\gamma$ and $\int_{I_2} k_2(v^+)^\beta$ in (6.250) are replaced by $j_1(v) = \int_{I_1} k_1[(v-\psi_1)^-]^\gamma$ and $j_2(v) = \int_{I_2} k_2[(v-\psi_2)^+]^\beta$, where ψ_1 and ψ_2 are given continuous functions defined respectively on I_1 and I_2 such that

$$\psi_1 \le 0 \ \text{on} \ I_1, \ \psi_2 \ge 0 \ \text{on} \ I_2.$$

Let

$$J_1(v) = \begin{cases} I_{K_1}(v) & \text{if} \ 1 < \gamma < 2, \\ \int_{N(\psi_1)} k_1(v^-)^2 & \text{if} \ \gamma = 2, \\ 0 & \text{if} \ \gamma > 2, \end{cases}$$

$$J_2(v) = \begin{cases} I_{K_2}(v) & \text{if } 1 < \beta < 2, \\ \displaystyle\int_{N(\psi_2)} k_2(v^+)^2 & \text{if } \beta = 2, \\ 0 & \text{if } \beta > 2, \end{cases}$$

where

$$K_1 = \{u \in V : u \geq 0 \text{ on } N(\psi_1)\}, \quad K_2 = \{u \in V : u \leq 0 \text{ on } N(\psi_2)\},$$

and

$$N(\psi_i) = \{x \in I_i : \psi_i(x) = 0\}, \quad i = 1, 2.$$

One easily shows that, as $\sigma \to 0^+$, j_1, j_2, and $j_1 + j_2$ tend to J_1, J_2, and $J_1 + J_2$, respectively, in the sense of (6.45) and (6.46). From this follows a result similar to Corollary 6.22.

Example 6.9 The following example is concerned with a bifurcation problem for plates subject to unilateral conditions on the boundary. With the settings in section 6.3, we consider the variational inequality (6.81) with a different convex functional j.

First, we consider some assumptions.

Let Ω, a, σ_{ij}, and σ_{ij}^0 be as in section 6.3, and let V be given by (6.86). We consider the functional j defined by

$$j(u) = \int_S k \left[\left(\frac{\partial u}{\partial n} \right)^- \right]^\gamma dS, \ u \in V. \tag{6.268}$$

Here S is a closed subset of $\partial\Omega$, $|S| > 0$ with the surface measure dS, k is a given function in $L^\infty(\Omega)$, $k(x) \geq k_0 > 0$ for a.e. $x \in S$, and $\gamma > 1$. For $u \in H^2(\Omega)$, $|\nabla u \cdot n| \in H^1(\Omega)$. Hence, $\partial_n u (= \partial u / \partial n) \in L^2(\partial\Omega)$, j is well defined on V, and $j(0) = 0$. Because $H^1(\Omega) \hookrightarrow H^{1/2}(\partial\Omega)$, in fact, $\partial_n u \in H^{1/2}(\partial\Omega)$, and, because $\partial\Omega$ is a one-dimensional manifold in $\mathbb{R}^2$, by the Sobolev embedding theorem, $H^{1/2}(\partial\Omega) \hookrightarrow L^q(\partial\Omega)$, $\forall q \in [1, \infty)$. Hence, $\partial_n u \in L^\gamma(\partial\Omega)$, and, then, $D(J) = V$, i.e., $j(v) < \infty$, $\forall v \in V$. We can also prove that j is convex and lower semicontinuous on V. Let

$$K = \{u \in V : \partial_n u \geq 0 \text{ a.e. on } S\}.$$

Because the mapping $H^2(\Omega) \to L^2(\partial\Omega), u \mapsto u|_{\partial\Omega}$, is compact, K is a closed, convex cone in V. Let $J : V \to [0, \infty]$ be given by

$$J(u) = \begin{cases} I_K(u) & \text{if } 1 < \gamma < 2, \\ j(u) & \text{if } \gamma = 2, \\ 0 & \text{if } \gamma > 2. \end{cases} \tag{6.269}$$

Then, using a proof similar to that of Lemma 6.18, we see that j_σ tends to J in the sense of (6.45) and (6.46), as $\sigma \to 0^+$. On the other hand, from the proof of Corollary 6.7, f, given by (6.99), is the homogenization of B $(B(u, \lambda) = \lambda Lu - C(u))$ in the sense of (6.39). Moreover, because A, given by (6.94), is linear and bounded, we see that the homogenized variational inequality of (6.81) is the following variational inequality:

$$\begin{cases} \langle Au, v - u \rangle - \lambda \langle Lu, v - u \rangle + J(v) - J(u) \geq 0, \ \forall v \in H^2(\Omega) \cap H_0^1(\Omega), \\ u \in H^2(\Omega) \cap H_0^1(\Omega), \end{cases}$$

$$(6.270)$$

where A and L are defined by (6.94) and (6.95), and J is given by (6.269).

Using Corollary 6.13 (b), we obtain the following result:

Corollary 6.23

(a) *Assume that $\gamma > 2$. Then, for every bifurcation point $(0, \lambda_0)$ of (6.81), λ_0 is an eigenvalue of the linear equation,*

$$\begin{cases} Au = \lambda Lu, \\ u \in H^2(\Omega) \cap H_0^1(\Omega). \end{cases}$$

$$(6.271)$$

Moreover, if λ_0 is an eigenvalue of (6.271) of odd multiplicity, then, $(0, \lambda_0)$ is a bifurcation point of (6.81) with an associated global bifurcation branch satisfying the alternative in Theorem 6.4.

(b) *Assume that $\gamma \leq 2$ and that λ_0 is an eigenvalue of (6.271) with a corresponding eigenvector u_0 satisfying*

$$\partial_n u_0 \geq \zeta > 0 \quad a.e. \ on \ S. \tag{6.272}$$

Then, there exists a global bifurcation branch of (6.81) that emanates from $[0, \lambda_0]$ and satisfies the alternative in Theorem 6.4.

Proof. First, we note, from (6.94) and (6.95), that A and $\lambda_0 L = f(\cdot, \lambda_0)$ are symmetric on $H^2(\Omega) \cap H_0^1(\Omega)$. Moreover, L is compact. Hence, (6.271) has a sequence of eigenvalues with finite multiplicities.

If $\gamma > 2$, then, by (6.269), $J \equiv 0$. Therefore, (6.270) is equivalent to the equation

$$\begin{cases} \langle Au, v \rangle - \lambda \langle Lu, v \rangle = 0, \ \forall v \in H^2(\Omega) \cap H_0^1(\Omega), \\ u \in H^2(\Omega) \cap H_0^1(\Omega), \end{cases}$$

which is the same as (6.271).

The conclusion of (a) now follows from Corollary 6.6.

Now, assume that $\gamma \leq 2$. If $\gamma = 2$, then $J = j$, and, from (6.268), $u \in \ker J$ if and only if $(\partial_n u)^- = 0$ a.e. on S, which means that $\partial_n u \geq$

0 a.e. on S, i.e., $u \in K$. Thus, by (6.269), we see that ker $J = K$ in both cases where $\gamma = 2$ and $\gamma < 2$. Let u_0 be a solution of (6.271) (with $\lambda = \lambda_0$) that satisfies (6.272), and let $D = C^2(\overline{\Omega}) \cap C_0^1(\overline{\Omega})$. For $v \in D$, $x \in S$,

$$\partial_n(u_0 + \epsilon v)(x) = \partial_n u_0(x) + \epsilon \partial_n v(x) \geq \zeta - \epsilon \|v\|_{L^\infty(\partial\Omega)}.$$

Hence, if ϵ is sufficiently small,

$$\partial_n(u_0 + \epsilon v) \geq 0 \text{ a.e. on } S,$$

which means that $u_0 + \epsilon v \in K$. Because D is dense in $H^2(\Omega) \cap H_0^1(\Omega)$, we find that

$$u_0 \in (\ker J)^I,$$

and, therefore, ker $(A - \lambda_0 L) \cap (\ker J)^I \neq \emptyset$. Because $|S| > 0$, it follows, from the definition of K, that K is a proper subset of $H^2(\Omega) \cap H_0^1(\Omega)$. Hence, all conditions in (6.203) are satisfied. Our conclusion now follows from Corollary 6.14 (b). ∎

We note that similar arguments can be applied to establish global bifurcation results for (6.81) with other choices of unilateral conditions. For example, we can choose V as the space in (6.89), and, instead of (6.268),

$$j(u) = \int_{\partial\Omega \backslash \Gamma_1} k(u^-)^\gamma dS, \ u \in V,$$

where $\gamma > 1$, and Γ_1 is as in section 6.3.

Example 6.10 In the following example, we consider the bifurcation for an obstacle problem for thin plates with an interface model. This problem is related to examples (a) and (b) above, and a particular case of it gives us a global bifurcation result for the buckling problem of a plate resting on a foundation, considered in [69].

We consider the buckling problem for a plate clamped on the edge and resting on a foundation whose base occupies a subdomain Ω_0 of Ω ($\Omega \subset \mathbb{R}^2$ is the domain occupied by the plate in its resting position) (see Figure 6.7). The contact between the plate and the foundation is described by an interface model.

This problem can be formulated by the variational inequality (6.81) with $V = H_0^2(\Omega)$ and $j : V \to \mathbb{R}$ given by

$$j(u) = \int_{\Omega_0} k(x, u(x))[u^-(x)]^\gamma dx, \ u \in V. \tag{6.273}$$

Here, $\gamma > 1$ and $k : \Omega_0 \times \mathbb{R} \to \mathbb{R}$ is a given Carathéodory function, such that

$$0 < m \leq k(x, u) \leq M, \text{ for a.e. } x \in \Omega_0, \text{ all } u \in \mathbb{R}, \tag{6.274}$$

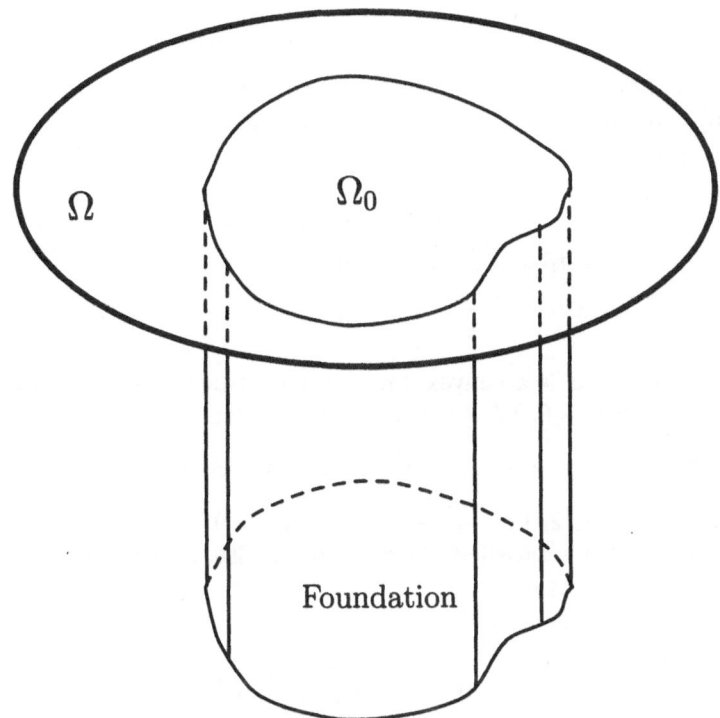

FIGURE 6.7. Example 6.10 - Buckling of plate with interface conditions.

and

$$\lim_{u \to 0} k(x, u) = k_0(x) \quad \text{a.e. on } \Omega_0. \tag{6.275}$$

We assume, furthermore, that the function

$$u \mapsto k(x, u)(u^-)^\gamma, \ u \in \mathbb{R}, \tag{6.276}$$

is convex for almost all $x \in \Omega_0$. Note that in the case $k(x, u) \equiv k_0(x)$ does not depend on u, (6.275) and (6.276) are immediately satisfied, and we obtain the functional representing an interface contact similar to those in Examples 6.6 and 6.7 above. The particular case, where $\gamma = 2$, gives us the problem in [69].

Because $H^2(\Omega) \hookrightarrow C(\overline{\Omega})$, we see, from (6.274), (6.276) that j is convex and nonnegative in V with $D(j) = V$, $j(0) = 0$. As above, by using Fatou's lemma, we can prove that j is lower semicontinuous on V.

We define j_σ by (6.44) with $p = 2$:

$$j_\sigma(u) = \sigma^{\gamma-2} \int_{\Omega_0} k(x, \sigma u(x))[u^-(x)]^\gamma dx,$$

$$j_0(u) = \int_{\Omega_0} k_0(x)[u^-(x)]^\gamma dx, \ u \in V,$$

and

$$K = \{u \in H_0^2(\Omega) : u(x) \geq 0, \ \forall x \in \Omega_0\}.$$

It is clear that j_0 is a convex, lower semicontinuous functional from V to $\mathbb{R}^+$, $j_0(0) = 0$, and K is a closed convex cone in V.

For $u \in V$, let

$$J(u) = \begin{cases} I_K(u) & \text{if } 1 < \gamma < 2, \\ j_0(u) & \text{if } \gamma = 2, \\ 0 & \text{if } \gamma > 2. \end{cases} \tag{6.277}$$

Then, in all cases, J is a convex, lower semicontinuous functional from V to $[0, \infty]$, and $J(0) = 0$. We prove the following lemma:

Lemma 6.24 *As $\sigma \to 0^+$, j_σ tends to J in the sense of (6.45)-(6.46).*

Proof. • $1 < \gamma < 2$. Let $v_n \rightharpoonup v$ in V, $\sigma_n \to 0^+$. If $v \geq 0$ on Ω_0, i.e., $v \in K$, then, (6.45) is satisfied. Suppose, otherwise, that $v < 0$ on a subset of Ω_0 of positive measure.

It follows from (6.274) that

$$\int_{\Omega_0} k_0(v^-)^\gamma dx \geq m \int_{\Omega_0} (v^-)^\gamma dx > 0. \tag{6.278}$$

Because the embedding $H_0^2(\Omega) \hookrightarrow C(\overline{\Omega})$ is compact,

$$v_n \to v \ \text{uniformly on } \overline{\Omega} \supset \Omega_0, \tag{6.279}$$

and, therefore,

$$\sigma_n v_n \to 0 \ \text{uniformly on } \Omega_0.$$

From (6.274) and (6.275), it follows that

$$\lim \int_{\Omega_0} k(\cdot, \sigma_n v_n)(v_n^-)^\gamma dx = \int_{\Omega_0} k_0(v^-)^\gamma dx > 0. \tag{6.280}$$

This implies that

$$\begin{aligned} \liminf j_{\sigma_n}(v_n) &= \lim \sigma_n^{\gamma-2} \lim \int_{\Omega_0} k(\cdot, \sigma_n v_n)(v_n^-)^\gamma dx \\ &= \infty = J(v). \end{aligned}$$

Hence, (6.45) holds. Let $v \in H_0^2(\Omega)$, $\{\sigma_n\} \subset \mathbb{R}$, $\sigma_n \to 0^+$, and choose $v_n = v$, $n = 1, 2, \ldots$. If $v \in K$, then, for all n, $\sigma_n v_n \geq 0$ on Ω_0. Hence, $j_{\sigma_n}(v_n) = 0$, $\forall n$, and

$$J(v) = 0 = \lim j_{\sigma_n}(v_n).$$

If $v \notin K$, then, as above, we have (6.278), and, therefore,

$$
\begin{aligned}
\lim j_{\sigma_n}(v_n) &= \lim \sigma_n^{\gamma-2} \left[\int_{\Omega_0} k_0(v^-)^\gamma \right] \\
&= \infty = J(v),
\end{aligned}
$$

proving (6.46).

• $\gamma = 2$. If $v_n \rightharpoonup v$ in V and $\sigma_n \to 0^+$, then, we have (6.279) and (6.280) (with $\gamma = 2$), which implies that

$$
\lim_{n \to \infty} j_{\sigma_n}(v_n) = j_0(v) = J(v).
$$

(6.45) and (6.46) follow directly from this equality.

• $\gamma > 2$. Because $J \equiv 0 \le j_\sigma$, we, clearly, have (6.45). To prove (6.46), we let $v \in V, \sigma_n \to 0^+$, and again choose $v_n = v, \forall n$.

Hence, (6.280) holds, and, therefore,

$$
\begin{aligned}
j_{\sigma_n}(v_n) &= \sigma_n^{\gamma-2} \int_{\Omega_0} k(\cdot, \sigma_n v_n)(v_n^-)^\gamma dx \\
&\to 0 = J(v),
\end{aligned}
$$

completing the proof.   ∎

From Lemma 6.24 and the arguments used in Example 6.9 and section 6.3, we see that, in this case, the homogenized variational inequality corresponding to (6.81) becomes

$$
\begin{cases}
\langle Au, v - u \rangle - \lambda \langle Lu, v - u \rangle + J(v) - J(u) \ge 0, \ \forall v \in H_0^2(\Omega), \\
u \in H_0^2(\Omega),
\end{cases}
\tag{6.281}
$$

with A, L, and J given by (6.94), (6.95), and (6.277). The linear equation associated with (6.281) is given by (6.92), which is the same as

$$
\begin{cases}
Au = \lambda Lu, \\
u \in H_0^2(\Omega).
\end{cases}
\tag{6.282}
$$

Hence, we have the following global bifurcation result for (6.81).

Corollary 6.25

(a) Assume that $\gamma > 2$. Then, every bifurcation point $(0, \lambda_0)$ of (6.81) (with j given by (6.273)) has λ_0 as an eigenvalue of (6.282). Conversely, if λ_0 is an eigenvalue of (6.282) of odd multiplicity, then, $(0, \lambda_0)$ is a bifurcation point of (6.81) with a global bifurcation branch emanating from $(0, \lambda_0)$ and satisfying the alternative in Theorem 6.4.

(b) Assume that $\gamma \leq 2$ and that λ_0 is an eigenvalue of (6.282) with an associated eigenvector u_0, such that

$$u_0(x) > 0, \ \forall x \in \overline{\Omega_0} \cap \Omega. \tag{6.283}$$

Then, there exists a global bifurcation branch of (6.81), which emanates from $\{0\} \times [0, \lambda_0]$ and satisfies the alternative in Theorem 6.4.

Proof. The proof of (a) is similar to that of Corollary 6.23 and is, therefore, omitted. If $\gamma = 2$, then, $u \in \ker J = \ker j_0$ if and only if $u^- = 0$, i.e., $u \geq 0$ on Ω_0. Hence, $\ker J = K$. From (6.277), we see that this is also true when $\gamma < 2$. Now we verify that, if u_0 satisfies (6.283), then,

$$u_0 \in K^I. \tag{6.284}$$

In fact, let $v \in C_0^\infty(\Omega)$. Because $\operatorname{supp} v \cap \overline{\Omega_0}$ is a compact subset of $\Omega \cap \overline{\Omega_0}$, from (6.283),

$$u_0(x) \geq m, \ \forall x \in \operatorname{supp} v \cap \overline{\Omega_0}, \tag{6.285}$$

for some $m > 0$ (depending on u_0, v). Let $\epsilon > 0$, and $x \in \Omega_0$. If $x \notin \operatorname{supp} v$, then,

$$(u_0 + \epsilon v)(x) = u_0(x) \geq 0.$$

If $x \in \operatorname{supp} v$, then, by (6.285),

$$(u_0 + \epsilon v)(x) \geq m - \epsilon \|v\|_{L^\infty(\Omega)}.$$

Choosing $\epsilon = m(2\|v\|_{L^\infty(\Omega)})^{-1} > 0$, we see that

$$(u_0 + \epsilon v)(x) \geq 0, \ \forall x \in \Omega_0,$$

i.e., $u_0 + \epsilon v \in K$. Because this holds for all $v \in C_0^\infty(\Omega)$ and because $C_0^\infty(\Omega)$ is dense in $H_0^2(\Omega)$, we have (6.284). It follows that

$$u_0 \in \ker(A - \lambda_0 L) \cap (\ker J)^I.$$

On the other hand, it can be seen from the definition of K that $\ker J = K \neq H_0^2(\Omega)$. Hence (6.203) is satisfied, and our conclusion in (b) follows from Corollary 6.14 (b). ■

Remark 6.4 The above buckling problem was considered in [69] with the assumption that k depends only on x and $\|k\|_{L^\infty}$ is small. The authors proved the existence of an infinite sequence of bifurcation points of (6.81). In [101], the smallness assumption of $\|k\|_{L^\infty}$ was removed, and the existence of at least one bifurcation point of (6.81) was proved. Corollary 6.25 is a global bifurcation result for (6.71) that corresponds to the existence results of McLeod, Turner, and Ridell in [69] and [101].

Example 6.11 [Bifurcation for a variational inequality containing the p-Laplacian] In this example, we consider bifurcation problems for the following variational inequality:

$$\begin{cases} \displaystyle\int_\Omega |\nabla u|^{p-2}\nabla u \nabla(v-u) - \int_\Omega [\lambda |u|^{p-2}u + g(x,u,\lambda)](v-u) + j(v) \\[2mm] \qquad - j(u) \geq 0, \ \forall v \in V, \\[2mm] u \in V. \end{cases}$$

$$(6.286)$$

Here $p > 1$, Ω is a bounded domain in $\mathbb{R}^N$ ($N \geq 1$) with a smooth boundary,

$$V = \{u \in W^{1,p}(\Omega) : v = 0 \text{ on } \Gamma\},$$

where Γ is a (relatively) open subset of $\partial\Omega$ with positive measure. $W^{1,p}(\Omega)$ is the usual Sobolev space, equipped with the norm,

$$\|u\|_{W^{1,p}(\Omega)} = \left[\int_\Omega (|u|^p + |\nabla u|^p)\right]^{1/p}, \ u \in W^{1,p}(\Omega).$$

$(V, \|\cdot\|_{W^{1,p}(\Omega)})$ is a closed (Banach) subspace of $W^{1,p}(\Omega)$. By Poincaré's inequality, we know that

$$\|u\| = \left(\int_\Omega |\nabla u|^p\right)^{1/p}, \ u \in V,$$

defines a norm on V, equivalent to $\|\cdot\|_{W^{1,p}(\Omega)}$. In the sequel, we will always consider V with this norm. We also define the pairing between V and V^* by $\langle \cdot, \cdot \rangle$. We assume that

$$g : \Omega \times \mathbb{R} \times \mathbb{R} \to \mathbb{R}$$

is a Carathéodory function, such that

$$g(x, u, \lambda) = o(|u|^{p-1}), \tag{6.287}$$

as $u \to 0$, uniformly a.e. with respect to $x \in \Omega$ and uniformly with respect to λ on bounded intervals, and, moreover, g satisfies the growth condition

$$|g(x, u, \lambda)| \leq C(\lambda)[m(x) + M|u|^{p-1}], \tag{6.288}$$

for a.e. $x \in \Omega$, all $u, \lambda \in \mathbb{R}$, where $C(\lambda) \geq 0$ is bounded on bounded sets, $m \in L^{\frac{p}{p-1}}(\Omega)$, and $M > 0$ is a constant.

We assume that j is given by

$$j(u) = \int_{\partial\Omega \backslash \Gamma} \psi(u(x))dS, \ u \in V, \tag{6.289}$$

where $\psi : \mathbb{R} \to [0, \infty]$ is a proper, convex, lower semicontinuous function satisfying (6.84) and (6.85). Let A be given by

$$\langle Au, v \rangle = \int_\Omega |\nabla u|^{p-2} \nabla u \nabla v \, dx, \quad u, v \in V. \qquad (6.290)$$

Because $|\nabla u| \in L^p(\Omega)$, $u \in V$, we see that A is well defined and is a mapping from V to V^*, and $\|Au\| \leq \|u\|^{p-1}, u \in V$. Hence, A is bounded. Moreover, $A(0) = 0$, and A is continuous on V. In fact, if $u_n \to u$ in V, then, $\nabla u_n \to \nabla u$ in $[L^p(\Omega)]^N$. Hence,

$$|\nabla u_n|^{p-2} \nabla u_n \to |\nabla u|^{p-2} \nabla u \text{ in } \left[L^{\frac{p}{p-1}}(\Omega)\right]^N.$$

Because

$$\|Au_n - Au\| \leq \left[\int_\Omega \left||\nabla u_n|^{p-2} \nabla u_n - |\nabla u|^{p-2} \nabla u\right|^{\frac{p}{p-1}}\right]^{\frac{p-1}{p}},$$

$Au_n \to Au$ in V^*, proving the continuity of A. Moreover, in Remark 6.1 (b) and (c), we have proved that

$$\langle Au - Av, u - v \rangle \geq (\|u\|^{p-1} - \|v\|^{p-1})(\|u\| - \|v\|), \; \forall u, v \in V, \quad (6.291)$$

for all $p > 1$, and

$$\langle Au - Av, u - v \rangle \geq C\|u - v\|^p, \; \forall u, v \in V,$$

for $p \geq 2$, and, therefore, (6.5) is satisfied. We have also observed in these remarks that (6.1) and (6.2) hold. Hence, A satisfies (A2).

Now, we define the mapping B by

$$\langle B(u, \lambda), v \rangle = \int_\Omega [\lambda |u|^{p-2} u + g(x, u, \lambda)] v \, dx, \quad u, v \in V. \qquad (6.292)$$

From (6.288), we see that B is well defined and is a mapping from $V \times \mathbb{R}$ to V^*. As usual, we verify by direct calculations that B is completely continuous.

Now, let f be given by

$$\langle f(u, \lambda), v \rangle = \lambda \int_\Omega |u|^{p-2} uv \, dx, \quad u, v \in V. \qquad (6.293)$$

Then, f is well defined on $V \times \mathbb{R}$. By arguments similar to those used to prove the complete continuity of B, we can check that f is a completely continuous mapping from $V \times \mathbb{R}$ to V^*, and, moreover, (6.287) implies that, if

$$v_n \rightharpoonup v \text{ in } V, \; \lambda_n \to \lambda \text{ and } \sigma_n \to 0^+ \text{ in } \mathbb{R},$$

then,

$$\frac{1}{\sigma_n^{p-1}} B(\sigma_n v_n, \lambda_n) \to f(v, \lambda) \text{ in } V^*,$$

proving (6.39). On the other hand, because A is homogeneous of degree $p-1$, i.e.,

$$A(\sigma u) = \sigma^{p-1} A u, \ u \in V, \sigma > 0,$$

it follows from the continuity of A that, if $v_n \to v$ in V and $\sigma_n \to 0^+$, then,

$$\frac{1}{\sigma_n^{p-1}} A(\sigma_n v_n) \to A(v) \text{ in } V^*. \tag{6.294}$$

This means that (6.37) is satisfied with $\alpha = A$. Because A satisfies (6.291), (6.43) also holds, implying that A satisfies (6.277). Now, let

$$J = I_{W_0^{1,p}(\Omega)}. \tag{6.295}$$

We prove that j_σ tends to J as $\sigma \to 0^+$ in the sense of (6.45) and (6.46). In fact, let $v_n \to v$ in V and $\sigma_n \to 0^+$. We know ([103], [66]) that the mapping

$$W^{1,p}(\Omega) \to L^q(\partial\Omega), \ u \mapsto u|_{\partial\Omega}, \tag{6.296}$$

is compact for all q with $1 \le q < \bar{p}$,

$$\bar{p} = \begin{cases} \dfrac{(N-1)p}{N-p} & \text{if } N > p, \\ \infty & \text{if } N \le p. \end{cases}$$

Hence, letting q satisfy this condition,

$$v_n|_{\partial\Omega} \to v|_{\partial\Omega} \text{ in } L^q(\partial\Omega). \tag{6.297}$$

Therefore, by passing to a subsequence,

$$v_n|_{\partial\Omega} \to v|_{\partial\Omega} \text{ a.e. on } \partial\Omega. \tag{6.298}$$

Now, if $v = 0$ a.e. on $\partial\Omega$, i.e., $v \in W_0^{1,p}(\Omega)$, then, $J(v) = 0$, and (6.45) immediately holds. Suppose that $v \neq 0$ on a subset of $\partial\Omega$ of positive measure. Noting that $v = 0$ on Γ, from (6.84),

$$0 < \int_{\partial\Omega} \psi(v(x)) dS = \int_{\partial\Omega \setminus \Gamma} \psi(v(x)) dS.$$

Hence, by (6.298), Fatou's lemma, and the lower semicontinuity of ψ,

$$\begin{aligned}
0 &< \int_{\partial\Omega \setminus \Gamma} \psi(v(x)) dS \\
&\le \int_{\partial\Omega \setminus \Gamma} \liminf \psi(v_n(x)) dS \\
&\le \liminf j(v_n).
\end{aligned}$$

Because ψ and, then, j are homogeneous of degree 1 ((6.85)), it follows that

$$\liminf \frac{j(\sigma_n v_n)}{\sigma_n^p} = \liminf \frac{j(v_n)}{\sigma_n^{p-1}} = \infty.$$

Hence, (6.45) also holds in this case. To prove (6.46), we let $v \in V, \{\sigma_n\} \subset \mathbb{R}_*^+, \sigma_n \to 0^+$, and choose $v_n = v, \forall n$. Because

$$j(v) = \begin{cases} 0 & \text{if } v \in W_0^{1,p}(\Omega), \\ > 0 & \text{if } v \notin W_0^{1,p}(\Omega), \end{cases}$$

$$\begin{aligned}
\lim \frac{j(\sigma_n v_n)}{\sigma_n^p} &= \lim_{n \to \infty} \frac{j(v)}{\sigma_n^{p-1}} \\
&= \begin{cases} 0 & \text{if } v \in W_0^{1,p}(\Omega), \\ \infty & \text{if } v \notin W_0^{1,p}(\Omega), \end{cases} \\
&= J(v).
\end{aligned}$$

Thus, (6.46) holds.

Now, using (6.293), (6.294), and (6.295), we see that the homogenized variational inequality of (6.286) is the following:

$$\begin{cases} \int_\Omega |\nabla u|^{p-2} \nabla u \nabla(v-u) - \lambda \int_\Omega |u|^{p-2} u(v-u) \geq 0, \ \forall v \in W_0^{1,p}(\Omega), \\ u \in W_0^{1,p}(\Omega), \end{cases}$$

(6.299)

which, in turn, is equivalent to

$$\begin{cases} \int_\Omega |\nabla u|^{p-2} \nabla u \nabla v - \lambda \int_\Omega |u|^{p-2} uv = 0, \ \forall v \in W_0^{1,p}(\Omega), \\ u \in W_0^{1,p}(\Omega). \end{cases}$$

(6.300)

Theorem 6.4 permits us to reduce the investigation of the global bifurcation of (6.286) to the study of eigenvalues and eigenvectors of (6.300), which was done in [4], [8], [26], [30], [31], [34], [84]. As in section 6.1, we know that, for each $f \in W^{-1,p'}(\Omega) = [W_0^{1,p}(\Omega)]^*$, the equation

$$\begin{cases} \int_\Omega |\nabla u|^{p-2} \nabla u \nabla v - \int_\Omega fv = 0, \ \forall v \in W_0^{1,p}(\Omega), \\ u \in W_0^{1,p}(\Omega), \end{cases}$$

(6.301)

has a unique solution

$$u = u_f = P_{A,J}(f) \equiv P_0(f).$$

Hence, (6.300) is equivalent to

$$
\begin{aligned}
u &= P_0(\lambda |u|^{p-2} u) \\
&= P_0[f(u, \lambda)].
\end{aligned}
\tag{6.302}
$$

From [4], [8], and [26], we know that the first eigenvalue λ_1 of (6.300), characterized by

$$
\lambda_1 = \inf \left\{ \int_\Omega |\nabla u|^p : u \in W_0^{1,p}(\Omega), \int_\Omega |u|^p = 1 \right\} > 0,
\tag{6.303}
$$

is a simple, isolated eigenvalue of (6.300). Hence,

$$
\overline{\lambda_2} \equiv \inf\{\lambda > \lambda_1 : \lambda \text{ is an eigenvalue of (6.300)}\} > \lambda_1.
$$

Using this result, del Pino and Manásevich ([31]) have proved the following result about the computation of the degree of the operator in (6.302) for λ passing through λ_1:

Theorem *(Proposition 2.2, [31], Theorem 14.18, [34]) For $r > 0$,*

$$
d(I - P_0[f(\cdot, \lambda)], B_r(0), 0) = \begin{cases} 1 & \text{if } \lambda < \lambda_1, \\ -1 & \text{if } \lambda_1 < \lambda < \overline{\lambda_2}. \end{cases}
$$

Using this result and Theorem 6.4, we obtain the following global bifurcation property of (6.286):

Corollary 6.26 $(0, \lambda_1)$ *is a bifurcation point of (6.286), and the associated global bifurcation branch satisfies the alternative in Theorem 6.4.*

Proof. We need only to apply Theorem 6.4 (II) with $a = \lambda_1 - \epsilon < \lambda_1 < b = \lambda_1 + \epsilon$, $\epsilon > 0$ small. Then, by the theorem cited above,

$$
d(I - P_0[f(\cdot, \lambda_1 - \epsilon)], B_r(0), 0) = 1 \neq -1 = d(I - P_0[f(\cdot, \lambda_1 + \epsilon)], B_r(0), 0).
$$

■ In the ODE case, i.e., $N = 1$, we can say more about bifurcation from higher eigenvalues. For example, we can choose $\Omega = (0, 1)$, $\gamma = \{0\}$, and, then,

$$
V = \{u \in W^{1,p}(0,1) : u(0) = 0\}.
$$

In this case, $\partial\Omega \setminus \Gamma = \{1\}$, and j becomes

$$
j(u) = k|u(1)|, \ u \in V,
$$

where $k > 0$ is a given constant.

Much more has been studied in this case than in the case of general dimension. In [84], the author proved that the spectrum of the p-Laplacian is very much similar to that of the Laplacian, i.e., the eigenvalues of (6.300) form a countable discrete sequence $\{\lambda_i\}$, $0 < \lambda_1 < \lambda_2 < \ldots$ with $\lim \lambda_n = \infty$, and, moreover, each λ_i is simple.

Furthermore, we have the same degree calculation as the usual calculations of degrees of linear, compact perturbations of the identity mapping, namely, the following theorem:

Theorem *(Theorem 4.1, [30], Theorem 14.9, [34]) Let $\lambda > 0$, $\lambda \neq \lambda_n$, $\forall n \in \mathbb{N}$. Then, for every $r > 0$,*

$$\mathrm{d}(I - P_0[f(\cdot, \lambda)], B_r(0), 0) = (-1)^{\beta(\lambda)},$$

where $\beta(\lambda)$ is the number of eigenvalues λ_n of (6.300) less than λ.

Using this theorem together with Theorem 6.4, we immediately obtain the following corollary:

Corollary 6.27 *Let $N = 1$. Then, the set of bifurcation points of (6.286) is*

$$\{(0, \lambda_n) : n \in \mathbb{N}\},$$

where λ_n, $n = 1, 2, \ldots$ are eigenvalues of (6.300). Moreover, each global bifurcation branch emanating from $(0, \lambda_n)$ satisfies the alternative in Theorem 6.4.

To prove this corollary, we just need to observe that, for $n \in \mathbb{N}$, $a = \lambda_n - \epsilon$, $b = \lambda_n + \epsilon$, $\epsilon > 0$ small,

$$\mathrm{d}(I - P_0[f(\cdot, \lambda_n \pm \epsilon)], B_r(0), 0) \in \{-1, 1\},$$

and

$$\mathrm{d}(I - P_0[f(\cdot, \lambda_n - \epsilon)], B_r(0), 0) = -\mathrm{d}(I - P_0[f(\cdot, \lambda_n + \epsilon)], B_r(0), 0).$$

Example 6.12 [Bifurcation for a variational inequality containing the anisotropic p-Laplacian] This example is concerned with bifurcation for a variational inequality similar to (6.286) with the anisotropic version of the p-Laplacian:

$$\begin{cases} \int_\Omega \sum_{i=1}^N \int_\Omega a_i(x)|\partial_i u|^{p-2}\partial_i u(\partial_i v - \partial_i u)dx - \int_\Omega [\lambda b|u|^{p-2}u \\ \qquad + g(x, u, \lambda)](v - u) + j(v) - j(u) \geq 0, \ \forall v \in V, \\ u \in V. \end{cases} \qquad (6.304)$$

Here, the coefficients a_i satisfy the conditions in Remark 6.1 (4), $b \in L^\infty(\Omega)$, and g is as in Example 6.11. Letting V be as in that example, one sees that (6.304) is of the form (6.8) with A given as in Remark 6.1 (4) and

$$\langle B(u, \lambda), v \rangle = \int_\Omega [\lambda b|u|^{p-2}u + g(\cdot, u, \lambda)]v.$$

A satisfies (A2), and, moreover, as A is homogeneous of degree $p-1$, (6.37) holds with $\alpha = A$.

As in Example 6.11, one can show that B is completely continuous and its homogenized mapping f is given by

$$\langle f(u, \lambda), v \rangle = \lambda \int_\Omega b|u|^{p-2}uv.$$

Thus, the homogenization of (6.304) is given by

$$
\begin{cases}
\displaystyle\int_\Omega \sum_{i=1}^N \int_\Omega a_i(x)|\partial_i u|^{p-2}\partial_i u(\partial_i v - \partial_i u)dx - \lambda \int_\Omega b|u|^{p-2}u(v - u) \\
+J(v) - J(u) \geq 0, \ \forall v \in V, \\
u \in V,
\end{cases}
$$

$$(6.305)$$

where J is the homogenized functional corresponding to j. The results in sections 6.2, 6.3, and 6.4 give us relationships between eigenvalues of (6.305) and bifurcation points of (6.304). For example, if $J = I_{W_0^{1,p}(\Omega)}$, then, (6.305) becomes a (nonlinear) equation on Ω. One can define multiplicities of the eigenvalues of (6.305) (cf. [84]) and relate them to degrees of the solution mapping of (6.305) (cf. [34]). Together with the above theorems, those results yield conditions for global bifurcation of (6.304).

Remark 6.5   (a) In the particular case where $\Gamma = \partial\Omega$, i.e., $V = W_0^{1,p}(\Omega)$, and $j \equiv 0$, the variational inequality (6.286) becomes the following nonlinear equation, considered in [31] and [34]:

$$
\begin{cases}
\displaystyle\int_\Omega |\nabla u|^{p-2}\nabla u \nabla v - \int_\Omega [\lambda|u|^{p-2}u + g(x, u, \lambda)]v = 0, \ \forall v \in W_0^{1,p}(\Omega), \\
u \in W_0^{1,p}(\Omega).
\end{cases}
$$

$$(6.306)$$

Corollaries 6.26 and 6.27 give global bifurcation results for (6.306) and are, therefore, the same as Theorem 1.1 [31] and Theorems 14.8 and 14.9, [34].

(b) Note that we have results similar to Corollaries 6.26 and 6.27 for other choices of the functional j or for variational inequalities containing some small perturbations of the p-Laplacian operator A.

(c) Bifurcation results for more general classes of quasilinear, elliptic, partial differential equations have recently become available. We refer to [41] and [42] and the references in these papers for such directions of investigation.

7

Bifurcation from Infinity in Banach Spaces

In this chapter, we develop results for bifurcation from infinity for variational inequalities defined in reflexive Banach spaces and containing nonlinear operators and convex functionals (that are not necessarily indicator functions of convex sets). These theorems generalize a number of results in Chapter 5 for variational inequalities in Hilbert spaces and are parallel to those in Chapter 6 about bifurcation from trivial solutions. General results are presented in Section 7.1. Some applications and examples are considered in Section 7.2.

7.1 Existence of global bifurcation branches

7.1.1 An asymptotic bifurcation theorem

Let $V, V^*, \|\cdot\|, \langle\cdot,\cdot\rangle$ and A, B, j be as in section 6.1. In particular, we assume that (A1) and (A2) are satisfied. We consider the problem of bifurcation from infinity for the variational inequality (6.8).

As in Chapter 5, we say that (∞, λ) ($\lambda \in \mathbb{R}$) is an (asymptotic) bifurcation point of (6.8), if there exists a sequence $\{(u_n, \lambda_n)\}$ of solutions of (6.8), such that $\lambda_n \to \lambda_0$ and $\|u_n\| \to \infty$, as $n \to \infty$. We say that bifurcation occurs at $\{\infty\} \times [a, b]$ ($a, b \in \mathbb{R}, a < b$) if there exists a sequence $\{(u_n, \lambda_n)\}$ of solutions of (6.8), such that $\lambda_n \in [a, b]$, $\forall n$, and $\|u_n\| \to \infty$ ($n \to \infty$).

We recall that $P = P_{A,j}$ is the solution mapping defined by (6.14) and that (6.8) can be written in the operator form (6.15).

Applying Theorem 2.6, Chapter 2, we have the following generalization of Theorem 5.1 in Chapter 5.

Theorem 7.1 *Suppose $a < b$ are such that (6.8) with $\lambda = a$ and $\lambda = b$ does not have any solution with a large norm (i.e., there exists $R_0 > 0$, such that (6.8) with $\lambda = a, b$ does not have solutions u with $\|u\| \geq R_0$).*
Furthermore, assume that

$$d(I - P[B(\cdot, a)], B_R(0), 0) \neq d(I - P[B(\cdot, b)], B_R(0), 0), \qquad (7.1)$$

$(R \geq R_0)$. Then, there exists a bifurcation point (λ, ∞) with $a < \lambda < b$, and there exists, moreover, a continuum

$$\mathcal{C} \subset \{(u, \lambda) \in V \times [a, b] : (u, \lambda) \text{ is a solution of (6.8)}\},$$

which is unbounded in $V \times [a, b]$, and either
(i) $\mathcal{C}$ is unbounded in the λ direction, or
(ii) there exists an interval $[c, d]$, such that $(c, d) \cap (a, b) = \emptyset$ and $\mathcal{C}$ bifurcates from infinity in $V \times [c, d]$.

7.1.2 Homogenization at infinity

In this section, we will homogenize (6.8) at infinity to obtain a homogeneous variational inequality, which, in turn, will aid in the calculation of the degrees and indices.

Again, we prove that asymptotic bifurcation points of (6.8) are related to eigenvalues of the asymptotically homogenized variational inequality and the degrees of the operators associated with these variational inequalities are equal. First, we state some necessary assumptions.

First, we assume that A and B are differentiable at infinity in the following sense:

(A9) There exist $\alpha_\infty : V \to V^*$, $f_\infty : V \times \mathbb{R} \to V^*$ such that

(a) α_∞ satisfies (A1), (A2), and for all sequences $\{v_n\} \subset V$, $\{\sigma_n\} \subset \mathbb{R}_*^+$ satisfying

$$v_n \to v \text{ in } V, \ \sigma_n \to \infty,$$

$$\frac{1}{\sigma_n^{p-1}} A(\sigma_n v_n) \to \alpha_\infty(v) \text{ in } V^*. \qquad (7.2)$$

(b) A_{σ_n} $(n \in \mathbb{N})$ belongs to class (S) uniformly with respect to n in the sense of (6.19) and (6.20), and A_σ is given by (6.38).

(A10) f_∞ is completely continuous on $V \times \mathbb{R}$, and, for all sequences

$$\{v_n\} \subset V, \ \{\sigma_n\} \subset \mathbb{R}_*^+, \ \{\lambda_n\} \subset \mathbb{R}$$

satisfying

$$v_n \rightharpoonup v \text{ in } V, \ \lambda_n \to \lambda, \ \sigma_n \to \infty,$$

$$\frac{1}{\sigma_n^{p-1}} B(\sigma_n v_n, \lambda_n) \to f_\infty(v, \lambda) \text{ in } V^*. \tag{7.3}$$

We see that α_∞ and f_∞ are uniquely determined by (7.2) and (7.3), because

$$\alpha_\infty(v) = \lim_{t \to \infty} \frac{A(tv)}{t^{p-1}}, \text{ and } f_\infty(v, \lambda) = \lim_{t \to \infty} \frac{B(tv, \lambda)}{t^{p-1}},$$

for all $v \in V, \lambda \in \mathbb{R}$.

Moreover, α and $f(\cdot, \lambda)$ are positive homogeneous of order $p - 1$ in the sense of (6.42):

$$\alpha_\infty(\sigma v) = \sigma^{p-1} \alpha_\infty(v), \ f_\infty(\sigma v, \lambda) = \sigma^{p-1} f_\infty(v, \lambda), \tag{7.4}$$

for all $\sigma \geq 0, v \in V, \lambda \in \mathbb{R}$. The proof of these properties is similar to that of (6.42).

Next, we consider the homogenization of j at infinity. For $\sigma > 0$, we denote by j_σ the functional from V to $[0, \infty]$, defined by (6.44).

(A11) We assume that there exists a proper, convex, lower semicontinuous functional

$$J_\infty : V \to [0, \infty],$$

such that j_σ tends to J_∞ (as $\sigma \to \infty$) in the following sense:

(a) If $v_n \rightharpoonup v$ in V and $\sigma_n \to \infty$, then,

$$J_\infty(v) \leq \liminf j_{\sigma_n}(v_n). \tag{7.5}$$

(b) For each $v \in V$ and each sequence $\{\sigma_n\} \subset \mathbb{R}_*^+$, such that $\sigma_n \to \infty$, we can choose a sequence $\{v_n\} \subset V$, such that

$$\begin{cases} v_n \to v \text{ in } V \text{ and} \\ j_{\sigma_n}(v_n) \to J_\infty(v), \text{ as } n \to \infty. \end{cases} \tag{7.6}$$

We note that $J_\infty(0) = 0$ and that J_∞, if it exists, is uniquely determined by (7.5) and (7.6). The proof is similar to the case of homogenization at 0 and is omitted. As is J, J_∞ is positive homogeneous of degree p:

$$J_\infty(\sigma u) = \sigma^p J_\infty(u), \ \forall u \in V, \sigma \geq 0. \tag{7.7}$$

This is clearly true for $\sigma = 0$. For $\sigma > 0$ fixed, we define

$$J_1 : V \to [0, \infty], \ J_1(v) = \sigma^{-p} J_\infty(\sigma v), \ v \in V.$$

Then, J_1 is a proper convex, lower semicontinuous functional on V. As in section 6.1, we can also check that J_1 satisfies (7.5) and (7.6). By the above uniqueness property of J_∞, $J_1 = J_\infty$, and (7.7) follows.

Using the assumptions (A9)-(A11), we can now homogenize (6.8) at infinity and obtain the following variational inequality:

$$\begin{cases} \langle \alpha_\infty(u) - f_\infty(u, \lambda), v - u \rangle + J_\infty(v) - J_\infty(u) \geq 0, \ \forall v \in V, \\ u \in V. \end{cases} \tag{7.8}$$

Let us consider the particular case of the functional j in Chapter 5 : $j = I_K$, where K is a closed, convex set in V containing 0, and $A = I$. In this case, $K = D(j) = \ker j$. Let rcK be the recession cone of K, defined in Chapter 5, and let J_∞ be the indicator function of rcK.

Then, we can prove that (7.5) and (7.6) are satisfied in the particular case in Chapter 5 where j and J_∞ are respectively the indicator functions of K and rcK.

We also note that, if A satisfies (6.9), then,

$$\begin{aligned} \langle \alpha_\infty(u) - \alpha_\infty(v), u - v \rangle &= \lim_{t \to \infty} \left\langle \frac{A(tu)}{t} - \frac{A(tv)}{t}, u - v \right\rangle \\ &\geq C\|u - v\|^p, \ \forall t > 0. \end{aligned}$$

Hence, α_∞ also satisfies (6.9) and, therefore, (A1)-(A2).

Because α_∞ satisfies (A1)-(A2) and J_∞ is convex, lower semicontinuous, and nonnegative on V, we see, as before (Theorem 2.1), that, for all $f \in V^*$, the variational inequality,

$$\begin{cases} \langle \alpha_\infty(u) - f, v - u \rangle + J_\infty(v) - J_\infty(u) \geq 0, \ \forall v \in V, \\ u \in V, \end{cases} \tag{7.9}$$

has a unique solution,

$$u = u_f = P_{\alpha_\infty, J_\infty}(f) := P_\infty(f).$$

With this notation, we see that (7.8) is equivalent to the following operator equation:

$$u = P_\infty[f_\infty(u, \lambda)]. \tag{7.10}$$

From (7.4) and (7.8),

$$P_\infty(tf) = t^{\frac{1}{p-1}} P_\infty(f)$$

for all $f \in V^*, t \geq 0$, and, thus, if u is a solution of (7.8) (or (7.10)), so is tu, for all $t \geq 0$. The definitions of eigenvalues and eigenvectors of (7.8) and (7.10) used in Chapter 6 (section 6.1) will be carried over to the present situation.

We have the following relationships between the asymptotic bifurcation problem for (6.8) and the eigenvalue problem of the asymptotically homogenized variational inequality (7.8).

Theorem 7.2 *(I) If (∞, λ) is an asymptotic bifurcation point of (6.8), then, λ is an eigenvalue of (7.8).*

(II) If a and b $(a < b)$ are not eigenvalues of (7.8) and if

$$d(I - P_\infty[f_\infty(\cdot, a)], B_R(0), 0) \neq d(I - P_\infty[f_\infty(\cdot, b)], B_R(0), 0), \quad (7.11)$$

for some $R > 0$, then, there exists a bifurcation point (∞, λ) with $a < \lambda < b$ (λ is an eigenvalue of (7.8)) and a continuum C of solutions of (6.8) which is unbounded in $V \times [a, b]$, and either

(i) C is unbounded in the λ direction, or

(ii) there exists an interval $[c, d]$, such that $(c, d) \cap (a, b) = \emptyset$ and C bifurcates from infinity in $V \times [c, d]$.

Proof. The ideas for proving this theorem are combinations of those used in the proofs of Theorem 6.4 and Theorem 5.2. Hence, we present, here, only the main steps of the proof. Note that, if a, b are not eigenvalues of (7.8), then, 0 is the only zero of $I - P_\infty[f_\infty(\cdot, a)]$ and $I - P_\infty[f_\infty(\cdot, b)]$, and the degrees in (7.11) are defined for all $R > 0$ (and do not depend on R). For $\sigma \in [0, 1]$, $u \in V$, $\lambda \in \mathbb{R}$, we define

$$A_\sigma^\infty(u) = \begin{cases} \sigma^{p-1} A(\sigma^{-1} u) & \text{if} \quad \sigma \in (0, 1], \\ \alpha_\infty(u) & \text{if} \quad \sigma = 0, \end{cases} \quad (7.12)$$

$$B_\sigma^\infty(u, \lambda) = \begin{cases} \sigma^{p-1} B(\sigma^{-1} u, \lambda) & \text{if} \quad \sigma \in (0, 1], \\ f_\infty(u, \lambda) & \text{if} \quad \sigma = 0, \end{cases} \quad (7.13)$$

$$j_\sigma^\infty(u) = \begin{cases} \sigma^p j(\sigma^{-1} u) & \text{if} \quad \sigma \in (0, 1], \\ J_\infty(u) & \text{if} \quad \sigma = 0. \end{cases} \quad (7.14)$$

Now, let $\{\sigma_n\}$ be a sequence in $[0, 1]$, $\sigma_n \to \sigma$. We can prove that

$$A_{\sigma_n}^\infty \to A_\sigma^\infty, \quad j_{\sigma_n}^\infty \to j_\sigma^\infty, \quad \text{as} \quad n \to \infty$$

in the sense of (A4) and (A5), i.e., $A_{\sigma_n}^\infty$ belong to class (S) uniformly for $n \in \mathbb{N}$, and if $v_n \to v$ (respectively, $v_n \rightharpoonup v$) in V, then,

$$A_\sigma^\infty(v) = \lim_{n \to \infty} A_{\sigma_n}^\infty(v_n) \text{ (respectively, } j_\sigma^\infty(v) \leq \liminf_{n \to \infty} j_{\sigma_n}^\infty(v_n)).$$

Moreover, for each $v \in V$, there exists a sequence $\{v_n\} \subset V$, such that $v_n \to v$ and

$$j_\sigma^\infty(v) = \lim_{n \to \infty} j_{\sigma_n}^\infty(v_n).$$

The (somewhat lengthy) proof of these properties is accomplished by using arguments similar to those used in the proof of Theorem 6.4, and we do not present it here. Now, by applying Lemma 6.1,

$$P_{A_{\sigma_n}^\infty, j_{\sigma_n}^\infty}(f_n) \to P_{A_\sigma^\infty, j_\sigma^\infty}(f) \text{ in } V, \quad (7.15)$$

whenever $\sigma_n \to \sigma$ in $[0,1]$ and $f_n \to f$ in V^* ($P_{A,j}$ is defined by (6.14)). Using this property, (7.3), and the complete continuity of B, one can show that the mapping

$$(\sigma, v, \lambda) \mapsto P_{A_\sigma^\infty, j_\sigma^\infty}[B_\sigma^\infty(v, \lambda)] \qquad (7.16)$$

is completely continuous from $[0,1] \times V \times \mathbb{R}$ to V.

To prove (I), we suppose that (∞, λ) is an asymptotic bifurcation point of (6.8) and that $\{u_n\}$ and $\{\lambda_n\}$ satisfy

$$\|u_n\| \to \infty, \ \lambda_n \to \lambda \ (n \to \infty),$$

and

$$\langle A(u_n) - B(u_n, \lambda_n), v - u_n \rangle + j(v) - j(u_n) \geq 0, \ \forall v \in V. \qquad (7.17)$$

Letting $v_n = \|u_n\|^{-1} u_n$ and dividing both sides of (7.17) by $\|u_n\|^p$,

$$\left\langle \frac{A(u_n)}{\|u_n\|^{p-1}} - \frac{B(u_n, \lambda_n)}{\|u_n\|^{p-1}}, \frac{v}{\|u_n\|} - \frac{u_n}{\|u_n\|} \right\rangle + \frac{j(v)}{\|u_n\|^p} - \frac{j(u_n)}{\|u_n\|^p} \geq 0,$$

for all $n \in \mathbb{N}, v \in V$. This implies that

$$\langle A_{1/\|u_n\|}^\infty(v_n) - B_{1/\|u_n\|}^\infty(v_n, \lambda_n), w - v_n \rangle + j_{1/\|u_n\|}^\infty(w) - j_{1/\|u_n\|}^\infty(v_n) \geq 0,$$

for all n, all $w \in V, w = v/\|u_n\|$. It follows from (7.12) and (7.14) that this variational inequality is equivalent to

$$v_n = P_{A_{1/\|u_n\|}^\infty, j_{1/\|u_n\|}^\infty}\left[B_{1/\|u_n\|}^\infty(v_n, \lambda_n)\right]. \qquad (7.18)$$

By passing to a subsequence, we may assume that

$$v_n \rightharpoonup v \text{ in } V.$$

Because $1/\|u_n\| \to 0$, we see, by the complete continuity of the mapping in (7.16), that

$$\lim_{n\to\infty} P_{A_{1/\|u_n\|}^\infty, j_{1/\|u_n\|}^\infty}[B_{1/\|u_n\|}^\infty(v_n, \lambda_n)] = P_{A_0^\infty, j_0^\infty}[B_0^\infty(v, \lambda)]$$
$$= P_{\alpha_\infty, J_\infty}[f_\infty(v, \lambda)]$$
$$= P_\infty[f_\infty(v, \lambda)] \text{ in } V.$$

Using (7.18), this implies that

$$v_n \to v \text{ in } V$$

and

$$v = P_\infty[f_\infty(v, \lambda)].$$

Hence, λ is an eigenvalue of (7.8) with a corresponding eigenvector v. (I) is proved.

To prove (II), we assume that a and b are not eigenvalues of (7.8). We prove that 0 is an isolated solution of (6.8) with $\lambda = a$ and, for $R > 0$ sufficiently large,

$$\mathrm{d}(I - P[B(\cdot, a)], B_R(0), 0) = \mathrm{d}(I - P_\infty[f_\infty(\cdot, a)], B_R(0), 0). \qquad (7.19)$$

To prove this, we need only to show that there exists $R > 0$ sufficiently large, such that, for all $\sigma \in [0, 1]$, the equation

$$u - P_{A_\sigma^\infty, j_\sigma^\infty}[B_\sigma^\infty(u, a)] = 0 \qquad (7.20)$$

has no solution in $V \setminus B_R(0)$.

Suppose that this is not the case and there exist sequences $\{u_n\} \subset V, \{\sigma_n\} \subset [0, 1]$, such that $\|u_n\| \to \infty$ ($n \to \infty$) and

$$u_n = P_{A_{\sigma_n}^\infty, j_{\sigma_n}^\infty}[B_{\sigma_n}^\infty(u_n, a)], \ \forall n,$$

or, in the variational inequality form,

$$\langle A_{\sigma_n}^\infty(u_n) - B_{\sigma_n}^\infty(u_n, a), v - u_n \rangle + j_{\sigma_n}^\infty(v) - j_{\sigma_n}^\infty(u_n) \geq 0, \ \forall v \in V,$$

or, by (7.12)-(7.14),

$$\sigma_n^{p-1} \langle A(\sigma_n^{-1} u_n) - B(\sigma_n^{-1} u_n, a), v - u_n \rangle + \sigma_n^p j(\sigma_n^{-1} v) - \sigma_n^p j(\sigma_n^{-1} u_n) \geq 0,$$
$$\forall v \in V.$$

As before, by setting $v_n = u_n / \|u_n\|$ and dividing this inequality by $\|u_n\|^p$,

$$\left\langle \frac{A(\sigma_n^{-1} \|u_n\| v_n)}{\sigma_n^{1-p} \|u_n\|^{p-1}} - \frac{B(\sigma_n^{-1} \|u_n\| v_n, a)}{\sigma_n^{1-p} \|u_n\|^{p-1}}, \frac{v}{\|u_n\|} - v_n \right\rangle$$
$$+ \frac{1}{\sigma_n^{-p} \|u_n\|^p} j\left(\frac{\|u_n\|}{\sigma_n} \frac{v}{\|u_n\|} \right) - \frac{1}{\sigma_n^{-p} \|u_n\|^p} j\left(\frac{\|u_n\|}{\sigma_n} v_n \right) \geq 0, \ \forall v \in V.$$

Letting $w = v / \|u_n\|$,

$$\langle A_{\sigma_n/\|u_n\|}^\infty(v_n) - B_{\sigma_n/\|u_n\|}^\infty(v_n, a), w - v_n \rangle + j_{\sigma_n/\|u_n\|}^\infty(w)$$
$$- j_{\sigma_n/\|u_n\|}^\infty(v_n) \geq 0, \quad \forall w \in V,$$

which is equivalent to

$$v_n = P_{A_{\sigma_n/\|u_n\|}^\infty, j_{\sigma_n/\|u_n\|}^\infty}\left[B_{\sigma_n/\|u_n\|}^\infty(v_n, a) \right]. \qquad (7.21)$$

Now, assume that $v_n \rightharpoonup v$ in V.

Because $\sigma_n/\|u_n\| \to 0$, we see that

$$P_{A^\infty_{\sigma_n/\|u_n\|}, j^\infty_{\sigma_n/\|u_n\|}} \left[B^\infty_{\sigma_n/\|u_n\|}(v_n, a) \right] \to P_\infty[f_\infty(v, a)] \text{ in } V.$$

Hence, (7.21) implies that $v_n \to v$ in V and

$$v = P_\infty[f_\infty(v, a)].$$

This contradicts the fact that a is not an eigenvalue of (7.8) and proves the existence of $R > 0$, such that (7.20) has no solutions with $\|u\| \geq R$.

By the homotopy invariance property of the Leray–Schauder degree, from (7.20),

$$
\begin{aligned}
d(I - P_\infty[f_\infty(\cdot, a)], B_R(0), 0) &= d(I - P_{A^\infty_0, j^\infty_0}[B^\infty_0(\cdot, a)], B_R(0), 0) \\
&= d(I - P_{A^\infty_1, j^\infty_1}[B^\infty_1(\cdot, a)], B_R(0), 0) \\
&= d(I - P[B(\cdot, a)], B_R(0), 0),
\end{aligned}
$$

and (7.19) is proved.

We have a similar equality with a replaced by b. Our conclusion follows from Theorem 7.1.   ∎

7.1.3 Some degree calculations

In this section, we apply Theorem 7.2 together with theorems in previous chapters about calculations of degrees of the operators associated with the homogenized variational inequalities to obtain results about the global behavior of the asymptotic bifurcation branches of (6.8).

First, we consider the case where $W = D(J_\infty)$ is a (closed) vector subspace of V and $J_\infty \equiv 0$ on W. In this case, (7.8) is equivalent to the following equation on W:

$$
\begin{cases}
\langle \alpha_\infty(u) - f_\infty(u, \lambda), v \rangle = 0, \ \forall v \in W, \\
u \in W.
\end{cases}
\tag{7.22}
$$

We assume, furthermore, that α_∞ is a linear operator from V to V^* and that B can be written as

$$B(u, \lambda) = \lambda \beta u + G(u, \lambda), \tag{7.23}$$

where $\beta \in L(V, V^*)$ is compact and $G : V \times \mathbb{R} \to V^*$ is a completely continuous (nonlinear) mapping, such that

$$\frac{\|G(u, \lambda)\|}{\|u\|} \to 0,$$

as $\|u\| \to \infty$, uniformly for λ in bounded intervals. By arguments similar to those used in section 5.1, in this case, the mapping $f_\infty : V \times \mathbb{R} \to \mathbb{R}$, given by

$$f_\infty(u, \lambda) = \lambda \beta u, \ u \in V, \lambda \in \mathbb{R}, \tag{7.24}$$

is the partial asymptotic derivative of B with respect to u in the sense of (A10).

(7.22) can therefore be written as

$$\begin{cases} \langle \alpha_\infty(u) - \lambda\beta(u), v \rangle = 0, \ \forall v \in W, \\ u \in W, \end{cases} \tag{7.25}$$

which is a linear equation in W. For $g \in W^*$, we denote by

$$P_{\alpha_\infty}(g) = u_g$$

the unique solution of the (linear) equation:

$$\begin{cases} \langle \alpha_\infty(u_g) - g, v \rangle = 0, \ \forall v \in W \\ u_g \in W. \end{cases}$$

Hence, P_{α_∞} is a continuous linear mapping from W^* to W, and (7.25) is equivalent to

$$\begin{cases} u = \lambda P_{\alpha_\infty}(\beta u) \\ u \in W. \end{cases} \tag{7.26}$$

Note that, because $V^* \subset W^*$, we can consider $\beta|_W$ as a mapping from W to W^* and $\beta|_W \in L(W, W^*)$. Moreover, because $\beta|_W$ is compact, $P_{\alpha_\infty}\beta = P_{\alpha_\infty}\beta|_W$ is a compact linear mapping from W into itself. Under these assumptions, we have the following consequence of Theorem 7.2.

Corollary 7.3   *(6.8) has, at most, a countable number of asymptotic bifurcation points. Moreover,*

(I) If (∞, λ) is an asymptotic bifurcation point of (6.8), then, λ is an eigenvalue of (7.26). In particular, if $W = \{0\}$, then, (6.8) has no finite asymptotic bifurcation point.

(II) If λ is an eigenvalue of (7.26) of odd multiplicity, then, (∞, λ) is an asymptotic bifurcation point of (6.8) corresponding to an asymptotic bifurcation branch that satisfies the alternative in Theorem 7.2.

The proof of this corollary is similar to that of Corollary 6.6, and it is, therefore, omitted.

We consider some analogs of Theorems 6.8 and 6.9 and their corollaries for bifurcation from infinity of (6.8). Again, we assume that α_∞ and $f_\infty(\cdot, \lambda)$ are linear bounded operators from V to V^*. Let α_∞^* and $f_\infty^*(\cdot, \lambda)$ be their adjoints, which are also linear, bounded operators from V to V^*. We assume the following:

- f_∞^* can be written as a sum

$$f_\infty^*(u, \lambda) = g(u, \lambda) + h(u), \quad u \in V, \lambda \in \mathbb{R}, \tag{7.27}$$

where g is homogeneous of degree γ ($\gamma > 0$) with respect to $\lambda \in \mathbb{R}^+$,

- 0 is not an eigenvalue of (7.8), i.e., $0 \in V$ is the unique solution of (7.8) with $\lambda = 0$ and (7.8) is monotone in the sense that

$$\langle \alpha_\infty u - f_\infty(u, 0), u \rangle \geq 0, \quad \forall u \in D(J_\infty). \tag{7.28}$$

We note that this condition is satisfied if $\alpha_\infty - f_\infty(\cdot, 0)$ is strictly monotone on $D(J_\infty)$, i.e.,

$$\langle \alpha_\infty u - f_\infty(u, 0), u \rangle > 0, \quad \forall u \in D(J_\infty) \setminus \{0\}. \tag{7.29}$$

In particular, if B and f_∞ are given by (7.23), (7.24), then, $f_\infty(\cdot, 0) = 0$, and (7.29) holds immediately.

The following result is the counterpart of Theorem 6.8 for bifurcation from infinity.

Theorem 7.4 *Suppose that λ_0 is an eigenvalue of (7.8). Let*

$$\begin{aligned}
K_\infty(\lambda_0) &= \ker\left[\alpha_\infty^* - f_\infty^*(\cdot, \lambda_0)\right] \cap \ker J_\infty \\
&= \{u \in V : \alpha_\infty^* u - f_\infty^*(u, \lambda_0) = 0, J_\infty(u) = 0\}.
\end{aligned}$$

Assume that either one of the following conditions is satisfied:

(a) λ_0 is a simple eigenvalue of (7.8) with a corresponding eigenvector u_1. Moreover, there exists $u_0 \in K_\infty(\lambda_0)$, such that

$$\langle \alpha_\infty^* u_0 - h(u_0), u_1 \rangle > 0. \tag{7.30}$$

(b) $K_\infty(\lambda_0)$ is not symmetric, and, for each eigenvector u_1 of (7.8) corresponding to λ_0, there exists $u_0 = u_0(u_1) \in K_\infty(\lambda_0)$, such that (7.30) holds. Then, there exists a global asymptotic bifurcation branch of solutions of (6.8), which emanates from $\{\infty\} \times [0, \lambda_0]$ and satisfies the alternative in Theorem 7.2.

Proof. The proof of this theorem follows the same lines as that of Theorems 6.8 and 5.7. Hence, we just sketch its outline here very briefly.

First, by using the assumption (7.28), and the fact that the family of completely continuous perturbations of the identity,

$$\{H(u, t) : 0 \leq t \leq 1, u \in V\},$$

given by

$$H(u, t) = u - P_\infty[t f_\infty(u, 0)], \quad u \in V, 0 \leq t \leq 1,$$

has the property that

$$H(u,t) \neq 0, \ \forall u \in \partial B_R(0), \ t \in [0,1] \ (R > 0),$$

one can show that

$$d(I - P_\infty[f_\infty(\cdot, 0)], B_R(0), 0) = 1, \tag{7.31}$$

for all $R > 0$.

On the other hand, using the family $\{H(t, u, \lambda) : t \in [0,1], u \in V, \lambda \in \mathbb{R}\}$, with

$$H(t, u, \lambda) = u - P_\infty[(1-t)f_\infty(u, \lambda) + tf_\infty(u, \lambda_0) + t\phi(u_0)]$$

(ϕ is a duality mapping on V satisfying (6.122)), we can show, as in Theorem 6.8, that there exist $R_0 > 0$, $\lambda_1 > \lambda_0$, such that

$$H(t, u, \lambda) \neq 0, \ \forall \lambda \in (\lambda_0, \lambda_1), \ \forall u \in V \setminus B_{R_0}(0) \tag{7.32}$$

and

$$H(1, u, \lambda_0) \neq 0, \ \forall u \in V. \tag{7.33}$$

These statements imply that, for $R > R_0$, $\lambda \in (\lambda_0, \lambda_1)$,

$$
\begin{aligned}
d(I - P_\infty[f_\infty(\cdot, \lambda)], B_R(0), 0) &\equiv d(I - H(0, \cdot, \lambda), B_R(0), 0) \\
&= d(I - H(1, \cdot, \lambda), B_R(0), 0) \\
&= d(I - H(1, \cdot, \lambda_0), B_R(0), 0) \\
&= 0.
\end{aligned}
\tag{7.34}
$$

Choosing $\lambda = \lambda_0 + \epsilon$, $\epsilon > 0$ small, it follows from (7.31) and (7.34) and Theorem 7.2 (II) that there exists an asymptotic bifurcation branch of solutions of (6.8) emanating from $\{\infty\} \times [0, \lambda_0]$, such that the alternative in Theorem 7.2 is satisfied.

For the proof of (b), we consider, instead, the homotopy family $\{H(t, u, \lambda) : t \in [0,1], u \in V, \lambda \in \mathbb{R}\}$, given by

$$H(t, u, \lambda) = u - P_\infty[(1-t)f_\infty(u, \lambda) + tf_\infty(u, \lambda_0) + t\psi],$$

where $\psi \in V^*$ is chosen, such that

$$
\begin{cases}
\langle \psi, u \rangle \geq 0, \ \forall u \in K_\infty(\lambda_0), \\
\langle \psi, x_0 \rangle < 0, \ \text{and} \ \langle \psi, \overline{u} \rangle > 0,
\end{cases}
$$

for some $x_0 \in \overline{K_\infty(\lambda_0) - K_\infty(\lambda_0)} \setminus K_\infty(\lambda_0)$ and $\overline{u} \in K_\infty(\lambda_0)$.

The remaining part of the proof can be carried out as in (a) and in the proof of Theorem 6.8 (b). ■

Another sufficient condition for the existence of an asymptotic bifurcation branch unbounded in the λ direction is given in the following theorem.

Theorem 7.5 *(a) Assume we have (7.27), and for $\lambda_0 > 0$, there exists $u_0 \in K_\infty(\lambda_0)$, such that*

$$\langle g(u_0, \lambda_0), u \rangle \geq 0, \ \forall u \in D(J_\infty, u_0, \lambda_0). \tag{7.35}$$

Then, if $\lambda > \lambda_0$ is not an eigenvalue of (7.8),

$$\mathrm{d}(I - P_\infty[f_\infty(\cdot, \lambda)], B_R(0), 0) = 0, \tag{7.36}$$

for all $R > 0$, and there exists a global asymptotic bifurcation branch of (6.8), which is unbounded in $V \times [0, \lambda]$ and satisfies the alternative in Theorem 7.2.

(b) Assume that all eigenvalues of (7.8) are positive and obey the following condition stronger than (7.35):

$$\langle g(u_0, \lambda_0), u \rangle > 0, \ \forall u \in D(J_\infty, u_0, \lambda_0) \setminus \{0\}. \tag{7.37}$$

Then, λ_0 is the greatest eigenvalue of (7.8), and we have (7.36) for all $\lambda > \lambda_0$. Moreover, the branch that bifurcates from $\{\infty\} \times [0, \lambda_0]$ is unbounded in the λ direction.

$(D(J_\infty, u_0, \lambda_0)$ is defined in section 6.4.) Before proving this theorem, we note that, if $h = 0$ and

$$\langle f_\infty(u, -1), u \rangle \leq 0, \ \forall u \in D(J_\infty), \tag{7.38}$$

then, all eigenvalues of (7.8) are positive. In fact, let (u, λ) satisfy (7.8). Hence, $u \in D(J)$, and, by choosing $v = tu$ in (7.8) with $t > 0$,

$$(t - 1)\langle \alpha_\infty u - f_\infty(u, \lambda), u \rangle + J_\infty(tu) - J_\infty(u) \geq 0.$$

Because J_∞ is homogeneous of degree 2,

$$(t - 1)[\langle \alpha_\infty u - f_\infty(u, \lambda), u \rangle + (t + 1)J_\infty(u)] \geq 0.$$

Dividing both sides of this inequality by $(t - 1)$ and letting $t \to 1^+$ and $t \to 1^-$,

$$\langle \alpha_\infty u - f_\infty(u, \lambda), u \rangle + 2J_\infty(u) = 0.$$

Now, assume that $\lambda \leq 0$. From the homogeneity of $f_\infty(u, \cdot)$, it follows that

$$
\begin{aligned}
\langle \alpha_\infty u, u \rangle + 2J_\infty(u) &= \langle f_\infty(u, \lambda), u \rangle \\
&= |\lambda|^\gamma \langle f_\infty(u, -1), u \rangle \\
&\leq 0.
\end{aligned}
$$

Hence, $\langle \alpha_\infty u, u \rangle = J_\infty(u) = 0$. The coerciveness of α_∞ implies that $u = 0$. Hence, (7.8) does not have solutions (u, λ) with $u \neq 0$, $\lambda \leq 0$, i.e., all eigenvalues of (7.8) are positive. Now, we sketch the proof of Theorem 7.5.

Proof of Theorem 7.5. The ideas of the proof of this theorem are based on those used in Theorem 6.9. We present, here, only the main steps. To prove (7.36), we use the family of mappings $\{H(u,t) : 0 \leq t \leq 1, u \in V\}$, given by

$$H(u,t) = u - P_\infty[f_\infty(u,\lambda) + t\phi(u_0)], \quad u \in V, 0 \leq t \leq 1,$$

where $\phi : V \to V^*$ is a duality mapping satisfying (6.122) and $u_0 \in K_\infty(\lambda_0)$ satisfies (7.35). One can show that the equation

$$H(u,1) \equiv u - P_\infty[f_\infty(u,\lambda) + \phi(u_0)] = 0 \tag{7.39}$$

has no solution in V and that

$$H(u,t) \neq 0, \; \forall t \in [0,1], \; \forall u \in V \setminus B_R(0).$$

Hence,

$$
\begin{aligned}
d(I - P_\infty[f_\infty(\cdot, \lambda)], B_R(0), 0) &= d(H(\cdot, 0), B_R(0), 0) \\
&= d(H(\cdot, 1), B_R(0), 0) \\
&= 0.
\end{aligned}
$$

Now, assume that (7.37) holds and that, for $\lambda > \lambda_0$, $u \in V$ is a solution of (7.8). Letting $v = u + u_0$ in (7.8),

$$\langle \alpha_\infty u - f_\infty(u,\lambda), u_0 \rangle + J_\infty(u + u_0) - J_\infty(u) \geq 0.$$

Consequently, we have the following estimates:

$$\langle \alpha_\infty u - f_\infty(u,\lambda), u_0 \rangle \geq 0,$$

$$\left[1 - \left(\frac{\lambda}{\lambda_0} \right)^\gamma \right] \langle g(u_0, \lambda_0), u \rangle \geq 0,$$

and

$$\langle g(u_0, \lambda_0), u \rangle \leq 0. \tag{7.40}$$

On the other hand,

$$\langle f_\infty(u,\lambda), u \rangle \geq \langle \alpha_\infty u, u \rangle + J_\infty(u) \geq 0,$$

by letting $v = 0$ in (7.8). These considerations imply that $u \in D(J_\infty, u_0, \lambda_0)$. However, because $u \neq 0$, we have (7.37), contradicting (7.40).

Hence, λ cannot be an eigenvalue of (7.8). Set $\lambda = \lambda_0 + \epsilon$ ($\epsilon > 0$ small). Because λ is not an eigenvalue of (7.8) and (7.36) holds, there exists, by (a), a branch $\mathcal{C}$ of solutions of (6.8) that bifurcates from $\{\infty\} \times [0, \lambda]$, and $\mathcal{C}$ is either unbounded in the λ direction or bifurcates from $\{\infty\} \times [c, d]$ for some interval $[c, d]$ with $(c, d) \cap (a, b) = \emptyset$, i.e., $(c, d) \subset (-\infty, 0] \cup [\lambda, \infty)$. Suppose that C is bounded in the λ direction. Then we have the second alternative, and there exists an asymptotic bifurcation point (∞, λ_1) with $\lambda_1 \in [c, d]$. Hence, $\lambda_1 > \lambda_0$ or $\lambda_1 \leq 0$, and λ_1 is an eigenvalue of (7.8) by Theorem 7.2 (a). However, this is impossible by the above proof and by the assumption that (7.8) does not have nonpositive eigenvalues. Thus, C is unbounded in the λ direction.   ∎

Now, we consider some consequences of the above theorems when the operators have some more specific features. First, we have the following lemma, whose proof is similar to that of Lemma 6.12.

Lemma 7.6 *(a) If*

$$\ker [\alpha_\infty^* - f_\infty^*(\cdot, \lambda_0)] \cap (\ker J_\infty)^I \neq \emptyset, \tag{7.41}$$

then, $u \in V$ is a solution of (7.8) with $\lambda = \lambda_0$ if and only if $u \in \ker J_\infty$ and u is a solution of the associated linear equation of (7.8):

$$\langle \alpha_\infty(u) - f_\infty(u, \lambda_0), v \rangle = 0, \ \forall v \in V. \tag{7.42}$$

(b) Suppose that λ_0 is an eigenvalue of (7.8) with an eigenvector $u_1 \in (\ker J_\infty)^I$ (respectively, $u_1 \in D(J_\infty)^I$ is also a solution of (7.42), and $D(J_\infty)$ is closed in V). Assume, furthermore, that (7.41) is satisfied and that $\ker J_\infty \neq V$ (respectively, $D(J_\infty) \neq V$).

Then, λ_0 is a simple eigenvalue of (7.8) whenever it is a simple eigenvalue of (7.42).

(c) Suppose that α_∞ and $f_\infty(\cdot, \lambda_0)$ are self-adjoint operators and that $J_\infty \not\equiv 0$ on V.

If λ_0 is a simple eigenvalue of (7.42) with an eigenvector $u_0 \in (\ker J_\infty)^I$, then, λ_0 is a simple eigenvalue of (7.8).

From this lemma and Theorem 7.4, we have the following result:

Corollary 7.7 *(a) Suppose that λ_0 is an eigenvalue of (7.8) satisfying the conditions in Lemma 7.6 (b), and (7.30). Then, we arrive at the conclusion of Theorem 7.4.*

(b) Assume that α_∞ and $f_\infty(\cdot, \lambda_0)$ are self-adjoint, $h \equiv 0$ in (7.27), and that λ_0 is a simple eigenvalue of (7.8) with an eigenvector $u_1 \in \ker J_\infty$, which also satisfies (7.42). Then, we arrive at the conclusion of (a).

(c) Assume that α_∞ and $f_\infty(\cdot, \lambda_0)$ are self-adjoint and that $h \equiv 0$. If λ_0 is a simple eigenvalue of (7.42) with an eigenvector $u_1 \in (\ker J_\infty)^I$, then, we arrive at the conclusion of (a).

The following result is a consequence of Lemma 7.6 and Theorem 6.8 (b), in which the simplicity of λ_0 is replaced by the nonsymmetry property of $(\ker J_\infty)^I \cap \ker [\alpha_\infty^* - f_\infty^*(\cdot, \lambda_0)]$:

Corollary 7.8 *(a) Suppose that*

$$[(\ker J_\infty)^I \setminus (-\ker J_\infty)] \cap \ker [\alpha_\infty^* - f_\infty^*(\cdot, \lambda_0)] \neq \emptyset, \tag{7.43}$$

and, for each $u_1 \in \ker J_\infty \setminus \{0\}$ satisfying the linear equation (7.42), there exists $u_0 \in K_\infty(\lambda_0)$, such that (7.30) holds. Then, we arrive at the conclusion of Theorem 7.4.

(b) Suppose that α_∞ and $f_\infty(\cdot, \lambda_0)$ are self-adjoint and $h \equiv 0$ in (7.27). If (7.43) holds, then, we arrive at the conclusion of Theorem 7.4.

Note that (7.43) is equivalent to the following condition:

$$\begin{cases} \ker [\alpha_\infty^* - f_\infty^*(\cdot, \lambda_0)] \cap (\ker J_\infty)^I \neq \emptyset, \\ \ker J_\infty \neq V. \end{cases} \tag{7.44}$$

7.2 Some applications

In this section, we apply the above abstract theorems to obtain global asymptotic bifurcation results for some particular examples, including variational inequalities containing second- and fourth-order, linear, elliptic, operators, quasilinear operators, or the p-Laplacian, and convex functionals of various types.

Example 7.1 In this example, we consider the bifurcation from infinity of the following semilinear, second-order, elliptic variational inequality:

$$\begin{cases} \int_\Omega \left[\sum_{i,j=1}^N a_{ij}\, \partial_i u\, \partial_j(v-u) + \sum_{i=1}^N a_i\, \partial_i\, u(v-u) + a_0 u(v-u) \right] \\ \quad -\lambda \int_\Omega g(\cdot, u, \lambda) u(v-u) + j(v) - j(u) \geq 0,\ \forall v \in V, \\ u \in V, \end{cases} \tag{7.45}$$

where $\Omega, V = H^1(\Omega)$ and a_{ij}, a_i, a_0 are as in Example 6.5, and

$$g : \Omega \times \mathbb{R} \times \mathbb{R} \to \mathbb{R}$$

is a Carathéodory function that is differentiable at infinity in the following sense:

There exists a Carathéodory function $F : \Omega \times \mathbb{R} \times \mathbb{R} \to \mathbb{R}$, such that g and F satisfy the following growth condition:

$$|g(x, u, \lambda)|, |F(x, u, \lambda)| \leq M(\lambda)[C(x)|u| + D(x)], \tag{7.46}$$

for a.e. $x \in \Omega$, all $u, \lambda \in \mathbb{R}$, where

$$C, D, M \geq 0 \ \text{ and } \ D \in L^r(\Omega), C \in L^q(\Omega), M \in L^\infty_{loc}(\Omega), \tag{7.47}$$

with

$$q > \left(\frac{2^*}{2}\right)' = \begin{cases} N/2 & \text{if } N > 2, \\ 1 & \text{if } N \leq 2, \end{cases}$$

and

$$r > (2^*)' = \begin{cases} \dfrac{2N}{N+2} & \text{if } N > 2, \\ 1 & \text{if } N \leq 2. \end{cases}$$

Moreover, for all sequences $\{\sigma_n\}, \{u_n\}$, and $\{\lambda_n\} \subset \mathbb{R}$, such that

$$\sigma_n \to \infty, u_n \to u, \lambda_n \to \lambda \ (n \to \infty),$$

we have (5.24) for almost all $x \in \Omega$.

We also assume that $j : V \to [0, \infty]$ is a convex, lower semicontinuous functional such that $j(0) = 0$. Let B and f_∞ be defined by

$$B, f_\infty : V \times \mathbb{R} \to \mathbb{R},$$

$$\langle B(u, \lambda), v \rangle = \int_\Omega g(x, u(x), \lambda) v(x) dx, \qquad (7.48)$$

$$\langle f_\infty(u, \lambda), v \rangle = \int_\Omega F(x, u(x), \lambda) v(x) dx, \qquad (7.49)$$

for $u, v \in V, \lambda \in \mathbb{R}$. By using arguments similar to those in Example 4.5, one can deduce from the assumptions (7.46), (7.47), and (5.24) and the compactness of the embedding

$$H^1(\Omega) \hookrightarrow L^p(\Omega), \ 1 \leq p < 2^* = \begin{cases} \dfrac{2N}{N-2} & \text{if } N > 2, \\ \infty & \text{if } N \leq 2, \end{cases}$$

that B and f_∞ are completely continuous mappings from $V \times \mathbb{R}$ to V and f_∞ is the derivative of B at infinity with respect to u in the sense of (A10).

Now, we define A from V to V^* by (6.210). In Example 6.5, we have seen that A satisfies (6.9) with $p = 2$. Hence, (A9) holds with $\alpha_\infty = A$. With these definitions, we see that (7.45) can be written in the form (6.8).

In what follows, we assume that g is of the form (5.25), for $x \in \Omega, u \in \mathbb{R}, \lambda \in \mathbb{R}$, where $b \in L^q(\Omega)$ and h satisfies the growth condition,

$$|h(x, u, \lambda)| \leq M(\lambda)[C(x)|u|^\gamma + D(x)], \ x \in \Omega, u, \lambda \in \mathbb{R}, \qquad (7.50)$$

with $0 \leq \gamma < 1$ and C, D as in (7.47). In this case,

$$\left| \frac{1}{\sigma_n} h(x, \sigma_n, \lambda_n) \right| \leq M(\lambda_n) \left[C(x) \frac{|u_n|^\gamma}{\sigma_n^{1-\gamma}} + \frac{D(x)}{\sigma_n} \right] \to 0 \text{ as } n \to \infty,$$

whenever $\sigma_n \to \infty, u_n \to u$, and $\lambda_n \to \lambda$ in $\mathbb{R}$. Moreover,

$$|g(x, u, \lambda)| \leq |\lambda| \, |b| \, |u| + M(\lambda) \, [C(x)(1 + |u|) + D(x)] \; (x \in \Omega, \; u, \lambda \in \mathbb{R}).$$

Hence, g satisfies (7.46) and (7.47). Thus, (5.24) holds with $F(x, u, \lambda) = \lambda bu$, and, therefore,

$$\langle f_\infty(u, \lambda), v \rangle = \lambda \int_\Omega buv dx, \; \forall u, v \in V. \tag{7.51}$$

In the case $N = 1$, note that (7.46), (7.47), and (7.50) reduce to the assumptions in Example 4.5. Now, we consider some asymptotic bifurcation results for (7.45) when j, given as in Example 6.5, represents a lower dimensional obstacle on the boundary: $j = I_K$ with K given by (6.209).

As observed in Section 7.1, in this case,

$$J_\infty = I_{rcK},$$

where rcK is the recession cone of K, given by

$$rcK = \{u \in V : u \geq 0 \text{ on } \partial\Omega\}. \tag{7.52}$$

The asymptotically homogenized variational inequality (7.8) corresponding to (7.45) in this case is given by

$$\begin{cases} \langle Au, v - u \rangle - \lambda \int_\Omega bu(v - u) \geq 0, \; \forall v \in rcK \\ u \in rcK. \end{cases} \tag{7.53}$$

The linear equation associated with (7.53) is equation (6.223), and its adjoint is given by (6.224).

Using the arguments in Corollaries 6.15, 6.16, and 6.17, we get the following consequence of Theorem 7.4, applied to (7.45).

Corollary 7.9 (a) *Suppose that λ_0 is a simple eigenvalue of (7.53) with an eigenvector u_1. Assume, furthermore, that there exists a solution u_0 of (6.224) (with $\lambda = \lambda_0$), such that*

$$u_0 \geq 0 \text{ on } \partial\Omega \tag{7.54}$$

and

$$\int_\Omega bu_0u_1 > 0. \tag{7.55}$$

Then, there exists a global asymptotic bifurcation branch of (7.45) emanating from $\{\infty\} \times [0, \lambda_0]$ and satisfying the alternative in Theorem 7.2.

(b) Suppose that λ_0 is an eigenvalue of (7.53) and that there exists a solution $\bar{u}$ of (6.224) (with $\lambda = \lambda_0$) satisfying (7.54), but

$$\bar{u} \not\equiv 0 \text{ on } \partial\Omega. \tag{7.56}$$

Assume, furthermore, that for each eigenvector u_1 of (7.53) corresponding to λ_0, there exists a solution of (6.224) such that (7.54), (7.55) are satisfied.

Then we arrive at the conclusion of (a).

(c) Suppose λ_0 is a simple eigenvalue of the linear equation (6.223) with an eigenvector u_1 satisfying (7.54). Assume, further, that there exists a solution u_0 of (6.224), such that

$$u_0 \geq \gamma > 0 \quad a.e. \text{ on } \partial\Omega, \tag{7.57}$$

for some $\gamma > 0$, and that (7.55) holds.

Then we arrive at the conclusion of (a).

(d) Suppose that A is self-adjoint. Then we arrive at the conclusion of (a) in the following cases:

(i) λ_0 is a simple eigenvalue of (7.53) with an associated eigenvector u_1 which is also a solution of (6.223).

(ii) λ_0 is an eigenvalue of (6.223) with an associated eigenvector u_1 satisfying (7.57).

The proof of this corollary is based on arguments similar to those in the results stated above.

We consider bifurcation from infinity for (7.45) with a different convex functional j, given by

$$j(u) = \int_{\partial\Omega} k|u|\,dS, \ u \in V, \tag{7.58}$$

where $k \in L^\infty(\partial\Omega)$ is given, $k \geq 0$ on $\partial\Omega$. Using the continuous mappings

$$H^1(\Omega) \to L^2(\partial\Omega) \hookrightarrow L^1(\partial\Omega), \ u \mapsto u|_{\partial\Omega} \mapsto u|_{\partial\Omega},$$

we see that j is a convex, nonnegative and continuous functional from V to $\mathbb{R}^+$ and $j(0) = 0$. We claim that the derivative of j at infinity is given by

$$J_\infty(u) \equiv 0, \ u \in V. \tag{7.59}$$

In fact, let $v_n \to v$ in $H^1(\Omega)$ and $\sigma_n \to \infty$. $v_n|_{\partial\Omega} \to v|_{\partial\Omega}$ in $L^2(\partial\Omega)$, and thus, in $L^1(\partial\Omega)$. Hence,

$$\int_{\partial\Omega} k|v_n|\,dS \to \int_{\partial\Omega} k|v|\,dS,$$

and, therefore,

$$\frac{j(\sigma_n v_n)}{\sigma_n^2} = \frac{1}{\sigma_n}\int_{\partial\Omega} k|v_n|\,dS \to 0 = J_\infty(v), \quad \text{as } n \to \infty,$$

(because $1/\sigma_n \to 0$). Thus, (7.5) holds. To prove (7.6), we let $v \in V, \sigma_n \to \infty$, and choose $v_n = v, \forall n$. Thus, using the above proof, again,

$$j_{\sigma_n}(v_n) = 0 = J_\infty(v),$$

proving (7.6). Thus, one obtains (7.59).

From (7.51) and (7.59), we see that the homogenized variational inequality (7.8) is the following:

$$
\begin{cases}
\langle Au, v - u \rangle - \lambda \displaystyle\int_\Omega bu(v - u) \geq 0, \ \forall v \in V, \\[2mm]
u \in V,
\end{cases}
\tag{7.60}
$$

which is equivalent to (6.223).

Because the homogenized variational inequality of (7.45) is a linear equation, we can apply Corollary 7.3 to get the following result:

Corollary 7.10 *The variational inequality (7.45), with j given by (7.58), has, at most, a countable number of asymptotic bifurcation points. Moreover, if (∞, λ) is an asymptotic bifurcation point of (7.45), then, λ is an eigenvalue of (6.223). If λ is an eigenvalue of (6.223) of odd multiplicity, then, (∞, λ) is an asymptotic bifurcation point of (7.45) corresponding to a asymptotic bifurcation branch that satisfies the alternative in Theorem 7.2.*

Note that we have similar results if j is given by $j(u) = \int_{\partial\Omega}(u)^\pm, u \in V$.

Example 7.2 In this example, we study the bifurcation from infinity for some variational inequalities containing fourth-order operators from the theory of plates and beams.

(a) First, consider the following variational inequality,

$$
\begin{cases}
\displaystyle\int_0^a u''(v - u)'' dx - \int_0^a g(x, u', \lambda)(v - u)' dx + j(v) - j(u) \geq 0, \\[2mm]
\quad \forall v \in V, \\[2mm]
u \in V,
\end{cases}
\tag{7.61}
$$

where $a > 0$, $V = H_0^2(0, a)$ or $H^2(0, a) \cap H_0^1(0, a)$, g is given in Example 5.3, and j is as in Example 6.7. For simplicity, we assume, here, that j consists only of the first integral in (6.250), i.e., $k_2 \equiv 0$ and

$$
j(u) = \int_{I_1} k_1(u^-)^\gamma dx, \ u \in V,
\tag{7.62}
$$

(I_1, k_1, and γ are as in Example 6.7). For $u \in V$, let

$$
J_\infty(u) =
\begin{cases}
0 & \text{if } \ 1 < \gamma < 2, \\[2mm]
\displaystyle\int_{I_1} k_1(u^-)^2 & \text{if } \ \gamma = 2, \\[2mm]
I_K(u) & \text{if } \ \gamma > 2,
\end{cases}
\tag{7.63}
$$

where $K = \{u \in V : u \geq 0 \text{ on } I_1\}$. We claim that j_σ tends to J_∞ as $\sigma \to \infty$ in the sense of (7.5) and (7.6). In fact, let $v_n \rightharpoonup v$ in V and $\sigma_n \to \infty$. By the compactness of the embedding $H^2(0, a) \hookrightarrow C^1[0, a]$, $v_n \to v$ uniformly on $[0, a]$, and, then,

$$k_1(v_n^-)^\gamma \to k_1(v^-)^\gamma \text{ uniformly on } I_1,$$

implying that

$$j(v_n) \to j(v), \ n \to \infty. \tag{7.64}$$

If $\gamma < 2$, then,

$$\frac{j(\sigma_n v_n)}{\sigma_n^2} = \frac{1}{\sigma_n^{2-\gamma}} j(v_n) \to 0 = J_\infty(v), \ \text{ as } n \to \infty,$$

(by (7.64) and $\lim \sigma_n^{\gamma-2} = 0$).
   If $\gamma = 2$, then, by (7.64),

$$\frac{j(\sigma_n v_n)}{\sigma_n^2} = j(v_n) \to j(v) = J_\infty(v).$$

Now, suppose that $\gamma > 2$. If $v \geq 0$ on I_1, then, $v \in K$ and $J_\infty(v) = I_K(v) = 0$, implying (7.5).
   If $v \notin K$, then, by (6.252),

$$j(v) \geq k_0 \int_{I_1} (v^-)^\gamma dx > 0. \tag{7.65}$$

Hence, by (7.64),

$$\begin{aligned} \lim \frac{j(\sigma_n v_n)}{\sigma_n^2} &= \lim [\sigma_n^{\gamma-2} j(v_n)] \\ &= \lim \sigma_n^{\gamma-2} \lim j(v_n) \\ &= \infty = J_\infty(v). \end{aligned} \tag{7.66}$$

Thus, (7.5) holds in all cases. Now, let $v \in V$, $\sigma_n \to \infty$, and choose $v_n = v$, $\forall n$. If $\gamma \leq 2$, then,

$$\begin{aligned} \lim \frac{j(\sigma_n v_n)}{\sigma_n^2} &= \left(\lim \frac{1}{\sigma_n^{2-\gamma}} \right) j(v) = \begin{cases} 0 & \text{if } \gamma < 2, \\ j(v) & \text{if } \gamma = 2, \end{cases} \\ &= J_\infty(v). \end{aligned}$$

If $\gamma > 2$ and $v \in K$, then, $j(\sigma_n v) = \sigma_n^\gamma j(v) = 0$, $\forall n$. If $v \notin K$, then, we have (7.65) and (7.66). In both cases, $j_{\sigma_n}(v_n) = j_{\sigma_n}(v) \to J_\infty(v)$, and (7.6) is proved.

Now, we define A, as in Example 6.7, and B and f_∞, as in Example 5.3. Because A is linear and bounded, and satisfies (6.9), $\alpha_\infty = A$. With these settings, we see that (7.61) is of the form (6.8) and its asymptotically homogenized variational inequality is given by

$$
\begin{cases}
\displaystyle \int_0^a u''(v-u)''dx - \lambda \int_0^a bu'(v-u)'dx + J_\infty(v) - J_\infty(u) \geq 0, \ \forall v \in V, \\
u \in V,
\end{cases}
$$

$$(7.67)$$

with J_∞ given by (7.63).

The linear equation associated with (7.67) is

$$
\begin{cases}
\displaystyle \int_0^a u''v''dx - \lambda \int_0^a bu'v'dx = 0, \ \forall v \in V, \\
u \in V.
\end{cases}
$$

Now assume that g is of the form (5.31). Then, by (5.32), $b = 1$, and the above equation becomes (6.261).

If $1 < \gamma < 2$, then, (7.67) is equivalent to (6.261). Because (6.261) has an unbounded sequence of eigenvalues, all of which are simple, by using Corollary 7.3, we get the following result:

Corollary 7.11 *If $1 < \gamma < 2$, then, (∞, λ) is a bifurcation point of (7.61) if and only if λ is an eigenvalue of (6.261). Moreover, each eigenvalue λ of (6.261) corresponds to a global asymptotic bifurcation branch that satisfies the alternative in Theorem 7.2.*

In the case where $\gamma \geq 2$, we find, from (7.63) and (7.62), that $\ker J_\infty = K$. Moreover, $u_0 \in (\ker J_\infty)^I$, whenever u_0 satisfies (6.263). Applying Theorem 7.4, one obtains the following result:

Corollary 7.12 *Assume that $\gamma \geq 2$. Let $\lambda_0 > 0$ be an eigenvalue of (6.261) with an eigenvector u_0 satisfying (6.263). Then, for some $\lambda \in (0, \lambda_0)$, (∞, λ) is an asymptotic bifurcation point of (7.61) with an asymptotic bifurcation branch emanating from $\{\infty\} \times [0, \lambda_0]$ that satisfies the alternative in Theorem 7.2.*

Note that similar arguments apply to the asymptotic bifurcation problem for (7.61) with two obstacles, i.e., j is given by (6.251) instead of (7.62).

(b) In this second example, we consider bifurcation from infinity for the variational inequalities in the theory of plates, considered in section 6.3.

We consider the variational inequality (6.81), where V and j are given by (6.82), (6.83), (6.86), and (6.87), or (6.89) and (6.90). Let A, a, B, and L be as in section 6.3. We assume that

$$\{\sigma_{ij}(u) : 1 \leq i, j \leq 2\}$$

satisfies (5.35) or, in particular, (5.37).

Because A is linear, bounded, and coercive in the sense of (6.9) (with $p = 2$), $\alpha_\infty = A$. As observed in Section 5.2, we see, from (5.35), that (A10) is satisfied with

$$f_\infty(u, \lambda) = \lambda Lu, \ u \in V, \lambda \in \mathbb{R}.$$

To compute J_∞, we consider, separately, the cases:

$$\begin{cases} \bullet \ \ \psi(\pm 1) < \infty, \\ \bullet \ \ \psi(-1) < \infty = \psi(1), \\ \bullet \ \ \psi(1) < \infty = \psi(-1). \end{cases}$$

First, we note that, if $\psi(\pm 1) < \infty$, then, for all three cases (i), (ii), and (iii), condition (A11) is satisfied with

$$J_\infty(u) = 0, \ \forall u \in V. \tag{7.68}$$

We check this, for instance, for the case (ii), i.e., for j given by (6.87). The cases where j is given by (6.83) and (6.90) are carried out in the same way. Because $J_\infty = 0 \le j$, (7.5) holds immediately. Now, let $v \in V$, and $\sigma_n \to \infty$. By the embedding (6.103), $\partial_n v \in L^2(\partial\Omega)$. For $x \in \partial\Omega$,

$$\begin{aligned} \psi(\partial_n(x)) &= |\partial_n v(x)| \, \psi(\text{sign}[\partial_n v(x)]) \\ &\le |\partial_n v(x)| \, [|\psi(-1)| + |\psi(1)|]. \end{aligned}$$

Hence, also, $\psi(\partial_n v(x)) \in L^2(\partial\Omega)$, i.e., $j(u) < \infty$. Letting $v_n = v, \ \forall n$,

$$\begin{aligned} \lim \frac{j(\sigma_n v)}{\sigma_n^2} &= \lim \frac{1}{\sigma_n} j(v) \\ &= 0 = J_\infty(v), \end{aligned}$$

proving (7.6).

Now, consider the case where $\psi(-1) < \infty = \psi(1)$. We prove that the functional J_∞ defined by

$$J_\infty(u) = I_K(u), \ u \in V, \tag{7.69}$$

with

$$K = \begin{cases} \{u \in V : u \le 0 \text{ on } \Omega\} & \text{in case} \quad \text{(i)}, \\ \{u \in V : \partial_n u \le 0 \text{ on } \partial\Omega\} & \text{in case} \quad \text{(ii)}, \\ \{u \in V : u \le 0 \text{ on } \partial\Omega \setminus \Gamma_1\} & \text{in case} \quad \text{(iii)}, \end{cases} \tag{7.70}$$

will satisfy (A11). In fact, let V, j be given by (6.82) and (6.83). To prove (7.5), we let $v_n \rightharpoonup v$ in $H_0^2(\Omega)$ and $\sigma_n \to \infty$. If $v \le 0$ on Ω, then, $J_\infty(v) = 0$, and (7.5) holds. Now, suppose $v(x_0) > 0$, for some $x_0 \in \Omega$. By the compact embedding (6.101), we see that $v_n \to v$ uniformly on $\overline{\Omega}$, and there exist m and $r > 0$, such that

$$v_n(x), v(x) \geq m, \ \forall x \in \overline{B_r(x_0)},$$

for all n sufficiently large. For such n,

$$\begin{aligned}
j(v_n) &\geq \int_{B_r(x_0)} \psi(v_n(x))dx \\
&= \int_{B_r(x_0)} v_n(x)\psi(1)dx \\
&= \infty.
\end{aligned}$$

Hence, $j_{\sigma_n}(v_n) = \dfrac{j(v_n)}{\sigma_n} = \infty$, for all n large, and

$$\liminf j_{\sigma_n}(v_n) = \infty = J_\infty(v).$$

Hence, (7.5) holds also. To prove (7.6), we let $v \in V$, and $\sigma_n \to \infty$. Choosing $v_n = v$ for all n,

$$j_{\sigma_n}(v_n) = \frac{j(\sigma_n v)}{\sigma_n^2} = \frac{1}{\sigma_n} j(v), \ \forall n.$$

If $v \leq 0$ on Ω, then,

$$j(v) = -\psi(-1) \int_\Omega v(x)dx < \infty.$$

Therefore,

$$\lim j_{\sigma_n}(v_n) = j(v) \lim \frac{1}{\sigma_n} = 0 = J_\infty(v).$$

If $v \not\leq 0$ on Ω, then, as in the above proof, $j(v) = \infty$, implying that

$$j_{\sigma_n}(v_n) = \infty, \ \forall n,$$

and, then,

$$\lim_{n \to \infty} j_{\sigma_n}(v_n) = \infty = J_\infty(v).$$

We have verified (7.6). The proofs for the other cases are similar. Also, the formula for J_∞ in the case where $\psi(1) < \infty = \psi(-1)$ is similar to (7.69) and (7.70).

From the discussion above, we see that, in the case where $\psi(\pm 1) < \infty$, (7.68) implies that the asymptotically homogenized variational inequality corresponding to (6.81) is the following linear equation:

$$\begin{cases}
a(u,v) - \lambda \int_\Omega \sum_{i,j=1}^{2} \sigma_{ij}^0 \, \partial_i u \, \partial_j v dx = 0, \ \forall v \in V, \\
\\
u \in V,
\end{cases} \tag{7.71}$$

where V is given by (6.82), (6.86), or (6.89), respectively. Because the asymptotically homogenized variational inequality, in this case, is a linear equation, we can apply Corollary 7.3 to obtain relationships between eigenvalues of odd multiplicity of (7.71) and asymptotic bifurcation points and bifurcation branches of (6.81).

Now, we consider the case where $\psi(-1) < \infty = \psi(1)$. Because J_∞ is given by (7.69), the asymptotically homogenized variational inequality (7.8) of (6.81), in this case, is the following:

$$
\begin{cases}
a(u, v - u) - \lambda \displaystyle\int_\Omega \sum_{i,j=1}^{2} \sigma_{ij}^0 \, \partial_i u \, \partial_j (v - u) dx \; \geq \; 0, \; \forall v \in K, \\[2ex]
u \in K,
\end{cases}
\tag{7.72}
$$

where K is given by (7.70).

Thus, we can apply Theorem 6.8 and its corollaries 6.13 and 6.14 to obtain global asymptotic bifurcation results for (6.81) from eigenvalues of (7.72). For example, we have the following result for case (i):

Corollary 7.13 *Suppose that j is given by (6.83) with $\psi(-1) < \infty = \psi(1)$. Assume, furthermore, that $\lambda_0 > 0$ is an eigenvalue of (7.72) with an eigenvector u_0, such that*

$$
u_0(x) > 0, \; x \in \Omega.
\tag{7.73}
$$

Then, there exists an asymptotic bifurcation branch of (6.81) which emanates from $\{\infty\} \times [0, \lambda_0]$ and satisfies the alternative in Theorem 7.2.

To prove this corollary, we apply Corollary 6.14 by noting that, if u_0 satisfies (7.73), then, $u_0 \in K^I = (\ker J_\infty)^I$. Hence,

$$
\ker \left[\alpha_\infty^* - f_\infty^*(\cdot, \lambda_0) \right] \cap (\ker J_\infty)^I = \ker \left[A - f_\infty(\cdot, \lambda_0) \right] \cap K^I \ni u_0.
$$

Thus, (7.44) follows and, consequently, also (7.43).

One has similar results for cases (ii) and (iii) and for the case where $\psi(1) < \infty = \psi(-1)$.

Example 7.3 In this example, we consider bifurcation from infinity for the following second-order, quasilinear, elliptic variational inequality:

$$
\begin{cases}
\displaystyle\int_\Omega \left[\sum_{i=1}^{N} a_i(x, u, \nabla u) \, \partial_i(v - u) + a_0(x, u, \nabla u) \, (v - u) \right] dx \\[2ex]
\quad + \displaystyle\int_\Omega g(x, u, \lambda) \, (v - u) dx + j(v) - j(u) \; \geq \; 0, \; \forall v \in H_0^1(\Omega), \\[2ex]
u \in H_0^1(\Omega).
\end{cases}
\tag{7.74}
$$

Here $\Omega, a_i, 0 \leq i \leq N$ are as in Example 6.3 and g is as in Example 7.1. In particular, we assume that g satisfies (7.46), (7.47), and (5.24). We assume that $j : V = H_0^1(\Omega) \to [0, \infty]$ is convex, lower semicontinuous, and $j(0) = 0$.

Suppose that the a_i satisfy the uniform monotonicity condition (6.171) and the following growth condition:

$$|a_i(x, u, \xi)| \leq K(|u| + |\xi|) + D(x), \ 0 \leq i \leq N, \qquad (7.75)$$

for a.e. $x \in \Omega$, all $u \in \mathbb{R}, \xi \in \mathbb{R}^N$, where $K > 0$ is a constant, and $D \in L^2(\Omega)$, $D \geq 0$ on Ω. Moreover, assume that the a_i are differentiable at infinity in the following sense:

For $0 \leq i \leq N$, there exist Carathéodory functions $A_i : \Omega \times \mathbb{R}^{N+1} \to \mathbb{R}$, such that

$$\frac{1}{\sigma_n} a_i(x, \sigma_n u_n, \sigma_n \xi_n) \to A_i(x, u, \xi), \qquad (7.76)$$

whenever $\sigma_n \to \infty, u_n \to u$ in $\mathbb{R}$, and $\xi_n \to \xi$ in $\mathbb{R}^N$.

Replacing u and p in (7.75) by $\sigma_n u_n$ and $\sigma_n p_n$, respectively, dividing the inequality, thus obtained, by σ_n, and letting $n \to \infty$, we see that A_i also satisfy the growth condition (7.75). By a similar observation, A_i satisfy (6.171).

Let A be defined by (6.179) in Example 6.3, and

$$\langle \alpha_\infty(u), v \rangle = \int_\Omega \left[\sum_{i=1}^N A_i(x, u, \nabla u) \, \partial_i v + A_0(x, u, \nabla u) \, v \right] dx, \ \forall u, v \in V.$$
$$(7.77)$$

Then, we can verify that α_∞ is the derivative of A at infinity in the sense of (A9).

Now, we consider the particular case where a_i can be written as

$$a_i(x, u, \xi) = a_{i0}^\infty(x) u + \sum_{j=1}^N a_{ij}^\infty(x) \xi_j + H_i(x, u, \xi), \ 0 \leq i \leq N, \quad (7.78)$$

for a.e. $x \in \Omega$, all $u \in \mathbb{R}$, $\xi \in \mathbb{R}^N$ with $a_{ij}^\infty \in L^\infty(\Omega), 0 \leq i, j \leq N$, and

$$|H_i(x, u, \xi)| \leq K_i(|u|^{\gamma_i} + |\xi|^{\gamma_i}) + D_i(x), \ x \in \Omega, u \in \mathbb{R}, \xi \in \mathbb{R}^N, \quad (7.79)$$

with $K_i \in \mathbb{R}^+, D_i \in L^2(\Omega)$, and $0 \leq \gamma_i < 1$. In this case, the a_i satisfy (7.75) and (7.76) with A_i linear with respect to u and ξ:

$$A_i(x, u, \xi) = a_{i0}^\infty(x) u + \sum_{j=1}^N a_{ij}^\infty(x) \xi_j, \ 0 \leq i \leq N, x \in \Omega, u \in \mathbb{R}, \xi \in \mathbb{R}^N.$$
$$(7.80)$$

In fact, for $\sigma_n \to \infty$, $u_n \to u$ in $\mathbb{R}$, and $\xi_n \to \xi$ in $\mathbb{R}^N$,

$$\frac{1}{\sigma_n} a_i(x, \sigma_n u_n, \sigma_n \xi_n) = a_{i0}^\infty(x) u_n + \sum_{j=1}^N a_{ij}^\infty(x) \xi_{nj} + \frac{1}{\sigma_n} H_i(x, \sigma_n u_n, \sigma_n \xi_n),$$

and, for n sufficiently large,

$$\frac{1}{\sigma_n}|H_i(x,\sigma_n u_n,\sigma_n\xi_n)| \leq \frac{K_i}{\sigma_n^{1-\gamma_i}}(|u_n|^{\gamma_i}+|\xi_n|^{\gamma_i}) + \frac{1}{\sigma_n}D_i(x).$$

Because $\{u_n\},\{\xi_n\}$ are bounded and $\gamma_i < 1$,

$$\frac{1}{\sigma_n}H_i(x,\sigma_n u_n,\sigma_n\xi_n) \to 0, \quad \text{as } n \to \infty,$$

implying (7.76) with A_i given by (7.80). In this case, α_∞ is of the form,

$$\langle\alpha_\infty(u),v\rangle = \int_\Omega \left\{\sum_{i,j=1}^N a_{ij}^\infty(x)\,\partial_j u(x)\,\partial_i v(x)\right.$$
$$\left.+\sum_{i=1}^N [a_{i0}^\infty(x)\,u(x)\,\partial_i v(x) + a_{0i}^\infty(x)\,\partial_i u(x)\,v(x)] + a_{00}^\infty u(x)v(x)\right\}dx.$$

$$(7.81)$$

Next, we define B and f_∞ by (7.48) and (7.51). Let J_∞ be the asymptotic homogenization of j at infinity, defined by (A11). Then, (7.74) is of the form (6.8) with the asymptotically homogenized variational inequality:

$$\begin{cases} \int_\Omega \left\{\sum_{i,j=1}^N a_{ij}^\infty\,\partial_j u\,\partial_i(v-u) + \sum_{i=1}^N [a_{i0}^\infty\,u\,\partial_i(v-u) + a_{0i}^\infty\,\partial_i u\,(v-u)] \right. \\ \left. + a_{00}^\infty\,u\,(v-u)\right\} - \lambda\int_\Omega bu(v-u) + J_\infty(v) - J_\infty(u) \geq 0,\ \forall v \in V, \\ u \in V. \end{cases}$$

$$(7.82)$$

We consider, for example, the obstacle problem on Ω, i.e., the problem with

$$j = I_K, \quad K = \{u \in H_0^1(\Omega) : u \geq \psi \text{ a.e. on } \Omega\}, \qquad (7.83)$$

where $\psi \in L^\infty(\Omega)$ is a given function with $\psi \leq 0$ on Ω. Using arguments as in Chapter 5 and section 7.1, we see that

$$J_\infty = I_{rcK}, \qquad (7.84)$$

with $rcK = \{u \in H_0^1(\Omega) : u \geq 0 \text{ a.e. on } \Omega\}$.

Applying Theorem 7.5, we get the following result:

Corollary 7.14 *Consider (7.74) with j given by (7.83). Assume that*

$$b(x) > 0, \quad \text{for a.e. } x \in \Omega, \qquad (7.85)$$

and that $\lambda_0 > 0$ is an eigenvalue of the linear equation ,

$$\begin{cases} \int_\Omega \left\{ \sum_{i,j=1}^N a_{ij}^\infty \, \partial_j u \, \partial_i v + \sum_{i=1}^N [a_{i0}^\infty \, u \, \partial_i v + a_{0i}^\infty \, \partial_i u \, v] + a_{00}^\infty \, u \, v \right\} dx \\ \quad - \lambda \int_\Omega buv = 0, \; \forall v \in V, \\ u \in V, \end{cases}$$

(7.86)

with an eigenvector u_0, such that

$$u_0(x) > 0 \; \text{ for a.e. } x \in \Omega. \tag{7.87}$$

Then, for all $\lambda > \lambda_0$, (∞, λ) is not an asymptotic bifurcation point of (7.74), and there exists a global asymptotic bifurcation branch of (7.74), which bifurcates from $\{\infty\} \times [0, \lambda_0]$ and is unbounded in the λ direction.

To prove this corollary, we, first, observe that $f_\infty(u, \lambda)$ given by (7.51) is symmetric with respect to u and homogeneous with respect to λ. Hence, $h = 0$ and $g = f_\infty^* = f_\infty$. For $u \in V$, $\lambda = -1$, from (7.51),

$$\langle f_\infty(u, -1), u \rangle = - \int_\Omega bu^2 \le 0.$$

Hence (7.38) is satisfied, and (7.82) does not have nonpositive eigenvalues. $D(J_\infty, u_0, \lambda_0) \subset D(J_\infty) = rcK$ by (7.84), and, then,

$$u(x) \ge 0 \; \text{ for a.e. } x \in \Omega, \; \forall u \in D(J_\infty, u_0, \lambda_0).$$

This implies that

$$\langle g(u_0, \lambda_0), u \rangle = \lambda_0 \int_\Omega bu_0 u \ge 0, \; \forall u \in D(J_\infty, u_0, \lambda_0),$$

by (7.85) and (7.87). Moreover, $\langle g(u_0, \lambda_0), u \rangle = 0$ only if $bu_0 u = 0$ a.e. on Ω. Again by (7.85) and (7.87), this happens only if $u = 0$ a.e. on Ω. This proves (7.37). Now, we see from (7.81) that (7.86) is the linear equation,

$$\begin{cases} \langle \alpha_\infty^*(u) - f_\infty^*(u, \lambda_0), v \rangle = 0, \; \forall v \in V, \\ u \in V. \end{cases}$$

Moreover, $u_0 \in K = \ker J_\infty$ by (7.87). Hence, $u_0 \in K_\infty(\lambda_0)$. All conditions in Theorem 7.5 (b) are satisfied, and our conclusion follows.

Another situation where we can apply Theorem 7.5 (b) is when j is given by

$$j(u) = \int_\Omega k(u^-)^\gamma dx, \; u \in V,$$

with $\gamma > 2$ and $k \in L^\infty(\Omega)$, ess $\inf_\Omega k > 0$. We can show that the recession functional J_∞ associated with j is also given by (7.84). By Theorem 7.5 (b), we see that, in the present case, (7.74) has an asymptotic bifurcation behavior similar to that of the problem in Corollary 7.14. We also note that, by applying Theorems 7.4, 7.5 and their corollaries, or Corollaries 5.9, 5.10 in Chapter 5, we can prove similar results for other choices of the convex functionals j, or for variational inequalities containing quasilinear elliptic operators of order higher than 2.

Example 7.4 In this example, we consider bifurcation from infinity for a variational inequality containing the p-Laplacian, i.e., a variational inequality of the form (6.286), where Ω, N, p, and Γ, V are as in Example 6.11. However, we assume that

$$g : \Omega \times \mathbb{R} \times \mathbb{R} \to \mathbb{R}$$

is a Carathéodory function such that $g(x, u, 0) = 0$ for a.e. $x \in \Omega$, all $u \in \mathbb{R}$, and g satisfies the growth condition,

$$|g(x, u, \lambda)| \leq M(\lambda)[C(x)|u|^{\gamma(p-1)} + D(x)] \tag{7.88}$$

for a.e. $x \in \Omega$, all $u, \lambda \in \mathbb{R}$, where $0 \leq \gamma < 1, M \in L^\infty_{loc}(\mathbb{R})$, and $D \in L^r(\Omega), C \in L^q(\Omega)$ with

$$r > (p^*)' = \begin{cases} \dfrac{Np}{Np - N + p} & \text{if } N > p, \\ 1 & \text{if } N \leq p, \end{cases} \tag{7.89}$$

$$q > \left(\dfrac{p^*}{1 + \gamma(p-1)}\right)' = \begin{cases} \dfrac{Np}{Np - (N-p)[1 - \gamma(p-1)]} & \text{if } N > p, \\ 1 & \text{if } N \leq p. \end{cases} \tag{7.90}$$

Here, $\beta' = \beta(\beta - 1)^{-1}$ denotes the conjugate exponent of $\beta \in [1, \infty]$, and $p^* = Np(N - p)^{-1}$ is the Sobolev conjugate exponent of p. We also assume that j is given by

$$j(u) = \int_{\partial\Omega \backslash \Gamma} k(x)|u(x)|^{p+1} dS, \quad u \in V, \tag{7.91}$$

where $k \in L^\infty(\partial\Omega \backslash \Gamma)$ and

$$k(x) \geq k_0 > 0 \text{ for a.e. } x \in \partial\Omega \backslash \Gamma. \tag{7.92}$$

Because $p > 1$, one can verify that j is a convex functional from V to $[0, \infty]$, and $j(0) = 0$. Moreover, by using Fatou's lemma and the continuity of the mapping

$$W^{1,p}(\Omega) \to L^p(\partial\Omega), \quad u \mapsto u|_{\partial\Omega},$$

one can show that j is lower semicontinuous on V. We define A and B by (6.290) and (6.292), and

$$\langle G(u, \lambda), v \rangle = \int_\Omega g(x, u, \lambda) v dx, \ \forall u, v \in V. \tag{7.93}$$

We see that G and B are well-defined mappings from $V \times \mathbb{R}$ to V^*. Moreover, G and B are completely continuous on $V \times \mathbb{R}$.

In fact, let $u_n \to u$ in V and $\lambda_n \to \lambda$ in $\mathbb{R}$. For $1 \le s < p^*$, by the compactness of the embedding

$$W^{1,p}(\Omega) \hookrightarrow L^s(\Omega), \tag{7.94}$$

$$u_n \to u \text{ in } L^s(\Omega). \tag{7.95}$$

From (7.89), (7.90), $r', [1 + \gamma(p - 1)]q' < p^*$. Hence, we can choose s such that

$$\max\{r', [1 + \gamma(p - 1)]q'\} \le s < p^*. \tag{7.96}$$

By Hölder's inequality and (7.94),

$$|\langle G(u_n, \lambda_n) - G(u, \lambda), v \rangle| \le \|g(\cdot, u_n, \lambda_n) - g(\cdot, u, \lambda)\|_{L^{s'}(\Omega)} \|v\|_{L^s(\Omega)}$$
$$\le C \|g(\cdot, u_n, \lambda_n) - g(\cdot, u, \lambda)\|_{L^{s'}(\Omega)} \|v\|,$$
$$\forall v \in W^{1,p}(\Omega)$$

($C > 0$ is a fixed constant).

It follows from these inequalities that

$$\|G(u_n, \lambda_n) - G(u, \lambda)\| \le C \|g(\cdot, u_n, \lambda_n) - g(\cdot, u, \lambda)\|_{L^{s'}(\Omega)}, \ \forall n \in \mathbb{N}. \tag{7.97}$$

From (7.95), by passing to a subsequence,

$$\begin{cases} u_n(x) \to u(x) & \text{a.e. in } \Omega, \text{ and} \\ |u_n| \le h & \text{a.e. in } \Omega, \end{cases} \tag{7.98}$$

for some $h \in L^s(\Omega)$. Therefore,

$$g(x, u_n(x), \lambda_n) \to g(x, u(x), \lambda) \text{ a.e. in } \Omega, \tag{7.99}$$

and, by (7.88),

$$|g(x, u_n, \lambda_n) - g(x, u, \lambda)|$$
$$\le \ M(\lambda_n)[C(x)|u_n|^{\gamma(p-1)} + D(x)] + M(\lambda)[C(x)|u|^{\gamma(p-1)} + D(x)]$$
$$\le \ [M(\lambda_n) + M(\lambda)][C(x)(h^{\gamma(p-1)} + |u|^{\gamma(p-1)}) + D(x)],$$
$$\tag{7.100}$$

a.e. in Ω. From (7.96), $r \geq s'$ and, then, $D \in L^{s'}(\Omega)$. Moreover, $h^{\gamma(p-1)}$, $|u|^{\gamma(p-1)} \in L^{s/[\gamma(p-1)]}(\Omega)$. Hence,

$$C(h^{\gamma(p-1)} + |u|^{\gamma(p-1)}) \in L^{\rho}(\Omega) \subset L^{s'}(\Omega),$$

because $\rho \equiv \left[\dfrac{1}{q} + \dfrac{\gamma(p-1)}{s} \right]^{-1} \geq s'$ by (7.96). Because $\{\lambda_n\}$ is bounded, $\{M(\lambda_n) + M(\lambda)\}$ is bounded. Hence, the function in the right-hand side of (7.100) actually belongs to $L^{s'}(\Omega)$.

From (7.97), (7.99), and (7.100) it follows that

$$G(u_n, \lambda_n) \to G(u, \lambda) \quad \text{in } V^*,$$

proving that G is completely continuous on $V \times \mathbb{R}$. The complete continuity of B follows from that of G, in view of (7.94) with $s = p$.

Now, letting $u_n \rightharpoonup u$ in V, $\lambda_n \to \lambda$ and $\sigma_n \to \infty$, one has (7.95) and (7.98). We show that

$$\lim \frac{G(\sigma_n u_n, \lambda_n)}{\sigma_n^{p-1}} = 0 \quad \text{in } V^*. \tag{7.101}$$

As in (7.97),

$$\frac{1}{\sigma_n^{p-1}} \|G(\sigma_n u_n, \lambda_n)\| \leq C \left\| \frac{1}{\sigma_n^{p-1}} g(\cdot, \sigma_n u_n, \lambda_n) \right\|_{L^{s'}(\Omega)}, \quad \forall n \in \mathbb{N}. \tag{7.102}$$

From (7.88), for a.e. $x \in \Omega$,

$$\frac{1}{\sigma_n^{p-1}} |g(x, \sigma_n u_n(x), \lambda_n)| \leq M(\lambda_n) \left[C(x) \frac{|u_n(x)|^{\gamma(p-1)}}{\sigma_n^{(p-1)(1-\gamma)}} + \frac{D(x)}{\sigma_n^{p-1}} \right]. \tag{7.103}$$

Because $\{M(\lambda_n)\}, \{u_n\}$ are bounded (by (7.98)) and $(p-1)(1-\gamma) > 0$, so that $\sigma_n^{p-1}, \sigma_n^{(p-1)(1-\gamma)} \to \infty$, as $n \to \infty$, we see that

$$\lim \frac{1}{\sigma_n^{p-1}} g(x, \sigma_n u_n(x), \lambda_n) = 0 \quad \text{for a.e. } x \in \Omega. \tag{7.104}$$

For n sufficiently large, $\sigma_n > 1$. Hence, as in (7.100), from (7.103),

$$\begin{aligned} \frac{1}{\sigma_n^{p-1}} |g(x, \sigma_n u_n, \lambda_n)| &\leq M(\lambda_n) \left[\frac{C(x) h^{\gamma(p-1)}}{\sigma_n^{(p-1)(1-\gamma)}} + \frac{D(x)}{\sigma_n^{p-1}} \right] \\ &\leq M[C h^{\gamma(p-1)} + D], \end{aligned} \tag{7.105}$$

and the function in the right-hand side is in $L^{s'}(\Omega)$. From (7.104), (7.105), and (7.102), (7.101) follows. Let $f_\infty : V \times \mathbb{R} \to V^*$ be given by

$$\langle f_\infty(u, \lambda), v \rangle = \lambda \int_\Omega |u|^{p-2} uv, \quad u, v \in V. \tag{7.106}$$

Because f_∞ is completely continuous on $V \times \mathbb{R}$ and homogeneous of degree $p - 1$ with respect to u, we conclude, from (7.101), that B satisfies (A10) with f_∞ given by (7.106).

Because A is continuous and homogeneous of degree $p - 1$ on V, and A satisfies (A2) (i.e. (6.9) holds for $N = 1$ and (6.291) for general N), (A9) is satisfied with

$$\alpha_\infty = A. \tag{7.107}$$

Now, we prove (A11) (a)-(b), with $J_\infty = I_{W_0^{1,p}(\Omega)}$. In fact, let $v_n \rightharpoonup v$ in V, and $\sigma_n \to \infty$. As in Example 6.11, we have (6.296), (6.297), and (6.298). If $v = 0$ on $\partial\Omega$, then, $J_\infty(v) = 0$, and (7.5) holds. If this is not the case, then, $v \not\equiv 0$ on $\partial\Omega \setminus \Gamma$ and

$$j(v) = \int_{\partial\Omega\setminus\Gamma} k(x)|v(x)|^{p+1} dS \geq k_0 \int_{\partial\Omega\setminus\Gamma} |v(x)|^{p+1} dS > 0.$$

Hence, by the weak lower semicontinuity of j,

$$\liminf j(v_n) \geq j(v) > 0,$$

and, therefore,

$$\liminf \frac{j(\sigma_n v_n)}{\sigma_n^{p-1}} = \liminf [\sigma_n j(v_n)]$$
$$= \infty = J_\infty(v).$$

We have (7.5). Now, let $v \in V$, $\sigma_n \to \infty$, and choose $v_n = v$, $\forall n$. Then,

$$\liminf \frac{j(\sigma_n v_n)}{\sigma_n^{p-1}} = \liminf [\sigma_n j(v)] = j(v) \lim \sigma_n$$

$$= \begin{cases} 0 & \text{if } j(v) = 0, \\ \infty & \text{if } j(v) > 0, \end{cases}$$

$$= \begin{cases} 0 & \text{if } v \equiv 0 \text{ on } \partial\Omega, \\ \infty & \text{if } v \not\equiv 0 \text{ on } \partial\Omega, \end{cases}$$

$$= J_\infty(v).$$

(7.6) and, then, (A11) are verified.

It follows from (7.106) and (7.107) that the asymptotically homogenized variational inequality (7.8) associated with (6.286) is (6.299) or, equivalently, (6.300).

Let λ_1 (defined by (6.303)) be the first eigenvalue of the p-Laplacian with the Dirichlet boundary condition. Combining Theorem 7.2 with the results by del Pino, Elgueta, Manásevich, Nečas, and Drábek (cf. [30], [31], [34], and [84]), we obtain the following corollary:

Corollary 7.15 *(a) (∞, λ_1) is an asymptotic bifurcation point of (6.286), and the associated asymptotic bifurcation branch satisfies the alternative in Theorem 7.2.*

(b) Assume that $N = 1$. Then, the set of asymptotic bifurcation points of (6.286) is $\{\infty\} \times \{\lambda_n : n \in \mathbb{N}\}$, where $\{\lambda_n : n \in \mathbb{N}\}$ is the set of all eigenvalues of (6.300). Moreover, each point (∞, λ_n) corresponds to a global asymptotic bifurcation branch of (6.286) that satisfies the alternative in Theorem 7.2.

Note that similar arguments can be applied to obtain global results for bifurcation from infinity of the variational inequality (6.286), with $V = W_0^{1,p}(\Omega)$, and j is of the form $j(u) = \int_{\Omega_0} k|u|dx$, or $j(u) = \int_{\Omega_0} k(u)^{\pm}dx$, $u \in V$, with $\Omega_0 \subset \Omega$. In this case, we can show that $J_{\infty}(u) = 0$, $u \in W_0^{1,p}(\Omega)$, and, then, the results in [30], [31], [34], and [84] about degree calculations are still valid. In Example 7.4, if $\Gamma = \partial\Omega$, then, $V = W_0^{1,p}(\Omega)$, $j \equiv 0$, and (6.286), therefore, reduces to the nonlinear equation (6.306). Corollary 7.15, thus, gives global asymptotic bifurcation results for (6.306).

References

[1] R. ADAMS, *Sobolev Spaces*, Academic, New York, 1975.

[2] J. C. ALEXANDER AND P. M. FITZPATRICK, *The homotopy of certain spaces of nonlinear operators and its relation with global bifurcation of fixed points of parametrized condensing operators*, J. Funct. Anal., 34 (1979), pp. 87–106.

[3] H. AMANN, *Fixed point equations and nonlinear eigenvalue problems in ordered Banach spaces*, SIAM Rev., 18 (1976), pp. 620–709.

[4] A. ANANE, *Simplicité et isolation de la première valeur propre du p-Laplacien*, C. R. Acad. Sci Paris, 305 (1987), pp. 725–728.

[5] S. ANTMAN, *The influence of elasticity on analysis: Modern developments*, Bull. Amer. Math. Soc., 9 (1983), pp. 267–291.

[6] ———, *Nonlinear Problems of Elasticity*, vol. 107 of Applied Mathematical Sciences, Springer, New York, 1995.

[7] C. BAIOCCHI AND A. CAPELO, *Variational and Quasivariational Inequalities: Applications to Free Boundary Problems*, John Wiley & Sons, New York, 1984.

[8] G. BARLES, *Remarks on uniqueness results of the first eigenvalue of the p-Laplacian*, Ann. Fac. Sci. Toulouse Math., 9 (1988), pp. 65–75.

[9] M. S. BERGER, *On von Karman's equations and the buckling of thin elastic plates, I, the clamped plates*, Comm. Pure Appl. Math., 20 (1967), pp. 687–719.

[10] M. S. BERGER AND P. C. FIFE, *On von Karman's equations and the buckling of thin elastic plates, II, plates with general edge conditions*, Comm. Pure Appl. Math., 21 (1968), pp. 227–241.

[11] H. BRÉZIS, *Equations et inéquations non linéaires dans les espaces vectoriels en dualité*, Ann. Inst. Fourier, 18 (1968), pp. 115–175.

[12] ———, *Problèmes unilatéraux*, J. Math. Pures Appl., 51 (1972), pp. 1–168.

[13] ———, *Operateurs Maximaux Monotones*, North-Holland, 1973.

[14] ———, *Analyse Fonctionnelle*, Masson, Paris, 1983.

[15] F. BROWDER, *Nonlinear monotone operators and convex sets in Banach spaces*, Bull. Amer. Math. Soc., 71 (1965), pp. 780–785.

[16] ———, *Existence and approximation of solutions of nonlinear varational inequalities*, Proc. Nat. Acad. Sci. USA, 56 (1966), pp. 1080–1086.

[17] ———, *Fixed point theory and nonlinear problems*, Bull. Amer. Math. Soc., 9 (1983), pp. 1–39.

[18] M. CHIPOT, *Variational Inequalities and Flow in Porous Media*, no. 52 in Applied Mathematical Sciences, Springer, New York, 1984.

[19] S. CHOW AND J. HALE, *Methods of Bifurcation Theory*, Springer, Berlin, 1982.

[20] P. G. CIARLET AND P. RABIER, *Les Equations de von Karman*, Springer, Berlin, 1980.

[21] F. CONRAD, F. ISSARD-ROCH, C.-M. BRAUNER, AND B. NICOLAENCO, *Nonlinear eigenvalue problems in elliptic variational inequalities: A local study*, Comm. Partial Differential Equations, 10 (1985), pp. 151–190.

[22] M. CRANDALL AND P. RABINOWITZ, *Nonlinear Sturm-Liouville problems and topological degree*, J. Math. Mech., 19 (1970), pp. 1083–1102.

[23] E. N. DANCER, *Global solution branches for positive mappings*, Arch. Rational Mech. Anal., 52 (1973), pp. 181–192.

[24] ———, *Global structure of the solutions on nonlinear real analytic eigenvalue problems*, Proc. London Math. Soc, 27 (1973), pp. 747–765.

[25] ———, *On the indices of fixed points of mappings in cones and applications*, J. Math. Anal. Appl., 91 (1983), pp. 131–151.

[26] F. DE THÉLIN, *Sur l'espace propre associé à la première valeur propre du pseudo-Laplacien*, C. R. Acad. Sci. Paris, 303 (1986), pp. 355–358.

[27] M. DEGIOVANNI, *Bifurcation problems for nonlinear elliptic variational inequalities*, Ann. Fac. Sci. Toulouse, 10 (1989), pp. 215–258.

[28] M. DEGIOVANNI AND A. MARINO, *Non-smooth variational bifurcation*, Atti. Acc. Lincei Rend. Fis., 81 (1987), pp. 259–270.

[29] K. DEIMLING, *Nonlinear Functional Analysis*, Springer, Berlin, 1985.

[30] M. A. DEL PINO, M. ELGUETA, AND R. MANÁSEVICH, *A homotopic deformation along p of a Leray–Schauder degree result and existence for* $(|u'|^{p-2}u')'+f(t,u) = 0$, $u(0) = u(T) = 0$, $p > 1$, J. Differential Equations, 80 (1989), pp. 1–13.

[31] M. A. DEL PINO AND R. MANÁSEVICH, *Global bifurcation from the eigenvalues of the p-Laplacian*, J. Differential Equations, 92 (1991), pp. 226–251.

[32] C. DO, *The buckling of a thin elastic plate subjected to unilateral conditions*, Lecture Notes Math., 503 (1976), pp. 307–316.

[33] ——, *Bifurcation theory for elastic plates subjected to unilateral conditions*, J. Math. Anal. Appl., 60 (1977), pp. 435–448.

[34] P. DRÁBEK, *Solvability and Bifurcations of Nonlinear Equations*, Longman Scientific and Technical Series, Essex, 1992.

[35] J. DUGUNDJI, *Topology*, Allyn and Bacon, Boston, 1966.

[36] G. DUVAUT AND J. L. LIONS, *Les Inéquations en Mécanique et en Physique*, Dunod, Paris, 1972.

[37] L. C. EVANS, *Partial Differential Equations*, Berkeley Lecture Notes, Berkeley, 1994.

[38] G. FICHERA, *Problemi elastostatici con vincoli unilaterali: il problema di signorini con ambigue condizionial contorno*, Mem. Accad. Naz. Lincei Ser. VII, 7 (1964), pp. 613–679.

[39] A. FRIEDMAN, *Variational Principles and Free Boundary Value Problems*, Wiley-Interscience, New York, 1983.

[40] M. FURI AND A. VIGNOLI, *Spectrum for nonlinear maps and bifurcation in the non differentiable case*, Ann. Mat. Pura ed Appl., 113 (1977), pp. 265–285.

[41] M. GARCÍA-HUIDOBRO, R. MANÁSEVICH, AND K. SCHMITT, *On principal eigenvalues of p−Laplacian like operators*, J. Differential Equations, 1996 (to appear).

[42] ——, *Some bifurcation results for a class of p−Laplacian like operators*, Differential Integral Equations, 1996 (to appear).

[43] D. GILBARG AND N. TRUDINGER, *Elliptic Partial Differential Equations of Second Order*, Springer, Berlin, 1983.

[44] D. GOELEVEN, V. H. NGUYEN, AND M. THÉRA, *Méthode du degré topologique et branches de bifurcation pour les inéquations variationnelles de von Karman*, C. R. Acad. Sci. Paris, 317 (1993), pp. 631–635.

[45] ——, *Nonlinear eigenvalue problems governed by a variational inequality of von Karman's type: a degree theoretic approach*, Topol. Methods Nonlinear Anal., 2 (1993), pp. 253–276.

[46] P. HESS AND T. KATO, *On some linear and nonlinear eigenvalue problems with an indefinite weight function*, Comm. Partial Differential Equations, 10 (1980), pp. 999–1030.

[47] I. HLAVÁČEK, J. HASLINGER, J. NEČAS, AND J. LOVIŠEK, *Solutions of Variational Inequalities in Mechanics*, no. 66 in Applied Mathematical Sciences, Springer, New York, 1988.

[48] R. H. W. HOPPE AND H. D. MITTELMANN, *A multi-grid continuation strategy for parameter-dependent variational inequalities. Continuation techniques and bifurcation problems.*, J. Comput. Appl. Math., 26 (1989), pp. 35–46.

[49] J. IZE, *Bifurcation Theory for Fredholm Operators*, Mem. Amer. Math. Soc., 174 (1976).

[50] D. KINDERLEHRER, *Remarks about Signorini's problem in linear elasticity*, Ann. Scuola Norm. Sup. Pisa, 8 (1981), pp. 605–645.

[51] D. KINDERLEHRER AND G. STAMPACCHIA, *An Introduction to Variational Inequalities*, Academic, New York, 1980.

[52] K. KIRCHGÄSSNER, *Bifurcation in nonlinear hydrodynamic stability*, SIAM Rev., 17 (1975), pp. 652–683.

[53] M. A. KRASNOSEL'SKII, *Topological Methods in the Theory of Nonlinear Integral Equations*, Pergamon, Oxford, 1963.

[54] ———, *Positive Solutions of Operator Equations*, Noordhoff, Groningen, 1964.

[55] R. S. KUBRUSLY, *Variational methods for nonlinear eigenvalue inequalities*, Differential Integral Equations, 3 (1990), pp. 923–932.

[56] R. S. KUBRUSLY AND J. T. ODEN, *Nonlinear eigenvalue problems characterized by variational inequalities with applications to the post-buckling analysis of unilaterally supported plates*, Nonlinear Anal., TMA, 5 (1981), pp. 1265–1284.

[57] A. KUFNER, O. JOHN, AND S. FUČIK, *Function Spaces*, Noordhoff, Leyden, 1977.

[58] M. KUČERA, *Bifurcation points of varational inequalities*, Czech. Math. J., 32 (1982), pp. 208–226.

[59] ———, *A new method for obtaining eigenvalues of variational inequalities: Operators with multiple eigenvalues*, Czech. Math. J., 32 (1982), pp. 197–207.

[60] ———, *A global continuation theorem for obtaining eigenvalues and bifurcation points*, Czech. Math. J., 38 (1988), pp. 120–137.

[61] M. KUČERA, J. NEČAS, AND J. SOUČEK, *The eigenvalue problem for variational inequalities and a new version of the Ljusternik-Schnirelmann theory*, in Nonlinear Analysis, New York, 1978, Academic, pp. 125–143.

[62] V. K. LE, *On Global Bifurcation for Variational Inequalities*, Ph.D. Thesis, University of Utah, 1995.

[63] ———, *Bifurcation from infinity for variational inequalities*, preprint, (1996).

[64] ———, *Global bifurcation results for variational inequalities*, J. Differential Equations, to appear, (1996).

[65] J. L. LIONS, *Quelques Méthodes de Résolution des Problèmes aux Limites Nonlinéaires*, Dunod, Paris, 1969.

[66] J. L. LIONS AND E. MAGENES, *Nonhomogeneous Boundary Value Problems and Applications*, Springer, Berlin, 1972.

[67] J. L. LIONS AND G. STAMPACCHIA, *Variational inequalities*, Comm. Pure Appl. Math., 20 (1967), pp. 493–519.

[68] R. MAGNUS, *A generalization of multiplicity and the problem of bifurcation*, Proc. London Math. Soc., 32 (1976), pp. 251–278.

[69] J. B. McLEOD AND R. E. L. TURNER, *Bifurcation for nondifferentiable operators with an application to elasticity*, Arch. Rational Mech. Anal., 63 (1976), pp. 1–45.

[70] E. MIERSEMANN, *Verzweigungsprobleme für Variationsungleichungen*, Math. Nachr., 84 (1975), pp. 187–210.

[71] ——, *Über nichtlineare Eigenwertaufgaben in konvexen Mengen*, Math. Nachr., 88 (1979), pp. 191–205.

[72] ——, *Über positive Lösungen von Eigenwertgleichungen mit Anwendungen auf elliptische Gleichungen zweiter Ordnung und auf ein Beulproblem für die Platte*, ZAMM, 59 (1979), pp. 189–194.

[73] ——, *Höhere Eigenwerte von Variationsungleichungen*, Beiträge Anal., 17 (1981), pp. 65–68.

[74] ——, *On higher eigenvalues of variational inequalities*, Comment. Math. Univ. Carolin., 24 (1983), pp. 657–665.

[75] ——, *Eigenvalue problems for variational inequalities*, Contemp. Math., 4 (1984), pp. 25–43.

[76] E. MIERSEMANN AND H. D. MITTELMANN, *Continuation for parametrized nonlinear variational inequalities. Continuation techniques and bifurcation problems*, J. Comput. Appl. Math., 26 (1989), pp. 23–34.

[77] ——, *On the continuation for variational inequalities depending on an eigenvalue parameter*, Math. Methods Appl. Sci., 11 (1989), pp. 95–104.

[78] ——, *Extension of Beckert's continuation method to variational inequalities*, Math. Nachr., 148 (1990), pp. 183–195.

[79] ——, *Stability in obstacle problems for the von Kármán plate*, SIAM J. Math. Anal., 23 (1992), pp. 1099–1116.

[80] H. D. MITTELMANN, *On continuation for variational inequalities*, SIAM J. Numerical Anal., 24 (1987), pp. 1374–1381.

[81] ——, *Continuation methods for parameter-dependent boundary value problems*, in Computational Aspects of VLSI Design with an Emphasis on Semiconductor Device Simulation, Providence, 1990, Amer. math. Soc., pp. 159–175.

[82] ——, *Nonlinear parametrized equations: New results for variational problems and inequalities*, in Computational Solution of Nonlinear Systems of Equations, Providence, 1990, Amer. Math. Soc., pp. 451–466.

[83] J. NAUMANN AND H. U. WENK, *On eigenvalue problems for variational inequalities*, Rend. Di Matematica, Univ. Roma, 6 (1976), pp. 439–463.

[84] J. NEČAS, *On the discreteness of the spectrum of a nonlinear Sturm-Liouville equation*, Sov. Math. Dokl., 12 (1971), pp. 1779–1783.

[85] R. NUSSBAUM, *The Fixed Point Index and Some Applications*, Montreal University Press, Montreal, 1985.

[86] J. T. ODEN AND J. A. C. MARTINS, *Models and computational methods for dynamic friction phenomena*, Comp. Methods Appl. Mech. Eng., 52 (1985), pp. 527–634.

[87] P. D. PANAGIOTOPOULOS, *Inequality Problems in Mechanics and Applications: Convex and Nonconvex Energy Functions*, Birkhäuser, Boston, 1985.

[88] D. PASCALI, *Elliptic eigenvalue variational inequalities*, Libertas Mathematica, 12 (1992), pp. 39–56.

[89] D. PASCALI AND J. SBURLAN, *Nonlinear Mappings of Monotone Type*, Sijthoff and Noordhoff, Bucharest, 1978.

[90] H. O. PEITGEN AND K. SCHMITT, *Global topological perturbations of nonlinear elliptic eigenvalue problems*, Math. Methods Appl. Sci., 5 (1983), pp. 376–388.

[91] ——, *Global analysis of two-parameter elliptic eigenvalue problems*, Trans. Amer. Math. Soc., 283 (1984), pp. 57–95.

[92] P. QUITTNER, *Spectral analysis of variational inequalities.*, Comment. Math. Univ. Carolin., 27 (1986), pp. 605–629.

[93] ——, *Bifurcation points and eigenvalues of inequalities of reaction-diffusion type*, J. Reine Angew. Math., 380 (1987), pp. 1–13.

[94] ——, *On the principle of linearized stability for variational inequalities*, Math. Ann., 289 (1989), pp. 257–270.

[95] ——, *Solvability and multiplicity results for variational inequalities.*, Comment. Math. Univ. Carolin., 30 (1989), pp. 281–302.

[96] P. J. RABIER AND J. T. ODEN, *Solution to Signorini-like contact problems through interface models, I, preliminaries and formulation of a variational equality*, Nonlinear Anal., TMA, 11 (1987), pp. 1325–1350.

[97] ——, *Solution to Signorini-like contact problems through interface models, II, existence and uniqueness theorems*, Nonlinear Anal., TMA, 12 (1988), pp. 1–17.

[98] P. RABINOWITZ, *Some global results for nonlinear eigenvalue problems*, J. Funct. Anal., 7 (1971), pp. 487–513.

[99] ——, *On bifurcation from infinity*, J. Differential Equations, 14 (1973), pp. 462–475.

[100] ——, *Some aspects of nonlinear eigenvalue problems*, Rocky Mtn. J. Math., 3 (1973), pp. 162–202.

[101] R. C. RIDELL, *Eigenvalue problems for nonlinear elliptic variational inequalities*, Nonlinear Anal., TMA, 3 (1979), pp. 1–33.

[102] T. R. ROCKAFELLAR, *Convex Analysis*, Princeton University Press, Princeton, NJ, 1970.

[103] J. F. RODRIGUES, *Obstacle Problems in Mathematical Physics*, North-Holland, Amsterdam, 1987.

[104] C. SACCON, *A global bifurcation result for variational inequalities*, Boll. Un. Mat. Ital., 7-A (1993), pp. 117–124.

[105] D. SATHER, *Branching of solutions of nonlinear equations*, Rocky Mtn. J. Math., 3 (1973), pp. 203–250.

[106] K. SCHMITT, *Positive solutions for semilinear elliptic boundary value problems*, in Topological Methods in Differential Equations, A. Granas and M. Frigon, eds., Kluwer, Doordrecht, 1995, pp. 447–500.

[107] K. SCHMITT AND H. SMITH, *On eigenvalue problems for nondifferentiable mappings*, J. Differential Equations, 33 (1979), pp. 294–319.

[108] ——, *Fredholm alternatives for nonlinear differential equations*, Rocky Mtn. J. Math., 12 (1982), pp. 817–841.

[109] K. SCHMITT AND Z. WANG, *On bifurcation from infinity for potential operators*, Differential Integral Equations, 4 (1991), pp. 933–943.

[110] F. SCHURICHT, *Bifurcation from minimax solutions by variational inequalities*, Math. Nachr., 154 (1991), pp. 67–88.

[111] ———, *Minimax principle for eigenvalue variational inequalities in the nonsmooth case*, Math. Nachr., 152 (1991), pp. 121–143.

[112] ———, *Minimax principle for eigenvalue variational inequalities in convex sets*, Math. Nachr., 163 (1993), pp. 117–132.

[113] ———, *Bifurcation from minimax solutions by variational inequalities in convex sets*, Nonlinear Anal., TMA, 26 (1996), pp. 91–112.

[114] A. SIGNORINI, *Questioni di elasticcità nonlinearizzata e semilinearizzata*, Rend. Mat., 18 (1959), pp. 1–45.

[115] G. STAMPACCHIA, *Formes bilinéaires coercitives sur les ensembles convexes*, C. R. Acad. Sci. Paris, 258 (1964), pp. 4413–4416.

[116] ———, *Le problème de Dirichlet pour les équations elliptiques du second ordre à coefficients discontinus*, Ann. Inst. Fourier, 258 (1965), pp. 189–258.

[117] C. A. STUART, *Some bifurcation theory for k-set contractions*, Proc. London Math. Soc., 27 (1973), pp. 531–550.

[118] A. SZULKIN, *Positive solutions of variational inequalities: A degree theoretic approach*, J. Differential Equations, 57 (1985), pp. 90–111.

[119] ———, *Existence and nonuniqueness of solutions of a noncoercive elliptic variational inequality*, Proc. Symp. Pure Math., 45 (1986), pp. 413–418.

[120] J. TOLAND, *Global bifurcation for k-set contractions without multiplicity assumptions*, Quart. J. Math., 27 (1976), pp. 199–216.

[121] G. M. TROIANIELLO, *Elliptic Differential Equations and Obstacle Problems*, Plenum, New York, 1987.

[122] S. L. TROJANSKI, *On locally uniformly convex and differentiable norms in certain non-separable Banach spaces*, Studia Math., 37 (1971), pp. 173–180.

[123] K. YOSIDA, *Functional Analysis*, Springer, New York, 1995.

[124] E. ZEIDLER, *Nonlinear Functional Analysis and its Applications, Vol.3: Variational Methods and Optmization*, Springer, Berlin, 1985.

[125] ———, *Nonlinear Functional Analysis and its Applications, Vol.1: Fixed-Point Theorems*, Springer, Berlin, 1986.

[126] ———, *Nonlinear Functional Analysis and its Applications, Vol.4: Applications to Mathematical Physics*, Springer, Berlin, 1988.

Index

Applied Mathematical Sciences

(continued from page ii)

(continued on next page)

Applied Mathematical Sciences

(continued from previous page)